THE NUMERICAL SOLUTION OF TWO-POINT BOUNDARY PROBLEMS IN ORDINARY DIFFERENTIAL EQUATIONS

BY

L. FOX, M.A., D.Sc.

MATHEMATICS DIVISION
NATIONAL PHYSICAL LABORATORY

DOVER PUBLICATIONS, INC.
NEW YORK

Published in Canada by General Publishing Company, Ltd., 30 Lesmill Road, Don Mills, Toronto, Ontario.

Published in the United Kingdom by Constable and Company, Ltd., 3 The Lanchesters, 162–164 Fulham Palace Road, London W6 9ER.

This Dover edition, first published in 1990, is an unabridged and unaltered republication of the work first published at the Clarendon Press, Oxford, in 1957 in the series *Monographs On Numerical Analysis*. It is reprinted by special arrangement with Oxford University Press, 200 Madison Avenue, New York, N.Y. 10016.

Manufactured in the United States of America

Dover Publications, Inc., 31 East 2nd Street, Mineola, N.Y. 11501

Library of Congress Cataloging-in-Publication Data

Fox, L. (Leslie)

The numerical solution of two-point boundary problems in ordinary differential equations / L. Fox.

p. cm.

Reprint. Originally published: Oxford : Clarendon Press, 1957.

Includes bibliographical references and index.

ISBN 0-486-66495-3 (pbk.)

1. Boundary value problems—Numerical solutions. I. Title.

QA379.F69 1990

515′.352—dc20 90-48640

CIP

PREFACE

FOR the purpose of numerical solution it is convenient to separate ordinary differential equations into two classes, according to the *position* of the associated prescribed conditions. If all such necessary conditions are given at one point in the range of the independent variable they are usually called initial conditions, and the differential system is of *initial-value* type. When more than one value of the independent variable is involved the conditions are called boundary conditions, and the differential system is of *boundary-value* type.

In this book I have concentrated on boundary-value problems, partly because much of my research has been in this field, and partly because this class has hitherto received only minor attention in published books. I have included most of the worthwhile methods known to me which are suitable for pencil-and-paper and desk-machine computation. Most of these methods are adaptable to automatic high-speed machines, but methods such as those of Monte-Carlo type, for which the use of such machines is essential, are deliberately excluded. For, as Professor Hildebrand remarks in the preface to his *Introduction to Numerical Analysis* (New York: McGraw-Hill, 1956), 'familiarity with computation on a desk calculator is a desirable preliminary to large-scale computation . . . and there is a continuing need for personnel sufficiently well-grounded in both the principles and the existent techniques of numerical analysis to be capable, not only of adapting available methods to a given machine, but also of comprehending the inherent limitations of each method and of devising error controls and modified techniques which are particularly appropriate to that machine'.

I have therefore given both methods and numerical techniques, and have tried to distinguish between good and bad computing principles. Each fresh point is illustrated by a worked example, and the reader is invited to perform again, for himself, the relevant computations. No unworked examples are listed, since the student can easily concoct suitable problems for additional practice.

The main part of the book describes various boundary-value methods for differential systems of orders one to four, including problems of eigenvalue type. Both linear and non-linear cases are considered, and special attention is given to the proper numerical treatment of the various possible types of boundary conditions. The precision of the solutions obtained is treated at some length in a separate chapter, and there is a discussion and warning on the use of initial-value techniques for boundary-value problems.

Throughout I have taken the view that truncation errors in finite-difference equations should not be tolerated, and have made full use of the 'difference-correction' method to eliminate such error without decreasing the finite-difference interval. I have also discussed in some detail, as an alternative device, the method of 'the deferred approach to the limit'.

Other relevant matters, such as the production and proper use of finite-difference equations, the solution of simultaneous algebraic equations, and the determination of the latent roots and vectors of arbitrary matrices, receive adequate but less detailed treatment; later volumes in this series will doubtless provide the necessary expansion.

There is little 'higher' mathematics or 'rigorous' argument. The book is intended for the practical computer and for the student seeking a career in computation.

On matters of arrangement, I have omitted references from the text and have collected them, and other additional notes relevant to the material of various sections, at the end of each chapter. On typography I have experimented, following the suggestion of Chaundy, Barrett, and Batey (*The Printing of Mathematics*, Oxford, 1954), by using ordinary italic type for matrices and vectors. In this, and in many other matters, the skill and advice of the Press have been invaluable. For any unexplained ambiguities, such as the use of y_0, y_r, and $y^{(r)}$ to mean different things in different contexts, I must take full responsibility.

In the preparation of this material I have benefited much from the conversations and writings of Dr. L. H. Thomas of Columbia University, an expert in this and many other computing fields. The first draft was read and criticized helpfully by the editors of this series and by several colleagues in the Mathematics Division, National Physical Laboratory, and all my arithmetic was checked, and often corrected, by Mrs. Joan Blake and Miss Pam Sivyer. To all these I give grateful thanks.

L. F.

Berkeley, California
12 *December* 1956

CONTENTS

NOTE. † indicates that there is a further mention of the section in *Additional notes and bibliography*, given at the end of each Chapter.

1

Introduction

Definitions

1. An *ordinary differential equation* is an equation involving one or more derivatives of the *dependent variable y* with respect to a single *independent variable x*. The *order* of the equation is that of the highest derivative contained in it, so that the general differential equation of order n can be written in the form

$$F\{y^{(n)}, y^{(n-1)},\dots, y', y, x\} = 0, \tag{1}$$

the symbol† $y^{(r)}$ denoting d^ry/dx^r. The *degree* of the equation is defined mathematically to be that of its highest-order derivative, when the equation has been made rational as far as the derivatives are concerned. The equation

$$y'' = (xy^{\frac{3}{2}}+y')^{\frac{1}{2}}, \tag{2}$$

for example, is of order 2 and degree 2, since the equation must be squared to rationalize the contributions from the derivatives.

This definition of order is adopted in numerical work also, but the definition of degree is not very relevant. We shall refer to an equation of the form

$$y^{(n)}+f_{n-1}(x)y^{(n-1)}+f_{n-2}(x)y^{(n-2)}+\dots+f_1(x)y'+f_0(x)y = g(x), \tag{3}$$

in which the $f_r(x)$ are functions of x only, as *linear*, and any other type of equation will be called *non-linear*. The reason for this is fairly clear: in many methods of solution we shall replace derivatives by linear combinations of several discrete values $y(x_r)$ and, if the f_r in (3) also contain y or any derivative, the resulting *algebraic* equation will contain non-linear terms y^s, $s \neq 1$.

2. *Simultaneous* ordinary differential equations involve several dependent variables y, z,... and their derivatives. The equations do not necessarily have the same order or degree, but there are usually the same number of equations as the number of unknown functions. General simultaneous equations in two variables will have the form

$$\left.\begin{aligned} F_1\{y^{(p)}, y^{(p-1)},\dots, y', y; z^{(q)}, z^{(q-1)},\dots, z', z\} = 0 \\ F_2\{y^{(r)}, y^{(r-1)},\dots, y', y; z^{(s)}, z^{(s-1)},\dots, z', z\} = 0 \end{aligned}\right\}. \tag{4}$$

† This is often replaced by y', y'', y''', y^{iv}, etc., with unbracketed roman superscripts, when a specific order is intended.

3. It can be proved that the most general solution of an ordinary differential equation of order n contains n *arbitrary constants*. This general solution is called the *Complete Primitive*, and a *Particular Integral* is obtained by giving specific values to these arbitrary constants. Non-linear equations may also have *singular solutions*, not obtainable from the complete primitive.

For the linear differential equation (3) the particular integral, as before, is any solution $y_0(x)$ which contains no arbitrary constants. The *Complementary Function* is of the form

$$y(x) = \sum_{1}^{n} A_r y_r(x), \tag{5}$$

in which the functions $y_r(x)$ are independent solutions of the homogeneous equation obtained from (3) by replacing $g(x)$ by zero, and the arbitrary constants A_r appear linearly. The general solution of (3) is then given by

$$y(x) = y_0(x) + \sum_{1}^{n} A_r y_r(x). \tag{6}$$

Analytical and numerical solutions

†4. With analytical methods we try to obtain a solution in the explicit form

$$y = f_1(x), \tag{7}$$

or implicit form

$$f_2(x, y) = 0, \tag{8}$$

containing the required number of arbitrary constants. The functions f_1 and f_2 may include only the elementary functions $\sin x$, e^x, etc., or involve higher mathematical functions such as those of Airy and Bessel. In practical work, however, the most important usual requirement is that y should be expressible quantitatively in terms of x, by means of a graph if no great precision is needed, or otherwise by a table of values of y calculated at various points in the range of x. In a useful analytical solution, therefore, the functions contained in it must be well-known and adequately tabulated.

Even in this case, however, much computation may be needed to produce numerical results from an analytical solution. For example, the simple differential equation

$$y' - \frac{2y}{1-x^4} = 0 \tag{9}$$

has the solution

$$y = A\left(\frac{1+x}{1-x}\right)^{\frac{1}{2}} e^{\tan^{-1}x}. \tag{10}$$

† See also additional note at the end of the chapter.

For effective tabulation of y we need tables of $x^{\frac{1}{2}}$, e^x, and $\tan^{-1}x$, and we shall generally have to interpolate, at least in the exponential table.

A further restriction on analytical methods is that they can only rarely produce closed solutions, of the required type, to differential equations which arise in practical problems. Even apparently innocuous equations are intractable. For example, the equation

$$y' - \frac{2y}{1-x^4} = x, \tag{11}$$

a seemingly trivial amendment of (9), has the solution

$$y = \left(\frac{1+x}{1-x}\right)^{\frac{1}{2}} e^{\tan^{-1}x} \left\{ \int x \left(\frac{1-x}{1+x}\right)^{\frac{1}{2}} e^{-\tan^{-1}x}\, dx + A \right\}, \tag{12}$$

and it is unlikely that the indefinite integral can be expressed in closed form in terms of known functions.

5. When a convenient analytical solution can be obtained, it has of course advantages over numerical methods. In particular, we have available the complete primitive containing arbitrary constants, and any particular integral is obtained easily by giving specific values to these constants. The latter are usually determined from a prescribed knowledge of some values of y or its derivatives, or both, at one or more points in the range of x. If several problems are presented with the same differential equation but different prescribed conditions, the complete primitive is the same in all cases, only the arbitrary constants needing recalculation. This advantage is not in general shared by numerical methods.

6. A numerical *solution* is a graph or table of y against x. A numerical *method*, from our point of view, is one which produces discrete values of y without the assistance of an analytical solution. By definition, a numerical solution contains no arbitrary constants, so that we always obtain particular integrals rather than complete primitives. Indeed the prescribed conditions are involved from the start in a numerical method, in direct contrast to analytical methods, and a change in these conditions usually presents a completely new problem, even for the same differential equation. Singular solutions, also, are seldom required.

Boundary conditions and initial conditions

7. We have mentioned the need for 'prescribed conditions' which fix the arbitrary constants in an analytical solution, and which are used directly in numerical methods. In most cases, and according to the order of the equation, these conditions specify the value of y or one or more of its derivatives, or some combinations of these quantities, at one or more points in the range of integration.

Consider in this respect a differential equation of second order, which needs two prescribed conditions to provide a definite solution in the range $x = x_0$ to x_n, say. The most general imposed conditions are given by

$$f_1(y, y') = 0, \qquad f_2(y, y') = 0, \tag{13}$$

at the respective specific points $x = x_r, x_s$. If x_r and x_s are both at x_0, so that both conditions are given at the same point, equations (13) permit the evaluation of both y and y' at this point. Conditions of this kind give rise to what we shall call an *initial-value* problem, only one point being involved in the prescribed conditions, in this case called *initial conditions*.

If x_r is at x_0, x_s at x_n, the other main possibility, we have one condition at each end or *boundary* of the range of integration. A problem of this kind we shall call a *boundary-value* problem, and we extend this definition to all problems in which two separate points are involved in the prescribed conditions, here called *boundary conditions*.

†8. The *position* of the prescribed conditions is of little importance in analytical methods, since the main part of the work is the formation of the complete primitive which depends only on the differential equation. In numerical work, however, the position of these conditions is more important, as regards choice of method, than the order, degree, or form of the differential equation.

We can illustrate this point with a simple example, involving the solution of the differential equation

$$y'' = f(x). \tag{14}$$

In many methods of solution we replace derivatives by simple linear combinations of pivotal values. The theory is given in the next chapter, but for our present purpose we anticipate the result that an approximate value for the second derivative y''_r at the point x_r is obtained from the equation

$$h^2 y''_r = y_{r+1} - 2y_r + y_{r-1}, \tag{15}$$

where $x_{r+1} = x_r + h$, $x_{r-1} = x_r - h$. We can then approximate to (14), at the point x_r, by the algebraic equation

$$y_{r+1} - 2y_r + y_{r-1} = h^2 f(x_r); \tag{16}$$

this equation holds for all values of r in the range of integration and depends only on the differential equation.

The method of solving the aggregate of equations typified by (16), however, depends on the prescribed conditions. If the latter are of initial-value type, enabling the first two values y_0 and y_1 to be calculated, we can calculate y_2, the value at the next pivotal point, directly from (16) with $r = 1$, and successive values also come from direct substitution in successive equations of the set (16). This is a *step-by-step* process, pivotal values being obtained successively; the value at x_r is determined from a knowledge of those at previous points but is independent of those at later points.

If the prescribed conditions are of boundary-value type, however, of which the simplest case is the specification of the two boundary values y_0 and y_n, the first and last equations of (16), at $r = 1$ and $n-1$ respectively, each contain two unknowns. All the other equations contain three unknowns, so that the pivotal values y_r can be obtained only by solving *simultaneous algebraic equations.*

This difference in technique between step-by-step solution and simultaneous solution is the main numerical distinction between initial-value and boundary-value problems.

Boundary-value problems can always be solved in theory, and often in practice, by a combination of initial-value problems, the extra initial conditions being chosen more or less arbitrarily but finally adjusted to satisfy the prescribed boundary conditions. This process often involves considerable labour, and it is therefore desirable to consider separately the two types of problem. In this book we concentrate on the solution of boundary-value problems, reserving for a second volume the treatment of initial-value problems.

Additional notes and bibliography

For the mathematical theory of ordinary differential equations see, for example,

INCE, E. L., 1944, *Ordinary differential equations.* Dover.

§ 4. A vast amount of information about mathematical tables is contained in

FLETCHER, A., MILLER, J. C. P., and ROSENHEAD, L., 1946, *An index of mathematical tables.* Scientific Computing Service. A revised up-to-date edition is being compiled.

§ 8. No published book has given extensive treatment of boundary-value methods for ordinary differential equations. Some mention is made in

HARTREE, D. R., 1952, *Numerical analysis.* Oxford, and

MILNE, W. E., 1953, *Numerical solution of differential equations.* Wiley, New York.

2

Finite Differences and Finite-difference Equations

Introduction

1. GIVEN the numerical values of a function $y(x)$ for a set of values of the independent variable, the theory of finite differences is concerned with the provision and use of formulae whereby we can calculate

(i) the value of y at any intermediate point (interpolation),

(ii) the derivative of y at any point (differentiation),

(iii) the definite integral of y between any two points (integration or quadrature).

The simplest functions in analysis are polynomials, and the general theory is based on this fact, providing expressions for interpolates, derivatives, and integrals which are exact for polynomials and which for other functions, subject to certain restrictions, have an error which can be estimated and often made as small as we please.

In this chapter we discuss some aspects of the finite-difference equations which we shall use in the rest of the book. The computer need not know how to establish the formulae, most of which are given in any book on numerical calculus. He does, however, need to know how to use them, and a knowledge of methods of production is of great value in stimulating confidence and facility in their application. We shall therefore first develop and illustrate a powerful operational method for the rapid and efficient derivation of such formulae, following this with a discussion of their use.

Differences

†2. The term 'finite differences' implies that the first thing to do with our column of pivotal values is to form their table of differences. For our purpose we take as constant the interval h between pivotal points.

The three main notations used for differences are indicated in Tables 1, 2, and 3.

Table 1 (Forward differences)

0	y_0				
		Δy_0			
1	y_1		$\Delta^2 y_0$		
		Δy_1		$\Delta^3 y_0$	
2	y_2		$\Delta^2 y_1$		etc.
		Δy_2		$\Delta^3 y_1$	
3	y_3		$\Delta^2 y_2$		
		Δy_3			
4	y_4				

Table 2 (Backward differences)

y_0				
	∇y_1			
y_1		$\nabla^2 y_2$		
	∇y_2		$\nabla^3 y_3$	
y_2		$\nabla^2 y_3$		etc.
	∇y_3		$\nabla^3 y_4$	
y_3		$\nabla^2 y_4$		
	∇y_4			
y_4				

Table 3 (Central differences)

y_0				
	$\delta y_{\frac{1}{2}}$			
y_1		$\delta^2 y_1$		
	$\delta y_{\frac{3}{2}}$		$\delta^3 y_{\frac{3}{2}}$	
y_2		$\delta^2 y_2$		etc.
	$\delta y_{\frac{5}{2}}$		$\delta^3 y_{\frac{5}{2}}$	
y_3		$\delta^2 y_3$		
	$\delta y_{\frac{7}{2}}$			
y_4				

The formation of the differences is indicated by the following typical equations:

$$y_1 - y_0 = \Delta y_0, \qquad y_1 - y_0 = \nabla y_1, \qquad y_1 - y_0 = \delta y_{\frac{1}{2}},$$
$$\Delta^2 y_1 - \Delta^2 y_0 = \Delta^3 y_0, \qquad \nabla^2 y_3 - \nabla^2 y_2 = \nabla^3 y_3, \qquad \delta^2 y_2 - \delta^2 y_1 = \delta^3 y_{\frac{3}{2}}.$$

It is conventional to label as 0, 1, 2,... the pivotal points x_0, x_0+h, x_0+2h,..., and as y_0, y_1, y_2,..., the corresponding pivotal values. Also the function is often omitted from the symbol for the difference so that, for example, we may sometimes write ∇_1 for ∇y_1, $\delta^3_{\frac{3}{2}}$ for $\delta^3 y_{\frac{3}{2}}$, etc.

The symbol Δ denotes a forward† difference: $\Delta^s y_r$ is the forward difference of order s of y_r. The symbol ∇ denotes a backward† difference: $\nabla^s y_r$ is the backward difference of order s of y_r. The symbol δ denotes a central† difference: here the matter is more complicated, because only the even central differences are associated with pivotal values of y_r, the odd central differences belonging apparently to points midway between the pivotal points.

Our purpose now is to establish formulae for interpolation, differentiation and integration which will involve the pivotal values y_r and their differences. Since the main concern is the solution of differential equations we can restrict ourselves to producing formulae for derivatives at or midway between pivotal points, and for definite integrals taken between such points. Interpolation is almost an incidental which will be used chiefly to illustrate the method.

Operators

3. We are accustomed to the operational representation of the Taylor series

$$y(x+h) = y(x) + hy'(x) + \frac{h^2}{2!}y''(x) + \dots, \tag{1}$$

expressed in the form

$$y(x+h) = e^{hD}y(x). \tag{2}$$

† For forward differences the suffix r of y_r is constant along lines sloping downwards from left to right, for backward differences r is constant along lines sloping upwards from left to right, and for central differences r is constant along horizontal lines.

In the theory of finite differences we carry this farther, introducing first the operator E, the effect of change of position, and defined by the equation

$$y(x+ph) = E^p y(x). \tag{3}$$

In particular the effect of moving from one pivotal point to the next is given by

$$y(x+h) = Ey(x). \tag{4}$$

We define similarly the difference operators by the equations

$$\left.\begin{aligned} y(x+h)-y(x) &= \Delta y(x) \\ y(x+h)-y(x) &= \nabla y(x+h) \\ y(x+h)-y(x) &= \delta y(x+\tfrac{1}{2}h) \end{aligned}\right\}. \tag{5}$$

From equations (2), (3), (4), and (5) we easily produce the operational equivalents

$$\left.\begin{aligned} E &= e^{hD} \\ E-1 &= \Delta \\ E-1 &= \nabla E \\ E-1 &= \delta E^{\frac{1}{2}} \end{aligned}\right\}. \tag{6}$$

With few restrictions these operators can be manipulated according to the rules of ordinary algebra. For example we can write the last of (6) in the form

$$E^{\frac{1}{2}}-E^{-\frac{1}{2}} = \delta, \tag{7}$$

and applying each operator to y_0 we have

$$y_{\frac{1}{2}}-y_{-\frac{1}{2}} = \delta y_0, \tag{8}$$

the equation defining the first central difference of y_0, missing from the difference table.

Finally we include the 'averaging' operator μ, defined by the equation

$$\mu = \tfrac{1}{2}(E^{-\frac{1}{2}}+E^{\frac{1}{2}}), \tag{9}$$

so that, for example,

$$\mu y_0 = \tfrac{1}{2}(y_{-\frac{1}{2}}+y_{\frac{1}{2}}),\quad \mu\delta y_0 = \tfrac{1}{2}(\delta y_{-\frac{1}{2}}+\delta y_{\frac{1}{2}}),\quad \mu\delta^2 y_{\frac{1}{2}} = \tfrac{1}{2}(\delta^2 y_0+\delta^2 y_1). \tag{10}$$

From the definitions we can produce many relations between operators, noting in particular the following:

$$\left.\begin{aligned} E &= e^{hD} \\ \Delta &= \nabla E = \delta E^{\frac{1}{2}} = E-1 = e^{hD}-1 \\ \nabla &= 1-E^{-1} = 1-e^{-hD} \\ \delta &= E^{\frac{1}{2}}-E^{-\frac{1}{2}} = e^{\frac{1}{2}hD}-e^{-\frac{1}{2}hD} = 2\sinh(\tfrac{1}{2}hD) \\ \mu &= \tfrac{1}{2}(E^{-\frac{1}{2}}+E^{\frac{1}{2}}) = \tfrac{1}{2}(e^{\frac{1}{2}hD}+e^{-\frac{1}{2}hD}) = \cosh(\tfrac{1}{2}hD) \\ E^{\frac{1}{2}} &= \mu+\tfrac{1}{2}\delta \\ \mu^2 &= 1+\tfrac{1}{4}\delta^2 \end{aligned}\right\}. \tag{11}$$

Using these operators, and the relations between them, we proceed to establish finite-difference formulae for interpolation, differentiation and integration, with forward, backward, or central differences. With central-difference formulae we must be careful to use only those central differences actually present in the difference table, and this can be effected by judicious use of the operator μ, which must, further, appear only in its first power.

Interpolation

†**4.** Consider, for example, the fundamental problem of interpolation, the determination of the quantity $y(x+ph)$ for a non-integer value of p. We have

$$\begin{aligned} y(x+ph) &= E^p y(x) = (1+\Delta)^p y(x) \\ &= y(x)+p\Delta y(x)+\frac{p(p-1)}{2!}\Delta^2 y(x)+..., \end{aligned} \tag{12}$$

which is the Gregory–Newton formula for interpolation with forward differences. The corresponding formula for use with backward differences is given by

$$\begin{aligned} y(x+ph) &= E^p y(x) = (1-\nabla)^{-p} y(x) \\ &= y(x)+p\nabla y(x)+\frac{p(p+1)}{2!}\nabla^2 y(x)+\dots. \end{aligned} \tag{13}$$

The Bessel formula uses the central differences $\delta y_{\frac{1}{2}}$, $\delta^2 y_0+\delta^2 y_1$, $\delta^3 y_{\frac{1}{2}},\dots$. We can produce this by suitable operations, remembering to replace μ^2, wherever it occurs, by its equivalent $1+\frac{1}{4}\delta^2$, and noting that

$$\delta y_{\frac{1}{2}} \equiv \delta E^{\frac{1}{2}} y_0, \qquad \delta^2 y_0+\delta^2 y_1 = 2\mu\delta^2 E^{\frac{1}{2}} y_0, \quad \text{etc.}$$

Then

$$\begin{aligned} E^p &= (1+\delta E^{\frac{1}{2}})^p \\ &= 1+\left\{p\delta+\frac{p(p-1)}{2!}\delta^2 E^{\frac{1}{2}}+\frac{p(p-1)(p-2)}{3!}\delta^3 E+\dots\right\}E^{\frac{1}{2}}. \end{aligned} \tag{14}$$

If in the braces on the right of (14) we substitute for $E^{\frac{1}{2}r}$ its equivalent $(\mu+\frac{1}{2}\delta)^r$, expand the latter and replace μ^{2r} by $(1+\frac{1}{4}\delta^2)^r$, we find

$$\begin{aligned} E^p = 1+\Big\{p\delta+\binom{p}{2}\delta^2(\mu+\tfrac{1}{2}\delta)+\binom{p}{3}\delta^3(1+\mu\delta+\tfrac{1}{2}\delta^2)+ \\ +\binom{p}{4}\delta^4(\mu+\tfrac{3}{2}\delta+\mu\delta^2+\tfrac{1}{2}\delta^3)+\dots\Big\}E^{\frac{1}{2}}. \end{aligned}$$

Collecting terms and interpreting the operations in terms of differences we have finally Bessel's formula

$$y(x+ph)$$
$$= y_0+p\delta y_{\frac{1}{2}}+\frac{p(p-1)}{2!}\tfrac{1}{2}(\delta^2 y_0+\delta^2 y_1)+\frac{p(p-1)(p-\frac{1}{2})}{3!}\delta^3 y_{\frac{1}{2}}+\dots. \quad (15)$$

Differentiation

5. Formulae for derivatives follow in like manner. We find a forward-difference formula for the first derivative by writing

$$hD = \log E = \log(1+\Delta) = \Delta-\tfrac{1}{2}\Delta^2+\tfrac{1}{3}\Delta^3-\dots,$$

so that

$$hy_0' = \Delta y_0-\tfrac{1}{2}\Delta^2 y_0+\tfrac{1}{3}\Delta^3 y_0-\dots. \quad (16)$$

The corresponding backward-difference formula is obtained from the equation

$$hD = \log E = -\log(1-\nabla) = \nabla+\tfrac{1}{2}\nabla^2+\tfrac{1}{3}\nabla^3+\dots,$$

so that

$$hy_0' = \nabla y_0+\tfrac{1}{2}\nabla^2 y_0+\tfrac{1}{3}\nabla^3 y_0+\dots. \quad (17)$$

For second derivatives we have

$$h^2D^2 = \{\log(1+\Delta)\}^2 = \Delta^2-\Delta^3+\tfrac{11}{12}\Delta^4-\tfrac{5}{6}\Delta^5+\tfrac{137}{180}\Delta^6-\dots,$$

so that

$$h^2y_0'' = \Delta^2 y_0-\Delta^3 y_0+\tfrac{11}{12}\Delta^4 y_0-\tfrac{5}{6}\Delta^5 y_0+\dots, \quad (18)$$

a corresponding backward-difference formula being obtained in like manner.

6. The central-difference formulae for derivatives are of prime importance in the solution of differential equations. We shall be concerned mainly with derivatives at a pivotal point or at a half-way point: in the former case the central differences involved will be $\mu\delta y_0$, $\delta^2 y_0$, $\mu\delta^3 y_0,\dots$, and in the latter $\delta y_{\frac{1}{2}}$, $\mu\delta^2 y_{\frac{1}{2}}$, $\delta^3 y_{\frac{1}{2}},\dots$, and in the production of the formulae we must arrange our operations accordingly.

For the first derivative at a pivotal point we write

$$hD = 2\sinh^{-1}\tfrac{1}{2}\delta = 2\left\{\left(\frac{\delta}{2}\right)-\frac{1}{2}\left(\frac{\delta}{2}\right)^3\frac{1}{3}+\frac{1.3}{2.4}\left(\frac{\delta}{2}\right)^5\frac{1}{5}-\dots\right\}. \quad (19)$$

Here only odd central differences are involved and they must therefore be associated with the operator μ. This we achieve by multiplying on the right by the identity operator $\mu(1+\frac{1}{4}\delta^2)^{-\frac{1}{2}}$, producing finally the result

$$hy_0' = \mu\delta y_0-\tfrac{1}{6}\mu\delta^3 y_0+\tfrac{1}{30}\mu\delta^5 y_0-\dots. \quad (20)$$

For the first derivative at a mid-point we have, directly from (19), the result

$$hy_{\frac{1}{2}}' = \delta y_{\frac{1}{2}}-\tfrac{1}{24}\delta^3 y_{\frac{1}{2}}+\tfrac{3}{640}\delta^5 y_{\frac{1}{2}}-\dots. \quad (21)$$

For the second derivative at a pivotal point we have

$$h^2D^2 = (2\sinh^{-1}\tfrac{1}{2}\delta)^2 = \delta^2-\tfrac{1}{12}\delta^4+\tfrac{1}{90}\delta^6-...,$$

giving directly the important result

$$h^2y_0'' = \delta^2y_0-\tfrac{1}{12}\delta^4y_0+\tfrac{1}{90}\delta^6y_0-... . \tag{22}$$

For the second derivative at a mid-point we must include the operator μ and write

$$h^2D^2 = \mu(1+\tfrac{1}{4}\delta^2)^{-\frac{1}{2}}(\delta^2-\tfrac{1}{12}\delta^4+\tfrac{1}{90}\delta^6-...),$$

giving

$$h^2y_{\frac{1}{2}}'' = \mu\delta^2y_{\frac{1}{2}}-\tfrac{5}{24}\mu\delta^4y_{\frac{1}{2}}+\tfrac{259}{5760}\mu\delta^6y_{\frac{1}{2}}-... . \tag{23}$$

7. It is interesting also to consider formulae for the derivatives at a point in terms of forward, backward, or central differences of some other point.

Formulae using sloping differences are obtained by use of the identity operators $(1+\Delta)E^{-1}$, $(1-\nabla)E$. Thus to find the formula expressing the derivative of y_0 in terms of the forward differences of y_{-1} we write

$$hD = (\Delta-\tfrac{1}{2}\Delta^2+\tfrac{1}{3}\Delta^3-...)(1+\Delta)E^{-1},$$

giving

$$hy_0' = \Delta y_{-1}+\frac{1}{1.2}\Delta^2y_{-1}-\frac{1}{2.3}\Delta^3y_{-1}+\frac{1}{3.4}\Delta^4y_{-1}-... . \tag{24}$$

The corresponding formula for backward differences is given by

$$hD = (\nabla+\tfrac{1}{2}\nabla^2+\tfrac{1}{3}\nabla^3+...)(1-\nabla)E,$$

so that

$$hy_0' = \nabla y_1-\frac{1}{1.2}\nabla^2y_1-\frac{1}{2.3}\nabla^3y_1-\frac{1}{3.4}\nabla^4y_1-... . \tag{25}$$

At a stage farther we can find

$$hD = (\Delta-\tfrac{1}{2}\Delta^2+\tfrac{1}{3}\Delta^3-...)(1+2\Delta+\Delta^2)E^{-2},$$

or

$$hy_0' = \Delta y_{-2}+\tfrac{3}{2}\Delta^2y_{-2}+\tfrac{1}{3}\Delta^3y_{-2}-\tfrac{1}{12}\Delta^4y_{-2}+\tfrac{1}{30}\Delta^5y_{-2}-... . \tag{26}$$

Similarly, for second derivatives, we use (18) and write

$$h^2D^2 = (\Delta^2-\Delta^3+\tfrac{11}{12}\Delta^4-\tfrac{5}{6}\Delta^5+...)(1+\Delta)E^{-1},$$

giving

$$h^2y_0'' = \Delta^2y_{-1}-\tfrac{1}{12}\Delta^4y_{-1}+\tfrac{1}{12}\Delta^5y_{-1}-..., \tag{27}$$

and

$$h^2D^2 = (\Delta^2-\Delta^3+\tfrac{11}{12}\Delta^4-\tfrac{5}{6}\Delta^5+...)(1+2\Delta+\Delta^2)E^{-2},$$

giving

$$h^2y_0'' = \Delta^2y_{-2}+\Delta^3y_{-2}-\tfrac{1}{12}\Delta^4y_{-2}+\tfrac{1}{90}\Delta^6y_{-2}-... . \tag{28}$$

Formulae of this kind using central differences are of little practical interest, and their production is as usual complicated by the necessity for the proper use of the operator μ. We illustrate the technique by finding the formula for the first derivative of y_0 in terms of the central differences of y_{-1}. Using (19), we write

$$hD = (\delta-\tfrac{1}{24}\delta^3+\tfrac{3}{640}\delta^5-...)$$
$$= (\delta-\tfrac{1}{24}\delta^3+\tfrac{3}{640}\delta^5-...)(\mu+\tfrac{1}{2}\delta)^2E^{-1}.$$

Now $(\mu+\frac{1}{2}\delta)^2 = 1+\frac{1}{2}\delta^2+\mu\delta$, but in the required formula μ must be associated with odd differences only, so that we must write

$$(\mu+\tfrac{1}{2}\delta)^2 = \mu(1+\tfrac{1}{2}\delta^2)(1+\tfrac{1}{4}\delta^2)^{-\frac{1}{2}}+\delta(1+\tfrac{1}{4}\delta^2)^{\frac{1}{2}}.$$

Then

$$hD = (\delta-\tfrac{1}{24}\delta^3+\tfrac{3}{640}\delta^5-...)\{\mu(1+\tfrac{3}{8}\delta^2-\tfrac{5}{128}\delta^4+...)+$$
$$+\delta(1+\tfrac{1}{8}\delta^2-\tfrac{1}{128}\delta^4+...)\}E^{-1},$$

leading to the final result

$$hy_0' = (\mu\delta+\tfrac{1}{3}\mu\delta^3-\tfrac{1}{20}\mu\delta^5+...)y_{-1}+(\delta^2+\tfrac{1}{12}\delta^4-\tfrac{1}{120}\delta^6+...)y_{-1}. \tag{29}$$

8. Another class of formulae connected with differentiation is that in which differences are expressed in terms of derivatives. We have, for example,

$$\delta^2 = 4\sinh^2(\tfrac{1}{2}hD) = 2(\cosh hD-1),$$

so that

$$\delta^2 y_0 = h^2y_0''+\frac{h^4}{12}y_0^{\text{iv}}+\frac{h^6}{360}y_0^{\text{vi}}+... . \tag{30}$$

Integration

9. We now turn to formulae for numerical integration. First, we can produce some useful formulae by inverting the results found for derivatives. Thus, from equation (16), we find

$$\Delta = (1-\tfrac{1}{2}\Delta+\tfrac{1}{3}\Delta^2-...)^{-1}hD,$$

so that

$$y_1-y_0 = (1+\tfrac{1}{2}\Delta-\tfrac{1}{12}\Delta^2+\tfrac{1}{24}\Delta^3-...)hy_0', \tag{31}$$

an expression for the definite integral of dy/dx taken between two adjacent pivotal points.

The corresponding formula for backward differences is obtained from (17) in the form

$$\nabla E = (1+\tfrac{1}{2}\nabla+\tfrac{1}{3}\nabla^2+...)^{-1}hDE,$$

so that

$$y_1-y_0 = (1-\tfrac{1}{2}\nabla-\tfrac{1}{12}\nabla^2-\tfrac{1}{24}\nabla^3-...)hy_1'. \tag{32}$$

We also have

$$\nabla E = (1+\tfrac{1}{2}\nabla+\tfrac{1}{3}\nabla^2+...)^{-1}(1-\nabla)^{-1}hD,$$

or

$$y_1-y_0 = (1+\tfrac{1}{2}\nabla+\tfrac{5}{12}\nabla^2+\tfrac{3}{8}\nabla^3+...)hy_0'. \tag{33}$$

Inversion of the central-difference formula (20) gives

$$\mu\delta = (1-\tfrac{1}{6}\delta^2+\tfrac{1}{30}\delta^4-\ldots)^{-1}hD,$$

so that

$$y_1-y_{-1} = 2(1+\tfrac{1}{6}\delta^2-\tfrac{1}{180}\delta^4+\ldots)hy_0', \tag{34}$$

an expression for the integral of dy/dx taken over two intervals.

Other useful formulae are obtained by inverting the formulae for a second derivative. From equation (22), for example, we have

$$\delta^2 = (1-\tfrac{1}{12}\delta^2+\tfrac{1}{90}\delta^4-\ldots)^{-1}h^2D^2,$$

giving

$$\delta^2 y_0 = (1+\tfrac{1}{12}\delta^2-\tfrac{1}{240}\delta^4+\ldots)h^2y_0'', \tag{35}$$

which can be regarded either as a formula for the second difference of a function in terms of its second derivative and its differences, or as a formula for the second difference of the double integral of y_0''.

Sums and sum operators

10. The results of the two previous sections could be extended to produce formulae for integration covering any number of intervals, but we shall use for this purpose a slightly different method, using the notion of a 'sum' analogous to a 'difference'.

In the calculus we frequently connect an indefinite integral with the operator D^{-1}. We retain this idea in the theory of finite differences and extend it to give meaning to inverse powers of Δ, ∇, and δ. From (6) we deduce the following relations:

$$\Delta^{-1}E-\Delta^{-1} = 1, \qquad \nabla^{-1}E-\nabla^{-1} = E, \qquad \delta^{-1}E^{\frac{1}{2}}-\delta^{-1}E^{-\frac{1}{2}} = 1, \tag{36}$$

which we interpret by equations similar to those for differences, given by

$$\Delta^{-1}y_1-\Delta^{-1}y_0 = y_0, \quad \nabla^{-1}y_1-\nabla^{-1}y_0 = y_1, \quad \delta^{-1}y_{\frac{1}{2}}-\delta^{-1}y_{-\frac{1}{2}} = y_0. \tag{37}$$

The difference tables 1, 2, and 3 can thus be extended backwards by the addition of columns of 'sums', as shown in Tables 4, 5, and 6. In the case of central sums the operator μ is used as for differences. For example, we write

$$\mu\delta^{-1}y_0 = \tfrac{1}{2}(\delta^{-1}y_{-\frac{1}{2}}+\delta^{-1}y_{\frac{1}{2}}). \tag{38}$$

TABLE 4 | TABLE 5 | TABLE 6

0					y_0					y_0					y_0
				Δ_1^{-1}					∇_0^{-1}					$\delta_{\frac{1}{2}}^{-1}$	
1			Δ_2^{-2}		y_1			∇_0^{-2}		y_1			δ_1^{-2}		y_1
		Δ_3^{-3}		Δ_2^{-1}			∇_0^{-3}		∇_1^{-1}			$\delta_{\frac{3}{2}}^{-3}$		$\delta_{\frac{3}{2}}^{-1}$	
2	etc.		Δ_3^{-2}		y_2	etc.		∇_1^{-2}		y_2	etc.		δ_2^{-2}		y_2
		Δ_4^{-3}		Δ_3^{-1}			∇_1^{-3}		∇_2^{-1}			$\delta_{\frac{5}{2}}^{-3}$		$\delta_{\frac{5}{2}}^{-1}$	
3			Δ_4^{-2}		y_3			∇_2^{-2}		y_3			δ_3^{-2}		y_3
				Δ_4^{-1}					∇_3^{-1}					$\delta_{\frac{7}{2}}^{-1}$	
4					y_4					y_4					y_4

It will be apparent that, unlike the differences, the first column of sums cannot be produced uniquely from the given pivotal values y_r: there is always a single arbitrary element, corresponding directly with the arbitrary constant attached to an indefinite integral. The same feature is present with sums of higher order: thus the second sums can be recorded uniquely only if one first sum and one second sum have prescribed values or if some equivalent extra conditions are imposed.

11. We now use this new concept to produce formulae of integration or numerical quadrature. First, for the formula using the forward operator Δ, we have

$$(hD)^{-1} = \{\log(1+\Delta)\}^{-1},$$

or

$$D^{-1}y_0 = h(\Delta^{-1}+\tfrac{1}{2}-\tfrac{1}{12}\Delta+\tfrac{1}{24}\Delta^2-\tfrac{19}{720}\Delta^3+...)y_0, \tag{39}$$

an expression for the indefinite integral of y at x_0. If we assign an arbitrary value, usually zero, to this quantity we can regard (39) as an equation defining the 'arbitrary constant' $\Delta^{-1}y_0$.

We can also write (39) in the form

$$\Delta D^{-1}y_0 = h(1+\tfrac{1}{2}\Delta-\tfrac{1}{12}\Delta^2+...)y_0,$$

which is effectively equation (31). Giving the difference between the indefinite integral at x_0 and x_1, it defines a definite integral in the form

$$\frac{1}{h}\int_{x_0}^{x_1} y\,dx = \tfrac{1}{2}(y_0+y_1)-\tfrac{1}{12}\Delta^2y_0+\tfrac{1}{24}\Delta^3y_0-\tfrac{19}{720}\Delta^4y_0+\dots. \tag{40}$$

For the definite integral between the pivotal points x_0 and x_n we have, from (39), the expression

$$D^{-1}y_n-D^{-1}y_0 = (E^n-1)D^{-1}y_0 = h(\Delta^{-1}+\tfrac{1}{2}-\tfrac{1}{12}\Delta+\tfrac{1}{24}\Delta^2-...)(y_n-y_0).$$

Now

$$(\Delta^{-1}+\tfrac{1}{2})(E^n-1) = \frac{E+1}{2(E-1)}(E^n-1) = \tfrac{1}{2}(E^n+1)+(E^{n-1}+...+E),$$

so that, finally,

$$\frac{1}{h}\int_{x_0}^{x_n} y\,dx = \tfrac{1}{2}y_0+y_1+...+y_{n-1}+\tfrac{1}{2}y_n+$$
$$+(-\tfrac{1}{12}\Delta+\tfrac{1}{24}\Delta^2-\tfrac{19}{720}\Delta^3+...)(y_n-y_0). \tag{41}$$

Equation (41) is not often convenient for integration, since in a normal difference table we have forward differences of y_0 but backward differences of y_n. We can produce the backward difference analogue of (39), and the equivalent of (32), by writing

$$(hD)^{-1} = -\{\log(1-\nabla)\}^{-1},$$

so that

$$\frac{1}{h}\int_{x_0}^{x_n} y\,dx = (hD)^{-1}(y_n-y_0) = (\nabla^{-1}-\tfrac{1}{2}-\tfrac{1}{12}\nabla-\tfrac{1}{24}\nabla^2-\ldots)y_n-$$
$$-(\Delta^{-1}+\tfrac{1}{2}-\tfrac{1}{12}\Delta+\tfrac{1}{24}\Delta^2-\ldots)y_0.$$

Now

$$(\nabla^{-1}-\tfrac{1}{2})y_n-(\Delta^{-1}+\tfrac{1}{2})y_0 = (E\Delta^{-1}-\tfrac{1}{2})y_n-(\Delta^{-1}+\tfrac{1}{2})y_0 = (\Delta^{-1}+\tfrac{1}{2})(y_n-y_0),$$

so that we have finally the Gregory quadrature formula, given by

$$\frac{1}{h}\int_{x_0}^{x_n} y\,dx = \tfrac{1}{2}y_0+y_1+\ldots+y_{n-1}+\tfrac{1}{2}y_n+(-\tfrac{1}{12}\nabla-\tfrac{1}{24}\nabla^2-\ldots)y_n+$$
$$+(\tfrac{1}{12}\Delta-\tfrac{1}{24}\Delta^2+\ldots)y_0. \quad (42)$$

†12. Turning now to central difference formulae of integration, we start with

$$(hD)^{-1} = \tfrac{1}{2}(\sinh^{-1}\tfrac{1}{2}\delta)^{-1} = \delta^{-1}+\tfrac{1}{24}\delta-\tfrac{17}{5760}\delta^3+\ldots. \quad (43)$$

Applying these operators at a half-way point, where odd differences exist, we have

$$D^{-1}y_{\frac{1}{2}} = h(\delta^{-1}+\tfrac{1}{24}\delta-\tfrac{17}{5760}\delta^3+\ldots)y_{\frac{1}{2}}, \quad (44)$$

a formula for the indefinite integral at a half-way point, from which, if this value is specified, we can compute the quantity $\delta^{-1}y_{\frac{1}{2}}$ and remove the uncertainty from the column of first sums.

For the definite integral between successive mid-points we can write

$$D^{-1}y_{\frac{1}{2}}-D^{-1}y_{-\frac{1}{2}} = h(\delta^{-1}+\tfrac{1}{24}\delta-\tfrac{17}{5760}\delta^3+\ldots)(y_{\frac{1}{2}}-y_{-\frac{1}{2}}),$$

or, remembering (8),

$$\frac{1}{h}\int_{x_{-\frac{1}{2}}}^{x_{\frac{1}{2}}} y\,dx = (1+\tfrac{1}{24}\delta^2-\tfrac{17}{5760}\delta^4+\ldots)y_0. \quad (45)$$

Similarly, for integer n, we have

$$(hD)^{-1}E^{n+\frac{1}{2}}-(hD)^{-1}E^{-\frac{1}{2}} = (\delta^{-1}+\tfrac{1}{24}\delta-\tfrac{17}{5760}\delta^3+\ldots)(E^{n+\frac{1}{2}}-E^{-\frac{1}{2}})$$
$$= \frac{E^{n+1}-1}{E-1}+(\tfrac{1}{24}\delta-\tfrac{17}{5760}\delta^3+\ldots)(E^{n+\frac{1}{2}}-E^{-\frac{1}{2}}),$$

giving

$$\frac{1}{h}\int_{x_{-\frac{1}{2}}}^{x_{n+\frac{1}{2}}} y\,dx = y_0+\ldots+y_n+(\tfrac{1}{24}\delta-\tfrac{17}{5760}\delta^3+\ldots)(y_{n+\frac{1}{2}}-y_{-\frac{1}{2}}). \quad (46)$$

To produce the indefinite integral at a pivotal point we must introduce the operator μ into equation (43), writing it in the form

$$(hD)^{-1} = \mu(1+\tfrac{1}{4}\delta^2)^{-\frac{1}{2}}(\delta^{-1}+\tfrac{1}{24}\delta-\tfrac{17}{5760}\delta^3+...),$$

so that

$$D^{-1}y_0 = h\mu(\delta^{-1}-\tfrac{1}{12}\delta+\tfrac{11}{720}\delta^3-...)y_0. \tag{47}$$

If this value is specified it enables us to find the quantity

$$\mu\delta_0^{-1} = \tfrac{1}{2}(\delta_{\frac{1}{2}}^{-1}+\delta_{-\frac{1}{2}}^{-1}).$$

Since we also know $\delta_{\frac{1}{2}}^{-1}-\delta_{-\frac{1}{2}}^{-1} = y_0$, both $\delta_{\frac{1}{2}}^{-1}$ and $\delta_{-\frac{1}{2}}^{-1}$ are then calculable.

For the definite integral between two pivotal points we have, from (47),

$$\begin{aligned}(E-1)(hD)^{-1} &= \mu(\delta^{-1}-\tfrac{1}{12}\delta+\tfrac{11}{720}\delta^3-...)(E-1)\\ &= \mu(1-\tfrac{1}{12}\delta^2+\tfrac{11}{720}\delta^4-...)E^{\frac{1}{2}},\end{aligned}$$

giving

$$\frac{1}{h}\int_{x_0}^{x_1} y\,dx = \tfrac{1}{2}(y_0+y_1)+(-\tfrac{1}{12}\mu\delta^2+\tfrac{11}{720}\mu\delta^4-...)y_{\frac{1}{2}}, \tag{48}$$

and for the integral over a number of intervals we have

$$\begin{aligned}\frac{1}{h}\int_{x_0}^{x_n} y\,dx &= \mu(\delta^{-1}-\tfrac{1}{12}\delta+\tfrac{11}{720}\delta^3-...)(E^n-1)y_0\\ &= \mu E^{\frac{1}{2}}(E^{n-1}+...+1)y_0+(-\tfrac{1}{12}\mu\delta+\tfrac{11}{720}\mu\delta^3-...)(y_n-y_0),\end{aligned}$$

so that

$$\begin{aligned}\frac{1}{h}\int_{x_0}^{x_n} y\,dx = \tfrac{1}{2}y_0+y_1+...+y_{n-1}+\tfrac{1}{2}y_n+&\\ +(-\tfrac{1}{12}\mu\delta+\tfrac{11}{720}\mu\delta^3-...)(y_n-y_0).&\end{aligned} \tag{49}$$

Finally, for the definite integral between a pivotal point and a half-way point, we write

$$\begin{aligned}\frac{1}{h}\int_{x_0}^{x_{n+\frac{1}{2}}} y\,dx &= (hD)^{-1}(E^{n+\frac{1}{2}}-1)y_0\\ &= \{(\delta^{-1}+\tfrac{1}{24}\delta-\tfrac{17}{5760}\delta^3+...)E^{n+\frac{1}{2}}-\mu(\delta^{-1}-\tfrac{1}{12}\delta+\tfrac{11}{720}\delta^3-...)\}y_0,\end{aligned}$$

using (43) and (47). Now

$$\delta^{-1}E^{n+\frac{1}{2}}-\mu\delta^{-1} = \frac{E^{n+1}-\frac{1}{2}(E+1)}{E-1} = E^n+E^{n-1}+...+E+\tfrac{1}{2}.$$

The required formula can then be written

$$\begin{aligned}\frac{1}{h}\int_{x_0}^{x_{n+\frac{1}{2}}} y\,dx = \tfrac{1}{2}y_0+y_1+...+y_n+(\tfrac{1}{24}\delta-\tfrac{17}{5760}\delta^3+...)y_{n+\frac{1}{2}}+&\\ +(\tfrac{1}{12}\mu\delta-\tfrac{11}{720}\mu\delta^3+...)y_0.&\end{aligned} \tag{50}$$

13. Formulae for repeated integrals can be obtained in a similar manner, or we can use the method of § 9, writing equation (35), for example, in the form

$$(hD)^{-2}y_0 = (\delta^{-2}+\tfrac{1}{12}-\tfrac{1}{240}\delta^2+...)y_0, \tag{51}$$

an expression for the indefinite double integral at the pivotal point x_0.

Relations between formulae

†14. We note in passing the existence of formulae which are effectively equivalent but slightly different in form.

For example, if we differentiate (48) we have

$$y_1-y_0 = \tfrac{1}{2}h(y_0'+y_1')+(-\tfrac{1}{12}\mu\delta^2+\tfrac{11}{720}\mu\delta^4-...)hy_{\frac{1}{2}}'. \tag{52}$$

On the other hand we have from (21) the equation

$$\tfrac{1}{2}h(y_0'+y_1') = h\mu y_{\frac{1}{2}}' = (1+\tfrac{1}{4}\delta^2)^{\frac{1}{2}}(\delta-\tfrac{1}{24}\delta^3+\tfrac{3}{640}\delta^5-...)y_{\frac{1}{2}},$$

so that
$$y_1-y_0 = \tfrac{1}{2}h(y_0'+y_1')+(-\tfrac{1}{12}\delta^3+\tfrac{1}{120}\delta^5-...)y_{\frac{1}{2}}. \tag{53}$$

Equations (52) and (53) are equivalent, but the former expresses a definite integral in terms of the integrand and its differences, while the latter expresses a definite integral in terms of the integrand and the differences of the integral.

From equations (35) and (22) we have a similar identity, given by

$$\left.\begin{aligned} y_1+y_{-1}-2y_0 &= h^2y_0''+(\tfrac{1}{12}\delta^2-\tfrac{1}{240}\delta^4+...)h^2y_0'' \\ y_1+y_{-1}-2y_0 &= h^2y_0''+(\tfrac{1}{12}\delta^4-\tfrac{1}{90}\delta^6+...)y_0 \end{aligned}\right\}, \tag{54}$$

of which the first member is properly regarded as an integration formula, the second as a formula of differentiation.

The difference correction

15. It is often convenient to regard finite-difference expressions for derivatives and integrals as consisting of two parts, a dominant term involving pivotal values of the function and a smaller term involving differences. In the case of equation (53), for example, we might write

$$\tfrac{1}{2}h(y_0'+y_1') = y_1-y_0+(\tfrac{1}{12}\delta^3-\tfrac{1}{120}\delta^5+...)y_{\frac{1}{2}}, \tag{55}$$

and regard the last term involving differences as a correction, which we shall call the *difference correction*, to the approximate formula resulting from the omission of this term.

The difference-correction operator will be denoted by the symbol C, so that equation (55) can be written in the form

$$\left.\begin{aligned} \tfrac{1}{2}h(y_0'+y_1') &= y_1-y_0+Cy_{\frac{1}{2}} \\ C &= \tfrac{1}{12}\delta^3-\tfrac{1}{120}\delta^5+... \end{aligned}\right\}. \tag{56}$$

As other examples we have, for a first derivative,

$$\left.\begin{aligned} hy_0' &= \tfrac{1}{2}(y_1-y_{-1})+Cy_0 \\ C &= -\tfrac{1}{6}\mu\delta^3+\tfrac{1}{30}\mu\delta^5-\dots \end{aligned}\right\}; \tag{57}$$

for a second derivative,

$$\left.\begin{aligned} h^2y_0'' &= y_1+y_{-1}-2y_0+Cy_0 \\ C &= -\tfrac{1}{12}\delta^4+\tfrac{1}{90}\delta^6-\dots \end{aligned}\right\}; \tag{58}$$

and for an integral,

$$\left.\begin{aligned} \frac{1}{h}\int_{x_0}^{x_n} y\,dx &= \tfrac{1}{2}y_0+y_1+\dots+y_{n-1}+\tfrac{1}{2}y_n+C(y_n-y_0) \\ C &= -\tfrac{1}{12}\mu\delta+\tfrac{11}{720}\mu\delta^3-\dots \end{aligned}\right\}. \tag{59}$$

Tables of formulae

In Table 7 we give some formulae for derivatives and integrals, many of which we shall use in following chapters and in the second volume. Additional formulae will be derived later as required.

TABLE 7

(1) *First derivatives*

$$\begin{aligned}
hy_0' &= (\Delta-\tfrac{1}{2}\Delta^2+\tfrac{1}{3}\Delta^3-\tfrac{1}{4}\Delta^4+\tfrac{1}{5}\Delta^5-\tfrac{1}{6}\Delta^6+\tfrac{1}{7}\Delta^7-\tfrac{1}{8}\Delta^8+\tfrac{1}{9}\Delta^9-\tfrac{1}{10}\Delta^{10}+\dots)y_0 \\
hy_0' &= (\Delta+\tfrac{1}{2}\Delta^2-\tfrac{1}{6}\Delta^3+\tfrac{1}{12}\Delta^4-\tfrac{1}{20}\Delta^5+\tfrac{1}{30}\Delta^6-\tfrac{1}{42}\Delta^7+\tfrac{1}{56}\Delta^8-\tfrac{1}{72}\Delta^9+\tfrac{1}{90}\Delta^{10}-\dots)y_{-1} \\
hy_0' &= (\Delta+\tfrac{3}{2}\Delta^2+\tfrac{1}{3}\Delta^3-\tfrac{1}{12}\Delta^4+\tfrac{1}{30}\Delta^5-\tfrac{1}{60}\Delta^6+\tfrac{1}{105}\Delta^7-\tfrac{1}{168}\Delta^8+\tfrac{1}{252}\Delta^9-\tfrac{1}{360}\Delta^{10}+\dots)y_{-2} \\
hy_0' &= (\Delta+\tfrac{5}{2}\Delta^2+\tfrac{11}{6}\Delta^3+\tfrac{1}{4}\Delta^4-\tfrac{1}{20}\Delta^5+\tfrac{1}{60}\Delta^6-\tfrac{1}{140}\Delta^7+\tfrac{1}{280}\Delta^8-\tfrac{1}{504}\Delta^9+\tfrac{1}{840}\Delta^{10}-\dots)y_{-3} \\
hy_0' &= (\Delta+\tfrac{7}{2}\Delta^2+\tfrac{13}{3}\Delta^3+\tfrac{25}{12}\Delta^4+\tfrac{1}{5}\Delta^5-\tfrac{1}{30}\Delta^6+\tfrac{1}{105}\Delta^7-\tfrac{1}{280}\Delta^8+\tfrac{1}{630}\Delta^9-\tfrac{1}{1260}\Delta^{10}+\dots)y_{-4}
\end{aligned}$$

(For the corresponding backward-difference formulae we merely alter the sign of *even* differences and of the suffix, and replace Δ by ∇.)

$$\begin{aligned}
hy_0' &= (\mu\delta-\tfrac{1}{6}\mu\delta^3+\tfrac{1}{30}\mu\delta^5-\tfrac{1}{140}\mu\delta^7+\tfrac{1}{630}\mu\delta^9-\dots)y_0 \\
hy_{\frac{1}{2}}' &= (\delta-\tfrac{1}{24}\delta^3+\tfrac{3}{640}\delta^5-\tfrac{5}{7168}\delta^7+\tfrac{35}{294912}\delta^9-\dots)y_{\frac{1}{2}}
\end{aligned}$$

(2) *Second derivatives*

$$\begin{aligned}
h^2y_0'' &= (\Delta^2-\Delta^3+\tfrac{11}{12}\Delta^4-\tfrac{5}{6}\Delta^5+\tfrac{137}{180}\Delta^6-\tfrac{7}{10}\Delta^7+\tfrac{363}{560}\Delta^8-\tfrac{761}{1260}\Delta^9+\tfrac{7129}{12600}\Delta^{10}-\dots)y_0 \\
h^2y_0'' &= (\Delta^2-\tfrac{1}{12}\Delta^4+\tfrac{1}{12}\Delta^5-\tfrac{13}{180}\Delta^6+\tfrac{11}{180}\Delta^7-\tfrac{29}{560}\Delta^8+\tfrac{223}{5040}\Delta^9-\tfrac{481}{12600}\Delta^{10}+\dots)y_{-1} \\
h^2y_0'' &= (\Delta^2+\Delta^3-\tfrac{1}{12}\Delta^4+\tfrac{1}{90}\Delta^6-\tfrac{1}{90}\Delta^7+\tfrac{47}{5040}\Delta^8-\tfrac{19}{2520}\Delta^9+\tfrac{17}{2800}\Delta^{10}-\dots)y_{-2} \\
h^2y_0'' &= (\Delta^2+2\Delta^3+\tfrac{11}{12}\Delta^4-\tfrac{1}{12}\Delta^5+\tfrac{1}{90}\Delta^6-\tfrac{1}{560}\Delta^8+\tfrac{1}{560}\Delta^9-\tfrac{37}{25200}\Delta^{10}+\dots)y_{-3} \\
h^2y_0'' &= (\Delta^2+3\Delta^3+\tfrac{35}{12}\Delta^4+\tfrac{5}{6}\Delta^5-\tfrac{13}{180}\Delta^6+\tfrac{1}{90}\Delta^7-\tfrac{1}{560}\Delta^8+\tfrac{1}{3150}\Delta^{10}-\dots)y_{-4}
\end{aligned}$$

(For the corresponding backward-difference formulae we merely alter the sign of *odd* differences and of the suffix, and replace Δ by ∇.)

$$h^2y_0'' = (\delta^2-\tfrac{1}{12}\delta^4+\tfrac{1}{90}\delta^6-\tfrac{1}{560}\delta^8+\tfrac{1}{3150}\delta^{10}-\dots)y_0$$

(3) *Higher derivatives*

$$h^3y_0''' = (\mu\delta^3-\tfrac{1}{4}\mu\delta^5+\tfrac{7}{120}\mu\delta^7-\tfrac{41}{3024}\mu\delta^9+\ldots)y_0$$
$$h^3y_{\frac{1}{2}}''' = (\delta^3-\tfrac{1}{8}\delta^5+\tfrac{37}{1920}\delta^7-\tfrac{3229}{967680}\delta^9+\ldots)y_{\frac{1}{2}}$$
$$h^4y_0^{\mathrm{iv}} = (\delta^4-\tfrac{1}{6}\delta^6+\tfrac{7}{240}\delta^8-\tfrac{41}{7560}\delta^{10}+\ldots)y_0$$

(4) *Reversion of derivative formulae*

$$y_1-y_0 = (1+\tfrac{1}{2}\Delta-\tfrac{1}{12}\Delta^2+\tfrac{1}{24}\Delta^3-\tfrac{19}{720}\Delta^4+\tfrac{3}{160}\Delta^5-\ldots)hy_0'$$
$$y_1-y_0 = (1-\tfrac{1}{2}\Delta+\tfrac{5}{12}\Delta^2-\tfrac{3}{8}\Delta^3+\tfrac{251}{720}\Delta^4-\tfrac{95}{288}\Delta^5+\ldots)hy_1'$$
$$y_1-y_0 = (1-\tfrac{1}{2}\nabla-\tfrac{1}{12}\nabla^2-\tfrac{1}{24}\nabla^3-\tfrac{19}{720}\nabla^4-\tfrac{3}{160}\nabla^5-\ldots)hy_1'$$
$$y_1-y_0 = (1+\tfrac{1}{2}\nabla+\tfrac{5}{12}\nabla^2+\tfrac{3}{8}\nabla^3+\tfrac{251}{720}\nabla^4+\tfrac{95}{288}\nabla^5+\ldots)hy_0'$$
$$y_1-y_{-1} = 2(1+\tfrac{1}{6}\delta^2-\tfrac{1}{180}\delta^4+\tfrac{1}{1512}\delta^6-\tfrac{23}{226800}\delta^8+\ldots)hy_0'$$
$$y_1+y_{-1}-2y_0 = (1+\tfrac{1}{1}\delta^2-{}_{2}\tfrac{1}{40}\delta^4+\tfrac{31}{60480}\delta^6-\tfrac{289}{3628800}\delta^8+\ldots)h^2y_0''$$

(5) *Indefinite integrals*

Sloping-difference formulae come immediately from the first four equations of group (4) above. For example

$$(1/h)D^{-1}y_0 = (\Delta^{-1}+\tfrac{1}{2}-\tfrac{1}{12}\Delta+\tfrac{1}{24}\Delta^2-\tfrac{19}{720}\Delta^3+\tfrac{3}{160}\Delta^4-\ldots)y_0$$

Central-difference formulae follow:

$$(1/h)D^{-1}y_{\frac{1}{2}} = (\delta^{-1}+\tfrac{1}{24}\delta-\tfrac{17}{5760}\delta^3+\tfrac{367}{967680}\delta^5-\tfrac{27859}{464486400}\delta^7+\ldots)y_{\frac{1}{2}}$$
$$(1/h)D^{-1}y_0 = (\mu\delta^{-1}-\tfrac{1}{12}\mu\delta+\tfrac{11}{720}\mu\delta^3-\tfrac{191}{60480}\mu\delta^5+\tfrac{2497}{3628800}\mu\delta^7-\ldots)y_0$$
$$(1/h^2)D^{-2}y_0 = (\delta^{-2}+\tfrac{1}{12}-\tfrac{1}{240}\delta^2+\tfrac{31}{60480}\delta^4-\tfrac{289}{3628800}\delta^6+\ldots)y_0$$

(6) *Definite integrals*

$$(1/h)\int_{x_0}^{x_n} y\,dx = \tfrac{1}{2}y_0+y_1+\ldots+y_{n-1}+\tfrac{1}{2}y_n+(-\tfrac{1}{12}\nabla-\tfrac{1}{24}\nabla^2-\ldots)y_n+ \\ +(\tfrac{1}{12}\Delta-\tfrac{1}{24}\Delta^2+\ldots)y_0$$

$$(1/h)\int_{x_0}^{x_n} y\,dx = \tfrac{1}{2}y_0+y_1+\ldots+y_{n-1}+\tfrac{1}{2}y_n+(-\tfrac{1}{12}\mu\delta+\tfrac{11}{720}\mu\delta^3-\ldots)(y_n-y_0)$$

$$(1/h)\int_{x_{-\frac{1}{2}}}^{x_{n+\frac{1}{2}}} y\,dx = y_0+\ldots+y_n+(\tfrac{1}{24}\delta-\tfrac{17}{5760}\delta^3+\ldots)(y_{n+\frac{1}{2}}-y_{-\frac{1}{2}})$$

$$(1/h)\int_{x_0}^{x_{n+\frac{1}{2}}} y\,dx = \tfrac{1}{2}y_0+y_1+\ldots+y_n+(\tfrac{1}{24}\delta-\tfrac{17}{5760}\delta^3+\ldots)y_{n+\frac{1}{2}}+ \\ +(\tfrac{1}{12}\mu\delta-\tfrac{11}{720}\mu\delta^3+\ldots)y_0$$

In all cases the coefficients of higher differences come from the corresponding equation of group (5).

The practical application of finite-difference equations

16. So far we have carried out certain operations, according to certain mathematical rules, and produced formulae of unspecified usefulness and accuracy. We must now consider how to use these formulae, in what circumstances they will give useful results, and the degree of accuracy they can attain. The general rigorous mathematical theory receives little attention; for the most part we shall give practical suggestions which will be satisfactory in nearly all circumstances.

†**17.** In Tables 8, 9, and 10 we give portions of the difference tables of a polynomial, the 'unit function', and a trigonometrical function respectively. In any column we follow the practice of putting signs on the first and last values and at a change of sign, unless the latter is frequent, when all signs are inserted. All differences are written without decimal points.

TABLE 8

x	x^4		δ^2		δ^4	
+0	+ 0		+ 2		+24	
		+ 1		+ 12		0
1	1		14		24	
		15		36		0
2	16		50		24	
		65		60		0
3	81		110		24	
		175		84		0
4	256		194		+24	
		369		+108		
5	625		+302			
		+671				
+6	+1296					

TABLE 9

y		δ^2		δ^4
0		0		0
	0		0	
0		0		+1
	0		+1	
0		+1		−4
	+1		−3	
+1		−2		+6
	−1		+3	
0		+1		−4
	0		−1	
0		0		+1
	0		0	
0		0		

TABLE 10

x	$\sin x$		δ^2		δ^4		δ^6		δ^8		δ^{10}
+0·0	+0·0000		0		0		0		0		0
		+1987		−80		+5		− 4		+ 7	
·2	·1987		− 80		+ 5		−4		+ 7		− 8
		1907		75		+1		+ 3		− 1	
·4	·3894		155		6		−1		+ 6		−30
		1752		69		0		+ 9		−31	
·6	·5646		224		6		+8		−25		+80
		1528		63		+8		−16		+49	
0·8	·7174		287		14		−8		+24		−82
		1241		49		0		+ 8		−33	
1·0	·8415		336		14		0		− 9		+53
		+ 905		−35		0		− 1		+20	
+1·2	+0·9320		−371		+14		−1		+11		−68

Table 8 confirms that the fifth differences of x^4 are zero. Table 9 shows the manner in which an error in any value of Table 8 would be propagated throughout the difference table, and shows that, when dealing with a polynomial, an examination of its differences would help to locate an isolated error. If the polynomial is of degree n, the differences of order $n+1$ should all be zero, but an error would cause an oscillation, of magnitude proportional to the binomial coefficients, and with the largest term (for even differences) on the same horizontal line as the offending pivotal value.

In the table of differences of $\sin x$ we note that the differences never disappear, but decrease steadily to orders five or six, then begin to increase and oscillate.

†18. Consider now the use of these differences for interpolation, differentiation or integration. For a polynomial there is no difficulty, since the formulae we have developed end at the term involving differences of order n, and the formulae are exact. For any other function each formula is an infinite series, and we have to decide whether the sum of this series has meaning, can be calculated and with what error.

We recall the theory of Taylor's series

$$f(x+t) = f(x)+tf'(x)+\frac{t^2}{2!}f''(x)+\dots. \tag{60}$$

For a polynomial the derivatives all vanish for orders greater than n, and the series terminates and gives an exact answer. For any other function the series is infinite, and the same questions arise. The series is usually summed by taking a finite number of terms and estimating the error so caused, expressed as a remainder term. Thus we have the well-known form

$$f(x+t) = f(x)+tf'(x)+\frac{t^2}{2!}f''(x)+\dots+\frac{t^n}{n!}f^{(n)}(x)+R_{n+1}, \tag{61}$$

where R_{n+1} is the remainder, given by

$$R_{n+1} = \frac{t^{n+1}}{(n+1)!}f^{(n+1)}(x+\theta t), \qquad 0 < \theta < 1. \tag{62}$$

If the series converges it can be summed to any accuracy by taking an n for which R_{n+1} is negligible. Sometimes, however, the derivatives increase at too fast a rate and the series diverges, a condition usually associated with a critical value of t (the radius of convergence). The series is also meaningless, of course, if any derivative of $f(x)$ is infinite.

In a similar way the rigorous theory of finite differences attaches a remainder term to every finite-difference formula and sometimes, when the interval between pivotal points is too large or the function or a derivative is infinite, the formula diverges and gives no useful result.

The remainder term is of the form (62), with t replaced by h. Thus, for example, it is shown in books on the theory of finite differences that the expression (20) for a first derivative is more properly written in the form

$$hy_0' = \mu\delta y_0 - \tfrac{1}{6}\mu\delta^3 y_0 + \tfrac{1}{30}\mu\delta^5 y_0 - \dots + \frac{h^{2n+1}(-1)^n(n!)^2 y^{(2n+1)}(\xi)}{(2n+1)!}, \quad (63)$$

where the last term involving differences is that containing $\mu\delta^{2n-1}y_0$ and ξ is some point in the range $x_{-n} \leqslant \xi \leqslant x_n$.

19. It is clearly desirable to know something about the function under investigation, and not to use the finite-difference formulae in an uncritical manner. Indeed, if a set of pivotal values of uncertain origin is presented, no definite statement can be made regarding finite-difference processes. This is easily seen by considering the effect of adding a function such as $\sin(\pi x/h)$ to the given pivotal values; the latter are unchanged, but the formulae will usually give wrong results.

In our problem, however, that of the solution of differential equations, the equations themselves will enable us to make pronouncements regarding the behaviour of the functions involved; our set of pivotal values is not of uncertain origin. We return later to this point in § 28.

Suppose for the moment that we have eliminated, by mathematical reasoning, the possibility that our pivotal values are contaminated by a function such as $\sin(\pi x/h)$. We then require some practical rule to enable the formulae to be used properly and with confidence, since the evaluation of derivatives of high order involved in the remainder term may not be a practicable undertaking.

We first consider the use of finite-difference formulae when central differences are everywhere available, and when the function and interval are such that the differences converge in the sense defined below.

Convergent differences

20. We turn again to Tables 8 and 9, and suppose that we add to x^4 a function which is not greater in magnitude than half a unit in the last figure. The effect on the differences can be calculated by consideration of Table 9, and it is not difficult to see that in the worst possible case, when the added function has the form $+\frac{1}{2}, -\frac{1}{2}, +\frac{1}{2}, -\frac{1}{2},\dots$, the differences of order n of the new function will be $\pm 2^{n-1}$ (for $n > 4$). Alternatively, if we round off our polynomial to one less figure, we see that the differences of order $n > 4$, instead of being zero, will oscillate between the limits $\pm 2^{n-1}$, but we would still expect our finite-difference formulae to be usable and to give an accuracy little inferior to that obtainable from the true polynomial.

Now all well-behaved functions exhibit similar difference behaviour at a small enough interval. The differences sooner or later decrease and finally reach a state of oscillation. If this oscillation has limits $\pm 2^{n-1}$ in the difference of order n, the differences are said to converge, and we can approximate to the true function by means of polynomials. In effect we regard the function as equivalent, within a rounding error, to a chain of polynomials. When we wish to use finite-difference formulae for interpolation, differentiation, or indefinite integration we use that polynomial whose origin is near the point concerned, and whose coefficients are determined by a nearly equal number of pivotal values on either side of this point. The degree of the polynomial is one less than the order of the first oscillating difference. In the corresponding finite-difference equations, of course, we curtail the infinite series at the first order of oscillating differences.

We note in passing that the amplitude of oscillation rarely reaches 2^{n-1}, and an oscillation of greater amplitude, in differences otherwise convergent, enables us to detect the position and rough magnitude of a blunder or blunders in one or more pivotal values.

21. In Table 10, a difference table of $\sin x$ to four decimals, we see that the sixth difference is the first which oscillates, so that at any point we can treat the function as a polynomial of degree five, ignoring differences of orders six and higher in finite-difference formulae.

TABLE 11

x	e^x		δ^2		x	e^x		δ^2	
−1·0	+0·368				+1·0	+2·718			
		+39					+286		
0·9	·407		+3		1·1	3·004		+30	
		42		+3			316		+3
·8	·449		6		1·2	3·320		33	
		48		−2			349		4
·7	·497		4		1·3	3·669		37	
		52		+2			386		4
·6	·549		6		1·4	4·055		41	
		58		−1			427		3
·5	·607		5		1·5	4·482		44	
		63		+3			471		6
·4	·670		8		1·6	4·953		50	
		71		−1			521		5
·3	·741		7		1·7	5·474		55	
		78		+1			576		5
·2	·819		8		1·8	6·050		60	
		86		+1			636		+7
·1	0·905		+9		1·9	6·686		+67	
		+95					+703		
−0·0	+1·000				+2·0	+7·389			

The degree of the 'equivalent' polynomials varies with the number of figures in the pivotal values, and may also vary with the argument x. For example the function $\sin x$, correctly rounded to three decimals in the range $x = 0(0{\cdot}2)1{\cdot}2$, has oscillating differences of fourth order, and the equivalent polynomials in this region are of degree three. In Table 11 we give some differences of the function e^x, and we notice that the third difference, oscillating in the early range, has become significant for larger values of the argument.

Numerical applications. Accuracy and precision

22. We now consider the numerical application of various formulae in our table of convergent differences. Two points call for discussion:

(i) do we use forward, backward, or central difference formulae;

(ii) what are their relative *accuracies*, and what is the maximum *precision* attainable?

If the function involved is a polynomial the questions present no difficulty. Since all pivotal values and differences are exact, and some order of differences is zero, all formulae for the same operation will give the same result, and this result will be exactly correct. For any other function the pivotal values and differences have rounding errors which get larger with successive orders of differences. We should therefore expect to get best accuracy with formulae whose coefficients are as small as possible.

A more accurate criterion is that the coefficient of the last difference retained in the relevant formula should be as small as possible. In this connexion it is interesting to examine the relations holding between the central-difference formula, say for the first derivative at a pivotal point, and the various sloping-difference formulae given in group 1 of Table 7. It is easy to verify the following relations:

$$\left.\begin{aligned} (\Delta+\tfrac{1}{2}\Delta^2)y_{-1} &= \mu\delta y_0 \\ (\Delta+\tfrac{3}{2}\Delta^2+\tfrac{1}{3}\Delta^3-\tfrac{1}{12}\Delta^4)y_{-2} &= (\mu\delta-\tfrac{1}{6}\mu\delta^3)y_0 \\ (\Delta+\tfrac{5}{2}\Delta^2+\tfrac{11}{6}\Delta^3+\tfrac{1}{4}\Delta^4-\tfrac{1}{20}\Delta^5+\tfrac{1}{60}\Delta^6)y_{-3} &= (\mu\delta-\tfrac{1}{6}\mu\delta^3+\tfrac{1}{30}\mu\delta^5)y_0 \end{aligned}\right\}. \qquad (64)$$

They show that to each formula, obtained by truncating the central-difference expression at different points, there corresponds exactly a forward-difference formula whose sloping line varies according to the number of differences retained.

In a similar way we find, for the second derivative, the relations

$$\left.\begin{aligned}\Delta^2 y_{-1} &= \delta^2 y_0\\ (\Delta^2+\Delta^3-\tfrac{1}{12}\Delta^4)y_{-2} &= (\delta^2-\tfrac{1}{12}\delta^4)y_0\\ (\Delta^2+2\Delta^3+\tfrac{11}{12}\Delta^4-\tfrac{1}{12}\Delta^5+\tfrac{1}{90}\Delta^6)y_{-3} &= (\delta^2-\tfrac{1}{12}\delta^4+\tfrac{1}{90}\delta^6)y_0\end{aligned}\right\}. \tag{65}$$

Now the fact that in equations (64) and (65) some of the earlier coefficients on the left are larger than those on the right does not mean that the central-difference formulae are more accurate. The differences are not independent, and the terms can be rearranged in many ways: only the last coefficient is crucial.

It is, of course, more convenient to use the central-difference formula. First, usually fewer terms appear, either odd or even differences being absent. Second, and more important, if it is desired at some stage in the work to use an extra term in the series, the central-difference amendment is trivial while the equivalent sloping-difference formula involves a complete new calculation along a different line.

The use, however, of a formula like

$$hy_0' = (\Delta-\tfrac{1}{2}\Delta^2+\tfrac{1}{3}\Delta^3-\tfrac{1}{4}\Delta^4+\tfrac{1}{5}\Delta^5)y_0,$$

for the first derivative at x_0, is not very satisfactory as regards accuracy. Rounding errors in the function could contribute as much as three units to the term $\tfrac{1}{5}\Delta^5 y_0$.

As a numerical illustration we use Table 10 to evaluate the first and second derivatives of $\sin x$ with the use of different formulae. With the forward-difference formula (16) for the first derivative at $x = 0{\cdot}6$, for example, we have the estimate

$$0{\cdot}2\cos 0{\cdot}6 = 0{\cdot}1528-\tfrac{1}{2}(-0{\cdot}0287)+\tfrac{1}{3}(-0{\cdot}0049)-\\ -\tfrac{1}{4}(+0{\cdot}0014)+\tfrac{1}{5}(0{\cdot}0000) = 0{\cdot}16517$$

for the unrounded value. Similar calculations at $x = 0{\cdot}0$, $0{\cdot}2$, $0{\cdot}4$, and the corresponding backward-difference formulae used at $x = 0{\cdot}8$ and $1{\cdot}0$, give estimates for $0{\cdot}2\cos x$ which have a maximum error of 2·6 units and an average error of 1·6 units in the fourth decimal. The central-difference formula (20) gives results with a maximum error of 0·6 units and an average error of 0·5 units. We might also try the more rapidly converging formula (24), applied at $x = 0{\cdot}0$, $0{\cdot}2$, $0{\cdot}4$, $0{\cdot}6$, and $0{\cdot}8$, and its equivalent (25) applied at $x = 1{\cdot}0$, which give results whose maximum error is 1·0 units and average error 0·7.

For the second derivative, the central-difference formula (22) gives estimates for $0{\cdot}04\sin x$ which have respective maximum and average errors of 1·4 and 0·6 units in the fourth decimal, while the sloping-difference formulae (18) and its equivalent have corresponding errors of amounts 7·9 and 4·1 units.

We note in passing that the respective formulae (21) and (44) for the first derivative and first indefinite integral at a half-way point have coefficients which, at least for the first few terms, decrease even faster than those in the formulae at a pivotal point. The converse is true for second derivatives and integrals.

Though central-difference formulae are more accurate than those using forward differences in formulae like (16), we shall find that the latter often play a valuable part in the solution of differential equations, particularly in step-by-step methods. As a general rule, we can say that such formulae are used to produce a first approximation to the required solution, central differences giving the final refined values.

23. In examining the precision of numerical differentiation, we note first that a derivative of order n cannot be estimated to more significant figures than are available in the corresponding difference of that order. Also the presence of rounding errors in the function, transmitted to its differences, further reduces the precision attainable in derivatives. These maximum errors are again of the order of 2^{n-1}, so that the maximum error in the first difference is 1, in the second 2, in the third 4, and so on, expressed in units of the last figure. In particular no estimate can be made for the derivative of the same order as the first oscillating difference. In the example of the last paragraph we should expect our central-difference formula for the derivative of $\sin x$ at $x = 0{\cdot}6$ to be correct within one part in 1600, the second derivative to one part in 100, and so on.

The differences generally become smaller as the interval of tabulation is reduced. To obtain maximum precision in the derivatives, therefore, it is desirable to use the largest possible interval in the tabulation. We shall see in the next section that there is an upper limit to the size of this interval.

For integration the situation is somewhat different, for here we are adding quantities instead of subtracting them, and the main source of error now lies in the statistical accumulation of a number of rounding errors. This is a much less serious limitation: integration is a 'smoothing' process.

In the case of interpolation, it is clear that best accuracy is attainable with an infinitesimally small interval.

Divergent differences

24. If the central differences at some points never begin to oscillate within the prescribed limits they are said to diverge. Divergent differences may arise for two reasons:

(i) the interval of tabulation is too large;

(ii) we are in the neighbourhood of an infinity in the function or one of its derivatives.

For a well-behaved function the divergence can be removed by reducing the interval, as illustrated for the function $(1+x^2)^{-1}$ in Tables 12 and 13.

TABLE 12

x	$(1+x^2)^{-1}$		δ^2		δ^4		δ^6		δ^8		δ^{10}
+0·0	+1·0000		−770		+322		−296		+444		−930
		− 385		+161		−148		+222		−465	
·2	0·9615		−609		+174		− 74		− 21		+291
		994		335		−222		+201		−174	
·4	·8621		−274		− 48		+127		−195		+244
		1268		287		− 95		+ 6		+ 70	
·6	·7353		+ 13		143		+133		−125		+144
		1255		144		+ 38		−119		+214	
0·8	·6098		+157		105		+ 14		+ 89		
		−1098		+ 39		+ 52		− 30			
+1·0	+0·5000		+196		− 53		− 16				

TABLE 13

x	$(1+x^2)^{-1}$		δ^2		δ^4		δ^6		δ^8
+0·0	+1·0000		−198		+22		−2		−14
		− 99		+11		− 1		− 7	
·1	0·9901		187		21		−9		+16
		286		32		10		+ 9	
·2	·9615		155		11		0		− 2
		441		43		10		+ 7	
·3	·9174		112		+ 1		+7		−20
		553		44		3		−13	
·4	·8621		68		− 2		−6		
		−621		+42		− 9			
+0·5	+0·8000		− 26		−11				

In the neighbourhood of an infinity of function or derivative the differences will always diverge, no matter how small the interval.

The divergence of differences does not necessarily prevent the accurate use of finite-difference formulae; it is necessary only that the series representing such a formula should converge. Thus in Table 12 we can estimate the second derivative at the origin to have the value

$$\left[\frac{d^2}{dx^2}(1+x^2)^{-1}\right]_{x=0} = 25[-0{\cdot}0770-\tfrac{1}{12}(0{\cdot}0322)+\tfrac{1}{90}(-0{\cdot}0296)-$$
$$-\tfrac{1}{560}(0{\cdot}0444)+\tfrac{1}{3150}(-0{\cdot}0930)+...],$$

the quantity in square brackets having the value $-0{\cdot}0801$ when we exclude the last term, contributing less than half a unit in the last figure. This figure compares well with the correct value $-0{\cdot}0800$.

Nevertheless divergent differences should be avoided where possible, for not only are we prevented from using the differences to detect an error, but also we are less confident of the accuracy of finite-difference formulae. Further, in many methods for the numerical solution of differential equations we require to be able to make reasonably accurate extrapolations from observations of the run of previous differences.

25. In the numerical solution of differential equations we can treat accurately only well-behaved functions, with no singularity in the range of integration. It will usually be possible to recognize from the given differential equations the existence, position, and type of singularity, and to remove it by transformations involving the dependent or independent variable or both.

We can therefore eliminate the possibility of divergent differences stemming from this source. The remaining possibility is the existence of a singularity outside the range, but sufficiently close to one end to affect the differences at points within the range. Consider, for example, some differences of the function x^{-1} shown in Table 14.

TABLE 14

x	x^{-1}		δ^2		δ^4		δ^6		δ^8		δ^{10}
+0·0	+∞										
		−∞									
·1	10·00		+∞								
		500		−∞							
·2	5·00		333		+∞						
		167		249		−∞					
·3	3·33		84		198		+∞				
		83		51		163		−∞			
·4	2·50		33		35		+136		+∞		
		50		16		27		−112		−∞	
·5	2·00		17		8		+24		+87		+∞
		33		8		3		−25		−57	
·6	1·67		9		5		− 1		+30		+17
		24		3		4		+ 5		−40	
·7	1·43		6		1		+ 4		−10		+57
		18		2		0		− 5		+17	
·8	1·25		4		1		− 1		+ 7		
		14		1		1		+ 2			
0·9	1·11		3		0		+ 1				
		−11		− 1		−0					
+1·0	+1·00		+ 2		+0						

It is clear that the forward differences at the origin are all infinite, so that at every point some central difference is infinite. This might suggest that finite-difference formulae would be meaningless for every point in the range, but it would be unreasonable to expect the behaviour at $x = 0$ to have such a catastrophic effect on the function or its derivatives at say $x = 1{\cdot}0$, ten intervals distant. We find in fact that it is perfectly safe to use our formulae at points sufficiently remote from the disturbance. Indeed for $x > 0{\cdot}5$ in Table 14 the sixth central differences are oscillating, and we can treat the function as a polynomial of degree five in each interval.

For the range $0 < x < 0{\cdot}5$ the issue is more complicated. If, for example, we want the first derivative at $x = 0{\cdot}3$, the central-difference formula is confronted with an infinite fifth difference, the contribution from third differences being still significant. In this case we are forced to use a forward-difference formula. Equation (16), for instance, gives

$$0{\cdot}1\left(\frac{dy}{dx}\right)_{x=0\cdot2} = -1{\cdot}67-0{\cdot}42-0{\cdot}17-0{\cdot}09-0{\cdot}05-$$
$$-0{\cdot}04-0{\cdot}04-0{\cdot}04-0{\cdot}04-0{\cdot}06\ldots .$$

What are we to say about this series? First, it is clear that the last four terms are most unreliable, since rounding errors in the function could make contributions of this size. Second, we cannot give an accurate result for this derivative; we are not even sure that the series is convergent. The rule often quoted is that the series should be truncated after its smallest term, with the expectation that the error committed is of the same order as this term. This rule would give the value $-2{\cdot}48$, as compared with the correct result $-2{\cdot}50$.

26. That this procedure is not without danger, however, can be seen by considering the function $x \ln x$, with infinite derivatives at the origin. Some differences are shown in Table 15.

TABLE 15

x	$y = x \ln x$		δ^2		δ^4		δ^6		δ^8	
+0·0	−0·000									
		−230								
·1	·230		+138							
		92		−85						
·2	·322		53		+65					
		39		20		−52				
·3	·361		33		+13		+41			
		− 6		7		−11		−33		
·4	·367		26		+ 2		+ 8		+35	
		+ 20		5		− 3		+ 2		−62
·5	·347		21		− 1		+10		−27	
		41		6		+ 7		−25		
·6	·306		15		+ 6		−15			
		+ 56		− 0		− 8				
+0·7	−0·250		+ 15		− 2					

If now we try to evaluate the first derivative at $x = 0$ we have the series

$$0{\cdot}1\left(\frac{dy}{dx}\right)_{x=0} = -0{\cdot}230-0{\cdot}069-0{\cdot}028-0{\cdot}016-0{\cdot}010-$$
$$-0{\cdot}007-0{\cdot}005-0{\cdot}004-0{\cdot}007-\ldots.$$

Again the last three terms might be expected to have little significance because of rounding errors in the function, and we are tempted to estimate the value of this series to be finite, with an error of about 0·005, though the true value is infinite.

27. On the other hand the forward-difference formulae are reliable, within the limits of accuracy determined by their coefficients, when the series are obviously converging. Thus from Table 14 we have, for the first derivative at $x = 0{\cdot}3$, the series

$$0{\cdot}1\left(\frac{dy}{dx}\right)_{x=0\cdot3} = -0{\cdot}83-0{\cdot}17-0{\cdot}05-0{\cdot}02-0{\cdot}01 = -1{\cdot}08,$$

further terms being truly negligible. Similarly from Table 15 we can find

$$0{\cdot}1\left(\frac{dy}{dx}\right)_{x=0\cdot1} = -0{\cdot}092-0{\cdot}026-0{\cdot}007-0{\cdot}003-0{\cdot}002-0{\cdot}001 = -0{\cdot}131.$$

In both cases the error is no greater than would be expected from the use of this formula in the most favourable case, and in the second case there is no possibility of using a central-difference formula.

Recapitulation

28. It will be clear from the foregoing that no golden rule is available for the guidance of the unwary computer. Experience in the use of differences is a considerable asset, and can be gained only by practice. We have nevertheless indicated some general rules for the successful application of finite-difference equations, and to this we add a final one: when in doubt from the behaviour of the differences as to the presence of singularities, consult the mathematician, and in other cases the interval can always be reduced.

Finally we illustrate by an example the necessity for having at all times an awareness of the underlying mathematics. The differential equation

$$y''+(314{\cdot}16)^2y = (314{\cdot}16)^2x, \quad \text{with} \quad y(0) = 0{\cdot}00001,\ y(1) = 1,$$

has the analytical solution

$$y = x-(0{\cdot}0136...)\sin(314{\cdot}16)x+(0{\cdot}00001)\cos(314{\cdot}16)x.$$

If an inexperienced computer is asked to solve this equation to four decimals, for example by one of the methods of Chapter 4, he is likely to use an interval of convenient size, say $h = 0{\cdot}1$. To four decimals he will produce the result $y = x$ at every pivotal point. Perhaps surprised at this strange answer, he takes an interval $h = 0{\cdot}05$, gets the same answer and is content. This answer is nevertheless quite useless for many purposes: it would, for example, give wildly erroneous values for the first derivative at all pivotal points.

The mathematician, on the other hand, would recognize the form of the complementary function, and he would know, if he were also a computer, that trigonometric functions do not have convergent differences at an interval of order 30 in x. He would therefore not fall into the trap of trying to produce an answer using this interval.

This example is trivial and artificial, but less obvious examples can easily be constructed, and it serves to illustrate the danger of divorcing mathematics completely from computation.

Lagrangian formulae

29. We end this chapter by discussing some aspects of Lagrangian formulae, which may be defined as formulae in which all significant differences have been replaced by pivotal values. In the equations of § 15, for example, we can regard the right-hand sides as the sum of some Lagrangian terms and a difference correction. Among other such formulae we quote the following, in which the term in brackets on the right is the first neglected difference.

Interpolation at a mid-point:

$$y_{\frac{1}{2}} = \tfrac{1}{2}(y_0+y_1), \qquad (\delta^2)$$

$$y_{\frac{1}{2}} = -\tfrac{1}{16}(y_2-9y_1-9y_0+y_{-1}); \qquad (\delta^4)$$

Differentiation:

$$hy_0' = \tfrac{1}{2}(y_1-y_{-1}), \qquad (\delta^3)$$

$$hy_0' = -\tfrac{1}{12}(y_2-8y_1+8y_{-1}-y_{-2}), \qquad (\delta^5)$$

$$h^2y_0'' = y_1-2y_0+y_{-1}, \qquad (\delta^4)$$

$$h^2y_0'' = -\tfrac{1}{12}(y_2-16y_1+30y_0-16y_{-1}+y_{-2}). \qquad (\delta^6)$$

Integration

We can turn the Gregory quadrature formula into Lagrangian form, but its coefficients depend not only on the neglected differences but also on the number of points in the range. Other Lagrangian quadrature formula are those of the Newton–Cotes type, whose most familiar member is that known as Simpson's rule, given by

$$\frac{3}{h}\int_{x_0}^{x_{2n}} y\,dx = y_0+4y_1+2y_2+4y_3+2y_4+...+2y_{2n-2}+4y_{2n-1}+y_{2n},$$

in which the neglected terms start with fourth differences.

30. Some workers use Lagrangian formulae throughout in preference to formulae involving differences and a difference correction. For work of high accuracy this procedure is not satisfactory. The only way of verifying the adequacy of the Lagrangian formula is to form the table of differences and verify that the neglected terms are really negligible, and if the differences must be formed they might as well be used. There is of course no danger in using such formulae if the results are checked by other means, and whenever they are used in this book their adequacy is invariably checked by a consideration of the resulting difference table.

Additional notes and bibliography

§ 2. Equal intervals are used almost always in numerical methods of solving differential equations. The general case of unequal intervals, involving the use of *divided differences*, is often made the starting-point of the theory of finite differences. See, for example,

MILNE-THOMSON, L. M., 1933, *The calculus of finite differences.* Macmillan.

§ 4. There are better ways of producing interpolation formulae which give directly the coefficients of the various differences. For Bessel's formula, for example, the problem is to determine the coefficients a_r in the expansion

$$y(x+ph) = y_0+a_1\,\delta y_{\frac{1}{2}}+a_2(\delta^2 y_0+\delta^2 y_1)+a_3\,\delta^3 y_{\frac{1}{2}}+a_4(\delta^4 y_0+\delta^4 y_1)+...,$$

or its equivalent

$$(1+\Delta)^p = 1+a_1\Delta+a_3\frac{\Delta^3}{1+\Delta}+...+a_{2s+1}\frac{\Delta^{2s+1}}{(1+\Delta)^s}+...+a_2\frac{\Delta^2}{1+\Delta}(2+\Delta)+$$
$$+a_4\frac{\Delta^4}{(1+\Delta)^2}(2+\Delta)+...+a_{2s}\frac{\Delta^{2s}}{(1+\Delta)^s}(2+\Delta)+....$$

Multiplying throughout with $(1+\Delta)^{s-1}$, expanding the remaining denominators, and equating coefficients of Δ^{2s} and Δ^{2s+1}, we find

$$2a_{2s} = \binom{p+s-1}{2s}, \qquad a_{2s+1}-a_{2s} = \binom{p+s-1}{2s+1},$$

or

$$a_{2s} = \frac{1}{2}\binom{p+s-1}{2s}, \qquad a_{2s+1} = \frac{p-\frac{1}{2}}{2s+1}\binom{p+s-1}{2s}.$$

For other symbolic methods see, for example,

HARTREE, D. R., 1952, *Numerical analysis.* Oxford.

MICHEL, J. G. L., 1946, Central-difference formulae obtained by means of operator expansions. *Journ. Inst. Actu.* **72,** 470–80.

§ 12. Formula (50) will rarely be needed except in the case $n = 0$, giving the definite integral between a pivotal point and the next half-way point, in the form

$$\frac{1}{h}\int_{x_0}^{x_{\frac{1}{2}}} y\,dx = \tfrac{1}{2}y_0+(\tfrac{1}{24}\delta-\tfrac{17}{5760}\delta^3+\ldots)y_{\frac{1}{2}}+(\tfrac{1}{12}\mu\delta-\tfrac{11}{720}\mu\delta^3+\ldots)y_0.$$

A rather more convenient formula, avoiding the formation of mean differences, is obtained by writing

$$\frac{1}{h}\int_{x_0}^{x_{\frac{1}{2}}} y\,dx = (hD)^{-1}(E^{\frac{1}{2}}-1)y_0 = \mu(\delta^{-1}-\tfrac{1}{12}\delta+\tfrac{11}{720}\delta^3-\ldots)(E^{\frac{1}{2}}-1)y_0.$$

The factor $\mu(E^{\frac{1}{2}}-1)$ can be written as $\frac{1}{2}\delta+(\mu-1)E^{\frac{1}{2}}$, and expansion of μ in powers of δ gives

$$\frac{1}{h}\int_{x_0}^{x_{\frac{1}{2}}} y\,dx = (\delta^{-1}-\tfrac{1}{12}\delta+\tfrac{11}{720}\delta^3-\ldots)\{\tfrac{1}{2}\delta+(\tfrac{1}{8}\delta^2-\tfrac{1}{128}\delta^4+\ldots)E^{\frac{1}{2}}\}y_0,$$

so that we have finally the result

$$\frac{1}{h}\int_{x_0}^{x_{\frac{1}{2}}} y\,dx = (\tfrac{1}{2}-\tfrac{1}{24}\delta^2+\tfrac{11}{1440}\delta^4-\ldots)y_0+(\tfrac{1}{8}\delta-\tfrac{7}{384}\delta^3+\ldots)y_{\frac{1}{2}}.$$

This result can also be obtained from (50), with $n = 0$, by substituting for a mean odd-difference $\mu\delta^{2s-1}y_0$ its equivalent $\delta^{2s-1}y_{\frac{1}{2}}-\frac{1}{2}\delta^{2s}y_0$.

§ 14. The different forms of the terms in brackets in (54) suggest important variations in techniques of integrating differential equations. In this book we use almost always formulae for differentiation. Many step-by-step methods, described in the second volume, use formulae of integration.

§ 17. In tables of functions and their differences we use a convention that positive and negative signs have equal value. The sign is inserted on the first and last entries in each column, whenever the sign changes, both before and after the change, and everywhere if sign changes are frequent. The same rule is observed if the figure preceding the decimal point is zero, but otherwise all leading figures are inserted.

§ 18. The rigorous theory of remainder terms in finite-difference equations is given by Milne-Thomson (see note on § 2).

3

The Solution of Algebraic Equations

Differential equations and algebraic equations

1. We have noted in Chapter 1 that the main feature of boundary-value methods is the necessity for solving simultaneous algebraic equations. The unknowns in these equations are the values of the wanted solution, at pivotal points, usually equidistant, in the range of integration. The equations arise in the replacement of derivatives by finite-difference expressions, and the representation of the dominant terms in these expressions by combinations of pivotal values. For example (equations (57) and (58) of Chapter 2) we have for the first and second derivatives the expressions

$$\left.\begin{aligned} hy'_r &= \tfrac{1}{2}(y_{r+1}-y_{r-1})+C_1 y_r \\ h^2y''_r &= (y_{r+1}-2y_r+y_{r-1})+C_2 y_r \end{aligned}\right\}, \tag{1}$$

where the 'difference-correction' operators C_1 and C_2 are respectively given by

$$\left.\begin{aligned} C_1 &= -\tfrac{1}{6}\mu\delta^3+\tfrac{1}{30}\mu\delta^5-\dots \\ C_2 &= -\tfrac{1}{12}\delta^4+\tfrac{1}{90}\delta^6-\dots \end{aligned}\right\}. \tag{2}$$

A second-order differential equation then reduces, at the pivotal point x_r, to an algebraic equation connecting y_{r+1}, y_r, and y_{r-1} and a difference correction Cy_r which may involve both C_1 and C_2. As r varies through all pivotal points the aggregate of such equations, and others derived from the boundary conditions, is a set of simultaneous algebraic equations, equal in number to that of unknown pivotal values.

2. Linear differential equations present the easier problem, since the simultaneous algebraic equations are then also linear. The general linear second-order equation, for example, leads to a set of algebraic equations typified by

$$a_r y_{r+1}+b_r y_r+c_r y_{r-1}+Cy_r = d_r. \tag{3}$$

There are many well-known techniques for solving simultaneous linear algebraic equations, and in general success is assured. There is usually a definite unique solution, and we can hardly fail to find it unless the number of equations is so large that the labour involved is prohibitive: the advent of high-speed automatic computing machines makes this contingency somewhat remote.

A non-linear differential equation is a more formidable proposition, since the resulting algebraic equations are non-linear and there is no certain method available for their solution. We shall see later that these problems are usually solved by methods of successive approximation, in which at each stage we solve linear algebraic equations with coefficients which are kept constant for that stage but varied from stage to stage.

The difference correction

†**3.** Our treatment of the difference correction in equations like (3) needs particular comment. Most writers neglect the difference correction, using more accurate formulae for derivatives, or a small enough interval, so that the error is either negligible or kept within reasonable bounds. In this book we shall work with a reasonably large interval and take full account of the difference correction. A large interval involves a correspondingly small number of algebraic equations and a proportionately small amount of labour: the inclusion of the difference correction yields high accuracy at this large interval.

The values of Cy_r are not known until the final solution is available, so that our process is one of successive approximation. For the linear case typified by equation (3), for example, we can write the set of equations in matrix form as

$$Ay+Cy = d, \tag{4}$$

where A is the matrix of coefficients, y the vector of pivotal values y_r, Cy the vector of difference corrections, and d the vector with components d_r. We first ignore Cy and obtain a first approximation $y^{(1)}$ from the equation

$$Ay^{(1)} = d. \tag{5}$$

The difference correction of $y^{(1)}$ will approximate to that of y, so that a first correction z to y is obtained from the equation

$$Az+Cy^{(1)} = 0, \tag{6}$$

and further corrections are obtained by repetitions of this cycle. Convergence is usually very rapid and the final solution $y = y^{(1)}+z+...$ can be checked and, if necessary, further corrected by insertion in (4).

The non-linear case is rather more involved, but the same general procedure is followed, of calculating a first approximation by ignoring the difference corrections, then computing approximations to the latter and using them to correct the approximate pivotal values.

The details of this process, applied to the solution of boundary-value problems of various orders, are described in succeeding chapters. For the rest of this chapter we shall discuss methods of solving simultaneous algebraic equations, with particular reference to the kind of algebraic equations with which we have to deal when solving differential equations.

Linear algebraic equations. Iterative methods

†4. Methods for the solution of linear equations are generally divided into two classes, the direct and the indirect. In a direct method all the computations are carried out with a fixed number of figures, and a single application of the process will give results of an accuracy determined by the number of figures retained throughout. If we could keep an infinite number of figures on our calculating machine a single application of a direct method would provide completely accurate solutions. An indirect method, on the other hand, is a process of successive approximation, more figures being added as trial solutions become steadily more accurate.

Most of the direct methods are refinements of simple elimination methods taught in schools, and can be used to solve any set of linear equations which has a solution. Indirect methods have a more restricted field of useful application, but we find that they are occasionally quicker and more convenient than direct methods for the kind of equations with which we are concerned.

The earliest indirect methods are generally attributed to Gauss and Seidel. Equations given by

$$\left.\begin{array}{l} a_{11}y_1+a_{12}y_2+\dots+a_{1n}y_n-b_1 = 0 \\ a_{21}y_1+a_{22}y_2+\dots+a_{2n}y_n-b_2 = 0 \\ \cdot\quad\cdot\quad\cdot\quad\cdot\quad\cdot\quad\cdot\quad\cdot\quad\cdot\quad\cdot \\ a_{n1}y_1+a_{n2}y_2+\dots+a_{nn}y_n-b_n = 0 \end{array}\right\} \tag{7}$$

are solved by iteration according to a fixed routine. If $y_r^{(s)}$ denotes the sth approximation to the unknown y_r, the two most common schemes of iteration are represented by the equations

$$\left.\begin{array}{l} a_{11}y_1^{(s+1)}+a_{12}y_2^{(s)}+\dots+a_{1n}y_n^{(s)} = b_1 \\ a_{21}y_1^{(s)}+a_{22}y_2^{(s+1)}+a_{23}y_3^{(s)}+\dots+a_{2n}y_n^{(s)} = b_2 \\ \cdot\quad\cdot\quad\cdot\quad\cdot\quad\cdot\quad\cdot\quad\cdot\quad\cdot\quad\cdot \\ a_{n1}y_1^{(s)}+\dots+a_{n,n-1}y_{n-1}^{(s)}+a_{nn}y_n^{(s+1)} = b_n \end{array}\right\}, \tag{8}$$

and

$$\left.\begin{array}{l} a_{11}y_1^{(s+1)}+a_{12}y_2^{(s)}+\dots+a_{1n}y_n^{(s)} = b_1 \\ a_{21}y_1^{(s+1)}+a_{22}y_2^{(s+1)}+a_{23}y_3^{(s)}+\dots+a_{2n}y_n^{(s)} = b_2 \\ \cdot\quad\cdot\quad\cdot\quad\cdot\quad\cdot\quad\cdot\quad\cdot\quad\cdot\quad\cdot \\ a_{n1}y_1^{(s+1)}+\dots+a_{nn}y_n^{(s+1)} = b_n \end{array}\right\}. \tag{9}$$

Of these processes the second, using at every stage the latest known information, is usually the more rapidly convergent.

In recent years Sir Richard Southwell has introduced a method of successive approximation, called by him *the relaxation method*. This we shall use in preference to the unrefined iterative processes given by equations (8) and (9), and we now proceed with its description.

The relaxation method

5. The accuracy of a trial solution can easily be tested by direct substitution in the given linear equations. If in equations (7), for example, we take arbitrary values for $y_1,..., y_n$ we shall obtain, for their left-hand sides, a set of numbers, called *residuals*, which give a measure of the error of the trial solution. The first residual, for example, is given by

$$R_1 = a_{11}y_1+a_{12}y_2+...+a_{1n}y_n-b_1. \tag{10}$$

Our aim is to find a set of y's for which the residuals are all zero, or so small that the y's constitute a correct solution, to some specified precision, to the given linear equations.

In the relaxation method we try to achieve this by making additions to the trial values of the unknowns so that the residuals are reduced, more or less systematically, to negligible proportions. The aggregate of the trial values and all subsequent additions then gives the required solution.

The increment in the residual R_s arising from a unit increment in the unknown y_r is easily seen to be the coefficient a_{sr} of y_r in the equation defining R_s, that is

$$\delta R_s = a_{sr}\delta y_r. \tag{11}$$

It is often convenient to incorporate the information contained in equation (11) into a table, showing the effect on any residual of an increment in any unknown. The act of making any change δy_r is often called an *operation*, and the table showing its effect on the residuals the *operations table*.

Any residual can be reduced momentarily to zero, at the expense of changing and perhaps increasing other residuals, by making an *appropriate* increment to *any* of the unknowns, and we have to decide at every step which residual to liquidate and which unknown is to be used for this purpose. If the process is to converge strongly, both increments and residuals must steadily decrease in absolute value. This suggests a good working rule, that at each stage *the largest residual should be liquidated and the unknown chosen for this purpose should be that which requires the smallest increment*; this unknown is therefore the one which has the largest coefficient in the equation defining the particular residual under consideration. In most problems of the type suitable for relaxation the equations can be arranged so that the biggest coefficient in any row lies in the diagonal, and the residual R_s is then associated with the variable y_s, and is completely liquidated by the operation

$$\delta y_s = -R_s/a_{ss}. \tag{12}$$

Convergence is somewhat faster if, instead of the rule given, we decide to choose as operation the greatest increment defined by (12). This does not necessarily correspond to the largest residual unless the diagonal terms are all equal. Choice of the largest residual is generally safe, however, and involves none of the arithmetic necessary to pick out the largest of the δy_s defined by (12).

Example

6. In illustration of these simple concepts, consider the solution of the following three simultaneous equations:

$$\left.\begin{aligned} 10y_1-2y_2+\ y_3-185 &= 0 \\ -3y_1-6y_2+2y_3-\ 93 &= 0 \\ y_1-2y_2-5y_3-\ 49 &= 0 \end{aligned}\right\}. \tag{13}$$

For arbitrary values of the unknowns the right-hand sides, the residuals, will have values R_1, R_2, and R_3. We wish to find a set of y's for which the R's have negligible values.

The computations are embodied in Tables 1 and 2, showing respectively the possible operations and details of the relaxation process.

TABLE 1

	δR_1	δR_2	δR_3
δy_1	+10	−3	+1
δy_2	− 2	−6	−2
δy_3	+ 1	+2	−5

TABLE 2

Row no.	y_1	y_2	y_3	R_1	R_2	R_3
1	0	0	0	−185	− 93	−49
2	+18			− 5	−147	−31
3		−24		+ 43	− 3	+17
4	− 4			+ 3	+ 9	+13
5			+2	+ 5	+ 13	+ 3
6		+ 2		+ 1	+ 1	− 1
7	+14	−22	+2	+ 1	+ 1	− 1

Accurate solution

$y_1 = +13{\cdot}95,\quad y_2 = -21{\cdot}89,\quad y_3 = +1{\cdot}75.$

In Table 2, row 1 gives the residuals corresponding to the initial guess $y_1 = y_2 = y_3 = 0$. If no good approximation is known from *a priori* considerations, the guess of zero initial values is probably the most convenient, the initial residuals being given immediately without computation.

R_1 is the biggest initial residual, and is almost completely liquidated by a change of $+18$ in y_1: row 2 records this change and the new residuals of which R_2 is now the largest. When this is liquidated (row 3) by means of a change -24 in y_2, R_1 again becomes the largest of the three residuals, and is in turn disposed of by a change of -4 in y_1 (row 4). At this stage R_3 becomes for the first time the largest residual, and is removed by a change of $+2$ in y_3 (row 5). Finally, in row 6, a change of $+2$ in y_2 reduces the residuals to $+1$, $+1$, and -1. In row 7 the contributions to the y's are accumulated, and the residuals checked by substitution in equations (13).

The residuals are now very small, and it is probable that we have obtained correctly the integer part of the true solutions. If higher accuracy is required we proceed farther, the easiest way being to multiply these residuals by 10 or 10^2, continuing the relaxation process and remembering that the new increments to the unknowns are in units of 10^{-1} or 10^{-2} respectively.

We can rarely by this means produce exact results in the form of rational fractions, unless these are also expressible as terminating decimals, but the error of approximate solutions can be made as small as we please. This is sufficient, for in all computational work an exact solution is rarely obtainable or required: our ambition is to be able to give an upper limit to the error, expressed in phrases such as 'the error in the last figure given will nowhere exceed m (specified) units'. Of the result given in Table 2, for example, we can say with fair confidence that the error in the last figure of y_1, y_2, and y_3 nowhere exceeds half a unit. It is not always easy, however, to decide the maximum error of an approximate solution from the size of the unliquidated residuals: we shall return later to this question.

Notes on relaxation

7. Facility in computing techniques can only be attained by practical work. This is perhaps more true of relaxation than of any other process, since there are many tricks a skilled relaxer can use to accelerate the convergence of the process and generally lighten his work. The following comments on the process include a few of these devices. They are intended only as a general guide to the computer, since he will appreciate them only in application to actual numerical examples: in the course of it he will indeed discover for himself these and many other labour-saving devices.

(i) It is clear that any operation of the kind shown in Table 1, when used for the complete liquidation of a particular residual, is equivalent to the satisfaction of the corresponding equation of the iterative set (9). By choosing at each stage the largest residual we are taking these equations, not in a fixed unvarying order, but in one which is likely to produce quicker convergence.

(ii) It is generally useless to make any residual exactly zero, by an exact change R_s/a_{ss} in the corresponding unknown y_s. This residual will almost certainly change again subsequently, and unnecessary arithmetic has been performed. One of the advantages of the method lies in the fact that 'easy' operations can be used, and much of the work done mentally without undue strain. In many problems the calculating machine need be used only for the calculation of residuals.

(iii) Very often the coefficients are not simple integers but may contain several figures, involving a little more work. The full number of figures is used for the calculation of residuals, but the coefficients can be rounded-off to convenient numbers for use in the operations table. When residuals are subsequently checked they will not agree exactly with those obtained from the relaxation process, but this merely requires a little extra relaxation.

(iv) In common with all methods of successive approximation, mistakes are less serious than in the corresponding direct methods. If some residual is incorrect the mistake will be found in the final check and further relaxation will be necessary. This final calculation of residuals is of course very important, and it also pays to check them fairly frequently in the course of the work, say whenever they have been reduced to a tenth of their previously calculated values.

(v) A further advantage over direct methods is that it is not necessary to keep a large number of figures throughout. We can start with a few figures and gradually add more as approximate solutions become more accurate.

(vi) It is sometimes possible, by observing the signs of successive increments to a particular y_r, to anticipate further additions by using devices often called *under-relaxation* or *over-relaxation*. If convergence is oscillatory each increment is clearly too large, and it would be advantageous subsequently to under-relax, that is to liquidate only part of the residual by a change in the corresponding unknown, expecting that some of the rest of this residual will be liquidated when other unknowns are changed. If convergence is monotonic it will pay to over-relax, that is to make a bigger change in the unknown than would be needed for the complete liquidation of the corresponding residual. The latter will now change its sign, and subsequent increments to other unknowns will then have a beneficial effect on it.

(vii) A little practice will convince the reader that the relaxation process is most powerful when the liquidation of any residual can be effected without, at the same time, materially changing the others. The best type of equations are therefore those in which the diagonal coefficients are large compared with the others. If this state of affairs does not already exist it may be possible to effect it by transforming the original equations into a new set with a dominant diagonal. Alternatively, and rather more easily, it is often possible to construct new operations from those given in the operations table. These operations, which might be called *group-operations*, and their employment *group-relaxation*, involve simultaneous increments by the same or different amounts to two or more of the unknowns. Their object is the liquidation of a particular residual without substantially changing the others.

With reference to Table 1, for example, two good group-operations for the liquidation of R_1 and R_3 are shown in Table 3.

TABLE 3

			δR_1	δR_2	δR_3
$\delta y_1 = 2,$	$\delta y_2 = -1,$	$\delta y_3 = 1$	$+23$	$+2$	-1
	$\delta y_2 = 1,$	$\delta y_3 = 3$	$+1$	0	-17

New operators of this kind would normally be incorporated in the operations table, and used whenever desirable.

(viii) Again it may happen, after some relaxation, that the remaining residuals are roughly proportional to their initial values. In this case a better solution can quickly be obtained by multiplying the approximate solution by a simple factor. In the problem of Table 2, for example, suppose that after some relaxation we had obtained an approximate solution

$$y_1 = +12, \qquad y_2 = -19, \qquad y_3 = +1{\cdot}5,$$

with corresponding residuals

$$R_1 = -25{\cdot}5, \qquad R_2 = -12, \qquad R_3 = -6{\cdot}5.$$

The ratios of these unliquidated residuals to their initial values are, to two decimals, respectively 0·14, 0·13, and 0·13. Our approximate solution has therefore removed about 0·87 of each of the initial residuals. If therefore we multiply the components of this approximate solution by $1/0{\cdot}87 = 1{\cdot}15$, we shall have values for which the residuals are all very small. This better solution is

$$y_1 = +13{\cdot}8, \qquad y_2 = -21{\cdot}8, \qquad y_3 = +1{\cdot}7,$$

and the corresponding residuals

$$R_1 = -1{\cdot}7, \qquad R_2 = -0{\cdot}2, \qquad R_3 = -0{\cdot}1.$$

This multiplying-factor device can clearly be regarded as a special group-operation.

These are some of the ways in which convergence can be accelerated, and we shall consider them again later in relation to the solution of the equations which arise from finite-difference theory. Meanwhile it should be emphasized that any technique is permissible, subject only to the criterion that the final solution gives negligible residuals.

Condition and convergence

8. We have not yet given any precise meaning to the phrase 'negligibility of residuals'. Computational standards suggest that the residuals can be considered negligible when our approximate solution is correct, that is correctly rounded-off, to the required precision. This, however, is merely an ambition, and gives no indication as to how it is to be attained.

The size of the unliquidated residuals, compared for example with the constant terms in the original equations, may not give a reliable indication of the accuracy of the approximate solution. In extreme cases a very poor approximation can give very small residuals, an example of which is provided by the following four simultaneous equations:

$$\left.\begin{aligned} 5x_1+7x_2+6x_3+5x_4-23 &= 0\\ 7x_1+10x_2+8x_3+7x_4-32 &= 0\\ 6x_1+8x_2+10x_3+9x_4-33 &= 0\\ 5x_1+7x_2+9x_3+10x_4-31 &= 0 \end{aligned}\right\}, \tag{14}$$

the solution of which is $x_1 = x_2 = x_3 = x_4 = 1$.

Table 4 shows three sets of residuals corresponding to three distinct approximate solutions.

TABLE 4

x_1	x_2	x_3	x_4	R_1	R_2	R_3	R_4
+14·6	−7·2	−2·5	+3·1	+0·1	−0·1	−0·1	+0·1
+ 2·36	+0·18	+0·65	+1·21	+0·01	−0·01	−0·01	+0·01
+ 1·136	+0·918	+0·965	+1·021	+0·001	−0·001	−0·001	+0·001

Here the approximate solutions have errors of 10 per cent. when the residuals have been reduced to less than 10^{-4} of their initial values. In the example of Table 2, on the other hand, the least accurate component of the solution had an error of only 15 per cent. when its associated residual was as much as one-fiftieth of the initial constant term.

Equations like (13) which have a fairly well-defined solution are often called *well-conditioned*: equations such as (14) are *ill-conditioned equations*. We can make the general definition that in a well-conditioned set of equations small residuals are relatively unimportant and can be liquidated by small increments to the unknowns: in an ill-conditioned set small residuals may require large changes in the unknowns for their liquidation.

9. We find the reason for the different behaviour of sets of equations in considering the reciprocal equations. In equations (13), for example, the equations defining the increments to y_1, y_2, and y_3 necessary for the liquidation of a set of residuals δR_1, δR_2, and δR_3 are given by

$$\left.\begin{aligned} 10\delta y_1-2\delta y_2+\delta y_3 &= -\delta R_1 \\ -3\delta y_1-6\delta y_2+2\delta y_3 &= -\delta R_2 \\ \delta y_1-2\delta y_2-5\delta y_3 &= -\delta R_3 \end{aligned}\right\}. \qquad (15)$$

These equations define the δR in terms of the δy. The reciprocal equations, defining the δy in terms of the δR, can be obtained by simple methods in the form

$$\left.\begin{aligned} -378\delta y_1 &= +34\delta R_1-12\delta R_2+2\delta R_3 \\ -378\delta y_2 &= -13\delta R_1-51\delta R_2-23\delta R_3 \\ -378\delta y_3 &= +12\delta R_1+18\delta R_2-66\delta R_3 \end{aligned}\right\}. \qquad (16)$$

For the ill-conditioned set (14) the reciprocal equations are given by

$$\left.\begin{aligned} -\delta x_1 &= +68\delta R_1-41\delta R_2-17\delta R_3+10\delta R_4 \\ -\delta x_2 &= -41\delta R_1+25\delta R_2+10\delta R_3-6\delta R_4 \\ -\delta x_3 &= -17\delta R_1+10\delta R_2+5\delta R_3-3\delta R_4 \\ -\delta x_4 &= +10\delta R_1-6\delta R_2-3\delta R_3+2\delta R_4 \end{aligned}\right\}. \qquad (17)$$

These reciprocal equations give the increments required to liquidate residuals. Equations (16) show, for example, that the largest effect on the δy caused by a single residual is only about one-sixth of the magnitude of that residual, while in the case of equations (17) this factor is as large as 68. In the latter case, further, a combination of residuals with signs $+ - - +$ reinforce each other, causing even larger increments to all the unknowns: in the former case no sign combination is bad for all the unknowns.

It will of course be realized that the reciprocal equations provide a complete solution to the original equations, when we replace the residuals by the initial constant terms. In the language of matrices, we have solved the equations

$$Ay = b \tag{18}$$

in the form

$$y = A^{-1}b. \tag{19}$$

The elements of the square matrix A are the coefficients of the unknowns in the original equations, the elements of the column matrix b the constant terms in those equations, and the elements of the reciprocal matrix A^{-1} are the coefficients in the inverse equations.

The determination of the inverse matrix is a more formidable undertaking than that of the solution of the simultaneous equations, so that this method of deciding the magnitude of negligible residuals is generally impracticable. The above illustration, however, shows that the problem exists and that its solution may be difficult.

10. In practice it will usually suffice to obtain approximate solutions to one or sometimes more figures than will be required in the final result. By means of the relaxation process, causing *steady and systematic* reduction in the size of the residuals, we can, moreover, see the way in which successive increments also get smaller. As the process goes on more and more significant figures become steady, showing no further change, and when the number of steady figures meets the prescribed requirement the work can stop.

†11. The question of convergence is closely allied with this matter of conditioning. It can be shown that the direct relaxation process, whereby at every stage the largest residual is liquidated by a change in its corresponding unknown, will always converge if the equations have been obtained from the condition that a positive definite quadratic function of the unknowns is a minimum. Many equations, for example in elasticity and statistics, arise naturally in this form, and any given set can be transformed to it by recording the conditions that the sum of the squares of the residuals shall be a minimum. In matrix notation, the equations

$$Ay = b$$

might be solved in the form

$$A'Ay = A'b,$$

where A' is the transpose of A. This is often called 'normalization'.

The question of theoretical convergence is not, however, of great importance: practical convergence is all-important. The set of equations

(14), for example, for which ultimate success is guaranteed, are almost impossible to solve by unrefined iteration, convergence being so slow as to make the labour prohibitive. For the unsymmetric set (13), on the other hand, convergence is rapid. The relaxation method, furthermore, by virtue of its flexibility, can be applied by a skilled computer to equations for which normal iterative methods would fail.

Relaxation is most useful when the set of equations has large coefficients down the principal diagonal, or when sufficient effective group-relaxation operators can be worked out to produce a similar effect. If the equations are ill-conditioned group-operators are hard to construct, particularly when the number of equations is large. It is found in the course of the work that small residuals require large changes in the unknowns for their liquidation, and the whole process is tedious and slow. This is a sure sign of ill-conditioning and provides a warning against accepting as accurate an approximate solution of the type shown in the first row of Table 4, if by chance this situation were ever reached.

Direct methods

†12. The method of relaxation requires for its efficient use a degree of skill which the casual computer may be unable to acquire. In some cases, moreover, it may be difficult to achieve a satisfactory rate of convergence when the equations are ill-conditioned. We therefore give a direct method for the solution of linear algebraic equations which is especially convenient when, as in our case, the coefficients of the equations are mainly grouped around the principal diagonal.

The method is based on the theorem that any square non-singular matrix can be expressed as the product of two triangular matrices, the first a lower triangle, with zero coefficients above the diagonal, and the second an upper triangle, with zero coefficients below the diagonal. The matrix equation

$$Ay = b \tag{20}$$

can then be written in the form

$$LUy = b, \tag{21}$$

where L and U are respectively the lower and upper triangles. If now we write

$$Uy = z, \tag{22}$$

then

$$Lz = b, \tag{23}$$

and z and y can be obtained successively from equations (23) and (22) respectively, the solution of such equations being relatively trivial.

The method of calculation of the coefficients of L and U can be seen most readily by considering a simple example. Given the coefficients a_{rs} of a third-order matrix A, we wish to determine the coefficients l_{rs} and u_{rs} in the equation

$$\begin{bmatrix} l_{11} & 0 & 0 \\ l_{21} & l_{22} & 0 \\ l_{31} & l_{32} & l_{33} \end{bmatrix} \begin{bmatrix} u_{11} & u_{12} & u_{13} \\ 0 & u_{22} & u_{23} \\ 0 & 0 & u_{33} \end{bmatrix} = \begin{bmatrix} a_{11} & a_{12} & a_{13} \\ a_{21} & a_{22} & a_{23} \\ a_{31} & a_{32} & a_{33} \end{bmatrix}. \tag{24}$$

From the rules for matrix multiplication we first have the three equations

$$\left.\begin{aligned} l_{11}u_{11} &= a_{11} \\ l_{11}u_{12} &= a_{12} \\ l_{11}u_{13} &= a_{13} \end{aligned}\right\}, \tag{25}$$

from which, if l_{11} or u_{11} is given an arbitrary non-zero value, the remaining three unknowns follow. Similar operations with the second row of L yield the equations

$$\left.\begin{aligned} l_{21}u_{11} &= a_{21} \\ l_{21}u_{12}+l_{22}u_{22} &= a_{22} \\ l_{21}u_{13}+l_{22}u_{23} &= a_{23} \end{aligned}\right\}, \tag{26}$$

from which, with l_{22} or u_{22} assigned an arbitrary value, the coefficients in the second row and column of L and U are obtained. The procedure can be continued and all the coefficients l_{rs} and u_{rs} are calculable.

The diagonal elements alone are arbitrary. It is common practice to put unit values in the diagonal elements of either L or U, the chosen triangle being then a *unit* triangle, or to make corresponding diagonal elements equal in magnitude. The latter method involves the computation of a square root, and is best reserved for the case when A is symmetric, the matrix U being then the transpose L' of L.

13. The next step in the solution of (20) is the determination of the auxiliary variable z from equation (23). We have the equation

$$\begin{bmatrix} l_{11} & 0 & 0 \\ l_{21} & l_{22} & 0 \\ l_{31} & l_{32} & l_{33} \end{bmatrix} \begin{bmatrix} z_1 \\ z_2 \\ z_3 \end{bmatrix} = \begin{bmatrix} b_1 \\ b_2 \\ b_3 \end{bmatrix}, \tag{27}$$

from which we can calculate z_1, z_2, and z_3 successively, using the equations

$$\left.\begin{aligned} l_{11}z_1 &= b_1 \\ l_{21}z_1+l_{22}z_2 &= b_2 \\ l_{31}z_1+l_{32}z_2+l_{33}z_3 &= b_3 \end{aligned}\right\}. \tag{28}$$

Finally, we obtain the required values of y from (22), which in this case represents the three equations

$$\left.\begin{aligned} u_{33}y_3 &= z_3 \\ u_{22}y_2+u_{23}y_3 &= z_2 \\ u_{11}y_1+u_{12}y_2+u_{13}y_3 &= z_1 \end{aligned}\right\}. \tag{29}$$

14. It is often desirable to know the precision obtainable in the solution of the linear equations. If the coefficients of the matrix A are assumed to be exact, but the constants b_r have uncertainties δb_r, then the uncertainties in the solution y are given by the equation

$$y = A^{-1}\delta b. \tag{30}$$

If the coefficients of A have small errors, so that the true matrix is $A+\delta A$, then we have

$$(A+\delta A)(y+\delta y) = b, \tag{31}$$

so that

$$\delta y = -A^{-1}(\delta A)y \tag{32}$$

is the correction to be added to y.

The evaluation of A^{-1} can be performed in many ways, of which the most obvious is the inversion of both triangles L and U and the calculation of A^{-1} from the equation

$$A^{-1} = U^{-1}L^{-1}. \tag{33}$$

The inversion of triangular matrices is effected by an obvious process, and needs no further description.

Non-linear equations

15. The solution of non-linear problems presents new difficulties. Linear differential equations almost always have a unique solution: the corresponding linear algebraic equations also have a unique solution, and this will correspond to that of the differential equation if the interval and finite-difference equations are well chosen.

Non-linear differential equations, however, may have no real solution or several real solutions. While there may be more than one mathematical solution to the given equation and associated boundary conditions, it is likely that only one is acceptable from physical considerations and this information, perhaps expressible mathematically in an inequality such as $|y| \not< Y$, a specified quantity, must be included in the specification of the problem. Moreover, a problem representing a physical system will almost certainly have at least one real solution, but it may happen that the equation or boundary conditions contain a parameter, whose value has to be adjusted to provide a real solution.

The corresponding non-linear algebraic equations have similar properties, with the additional complication that complex solutions may be produced because the chosen interval is too large. Consider, for example, the simple equation and boundary conditions given by

$$y''-y^2 = 0, \qquad y(0) = y(1) = -2{\cdot}1. \tag{34}$$

The simplest finite-difference equations are typified by

$$y_{r+1}-(2y_r+h^2y_r^2)+y_{r-1} = 0. \tag{35}$$

At the interval $h = \frac{1}{2}$ there is just one equation for calculating y at $x = \frac{1}{2}$, given by

$$y^2+8y+16{\cdot}8 = 0. \tag{36}$$

This equation has no real root, and we must conclude either that the interval is too large or that the differential equation has only complex solutions.

At the interval $h = \frac{1}{3}$ we find the results $y(\frac{1}{3}) = y(\frac{2}{3}) = -3{\cdot}34$ or $-5{\cdot}66$, and if these answers have any significance, that is if the approximate finite-difference equations are reasonable representations of the differential equation, we need an extra criterion to determine the physically acceptable solution.

16. The solution of the non-linear simultaneous algebraic equations may also be difficult, quite apart from questions of reality and uniqueness; there is no established numerical technique for this problem. The only practical method of general application seems to be to 'linearize' the equations and solve them by an iterative process.

Suppose, for example, that the equations are given by

$$f_r(y_1, y_2,..., y_n, x) = 0 \qquad (r = 1, 2,..., n). \tag{37}$$

If at any stage approximate solutions y_s are known, then corrections η_s are given, by Newton's rule, from the equations

$$\sum_{s=1}^{n} \eta_s \frac{\partial f_r}{\partial y_s} = -f_r \qquad (r = 1, 2,..., n). \tag{38}$$

Equations (38) constitute a set of linear equations from which the corrections η_s can be computed by methods already outlined. The new approximations $y_s+\eta_s$ are then used instead of the previous y_s, and the process carried out until further corrections are negligible.

Consider in this connexion the two equations corresponding to (35), with the interval $h = \frac{1}{3}$, given by

$$\left.\begin{aligned} -(2y_{\frac{1}{3}}+\tfrac{1}{9}y_{\frac{1}{3}}^2)+y_{\frac{2}{3}} &= +2{\cdot}1 \\ y_{\frac{1}{3}}-(2y_{\frac{2}{3}}+\tfrac{1}{9}y_{\frac{2}{3}}^2) &= +2{\cdot}1 \end{aligned}\right\}. \tag{39}$$

The equations corresponding to (38) are

$$\left.\begin{aligned} -(2+\tfrac{2}{9}y_{\frac{1}{3}})\eta_{\frac{1}{3}}+\eta_{\frac{2}{3}} &= (2y_{\frac{1}{3}}+\tfrac{1}{9}y_{\frac{1}{3}}^2)-y_{\frac{2}{3}}+2{\cdot}1 \\ \eta_{\frac{1}{3}}-(2+\tfrac{2}{9}y_{\frac{2}{3}})\eta_{\frac{2}{3}} &= -y_{\frac{1}{3}}+(2y_{\frac{2}{3}}+\tfrac{1}{9}y_{\frac{2}{3}}^2)+2{\cdot}1 \end{aligned}\right\}. \tag{40}$$

If we take as a first approximation $y_{\frac{1}{3}} = y_{\frac{2}{3}} = 0$ we find successive approximations to $y_{\frac{1}{3}} = y_{\frac{2}{3}}$ as follows:

$$0{\cdot}000, \quad -2{\cdot}100, \quad -3{\cdot}019, \quad -3{\cdot}304, \quad -3{\cdot}338.$$

The fifth value is correct to three decimals, and convergence is fairly rapid.

In general, however, convergence to any real solution is by no means certain, and it may be still more difficult to find the solution required in the physical problem. An *a priori* knowledge of a good approximation is therefore extremely useful in non-linear equations: convergence is then more certain, since the neglected terms in (38) are small, and there is a much better chance of securing the right solution.

17. It is interesting to compare Newton's method with the adaptation of the relaxation procedure for non-linear equations. With any guess at the solution, the residuals are given from (37) by

$$R_r = f_r(y_1, y_2,..., y_n, x) \qquad (r = 1, 2,..., n), \tag{41}$$

and the effect on R_r of a change in the unknown y_s is given by

$$\delta R_r = \frac{\partial f_r}{\partial y_s}\,\delta y_s, \tag{42}$$

provided that terms like $\delta y_s\,\delta y_r$, $(\delta y_s)^2$,... can be neglected. If we make simultaneous changes in the unknowns such that $\delta R_r = -R_r$ at all points, the magnitudes of the increments are clearly given by the solution of equations (38). In effect, therefore, the relaxation method uses the same equations as the 'direct' method, whether the differential equation is linear or non-linear. We may, however, find it convenient to solve the equations by relaxation, using the devices already explained for speeding convergence.

Solution of simultaneous finite-difference equations

18. We now consider the application of the relaxation and direct methods to the solution of the type of algebraic equations which are of common occurrence in our boundary-value techniques for solving ordinary differential equations, with particular reference to equations of orders two and four.

Relaxation: a computational arrangement for difference equations

19. A good computational arrangement is of some importance as regards efficiency and the prevention of errors. In this section we give a possible arrangement for the relaxational solution of finite-difference equations arising from differential equations of the second order. The adaptation to equations of higher order is obvious.

The computer is recommended to find and use an arrangement which best suits his convenience and temperament. The method favoured by the writer is illustrated in Fig. 1.

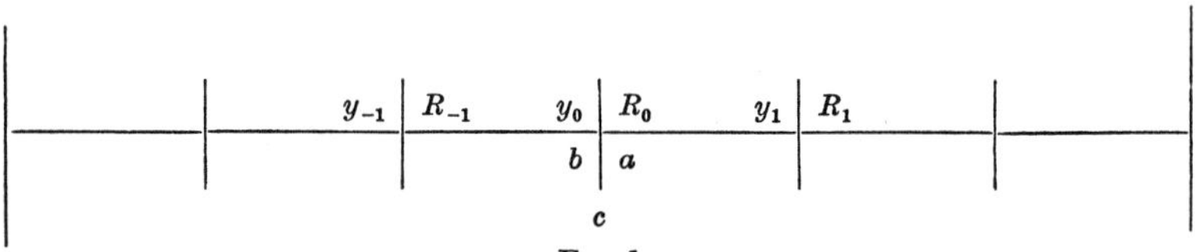

FIG. 1

In this diagram the horizontal line represents the range of the independent variable. Short vertical lines are drawn through the pivotal points, longer verticals through the end points of the range. Values of the unknowns are written to the left of the vertical lines, successive increments being recorded one above the other. Current residuals are recorded to the right of the vertical lines. Beneath the horizontal lines are written the coefficients in the equations. The part of the equation involving the unknowns to be satisfied at the typical pivotal point x_0, for example, is $ay_1+by_{-1}+cy_0$. The coefficients are often omitted, for clarity, in later diagrams.

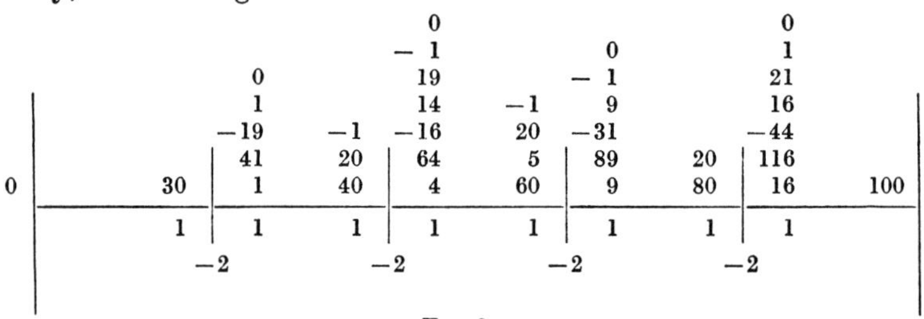

FIG. 2

Fig. 2 shows a typical piece of computation. The difference equations are typified by

$$y_1+y_{-1}-2y_0+x_0^2 = R_0, \tag{43}$$

with $y = 0$ at $x = 0$, $y = 100$ at $x = 5$. We wish to find y at $x = 1, 2, 3,$ and 4 for which the R's are negligibly small.

The given end values are written in ink to the left of their vertical lines, and the values of x^2 are recorded in ink to the right of the pivotal points. The initial guess is taken to be zero at all internal points, and the initial residuals are calculated. The latter are equal to the recorded values of x^2 except at the point $x = 4$, where the boundary value 100 is added to x^2 to produce the recorded value 116. The relaxation process is performed in pencil, successive increments being recorded to the left of the vertical lines. At any stage the *aggregate* of the quantities on the left of a line is the latest approximation to the unknown at that point, whereas the last or topmost figure to the right is the residual still unliquidated. This is of course analogous to the arrangement of Table 2 for the more general type of equations.

A rubber is essential. When the sheet is becoming uncomfortably crowded, or after the residuals have been substantially reduced, the increments to the y are summed, the column erased, and the new value inserted to the left of the vertical line. On the right the column of residuals is erased, the top one being placed in the bottom position. The residuals are then checked and more relaxation performed if necessary.

Modifications of this arrangement, which may be necessary in more complicated problems, will readily suggest themselves to the computer.

Relaxation: second-order equations

20. If the equation and boundary conditions are linear, the general second-order equation may be represented by an approximate finite-difference equation of the form

$$y_1(1+\tfrac{1}{2}hf_0)+y_{-1}(1-\tfrac{1}{2}hf_0)-y_0(2-h^2g_0)-h^2k_0 = 0, \qquad (44)$$

where f, g, and k are functions of x, and the suffix 0 indicates that the function is to be evaluated at the pivotal point x_0. The first step in the solution of a second-order differential equation is the solution of a set of algebraic equations typified by (44). Associated with them there will be two other equations, one at each end of the range of integration, expressing some condition between one or more adjacent points.

Consider a portion of the range of integration, shown in Fig. 3.

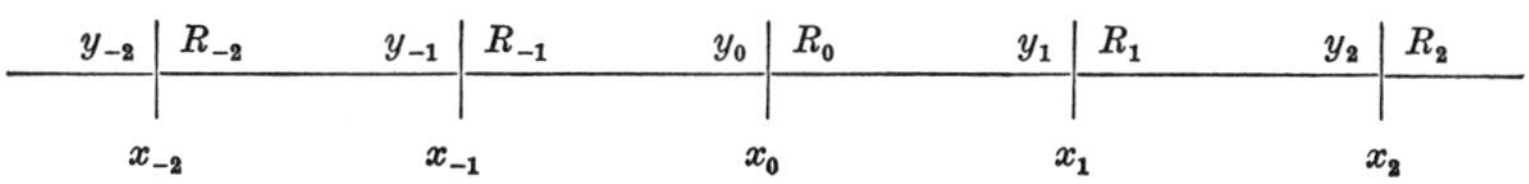

FIG. 3

The residual R_r at any point x_r is defined to be the left-hand side of equation (44), that is

$$R_r = y_{r+1}(1+\tfrac{1}{2}hf_r)+y_{r-1}(1-\tfrac{1}{2}hf_r)-y_r(2-h^2g_r)-h^2k_r. \qquad (45)$$

With any assumption of the y values at pivotal points we can calculate the residuals at pivotal points from (45). The next step is to alter the y values systematically so that the residuals are steadily reduced and ultimately become negligible. It is clear from the set of equations (45) that the effects on the residuals of a change δy_r in the quantities y_r are given by

$$\left.\begin{aligned} \delta R_r &= -(2-h^2g_r)\delta y_r \\ \delta R_{r-1} &= (1+\tfrac{1}{2}hf_{r-1})\delta y_r \\ \delta R_{r+1} &= (1-\tfrac{1}{2}hf_{r+1})\delta y_r \end{aligned}\right\}, \tag{46}$$

only three residuals, at three adjacent points, being affected by the change δy_r. This suggests that relaxation will be a useful and simple technique for the solution of problems of this kind, provided that the process converges reasonably quickly, that is if the equations are not badly conditioned.

Not only is each operation a simple one, but also the construction of group-relaxation operators should present no difficulty. It is, moreover, easy to decide when to over-relax and when to under-relax, a matter of some difficulty in the solution of a general set of algebraic equations.

21. The simplest case is that for which both $f(x)$ and $g(x)$ are everywhere zero, and the boundary values are specified. Equation (45) need then be applied only at internal points; at the boundaries we record the given values. Equations (46) then become

$$\left.\begin{aligned} \delta R_r &= -2\delta y_r \\ \delta R_{r-1} &= \delta y_r \\ \delta R_{r+1} &= \delta y_r \end{aligned}\right\} \tag{47}$$

at a general point, and at the points adjacent to the boundaries only two residuals are affected.

We can easily construct relaxation patterns as shown in Fig. 4.

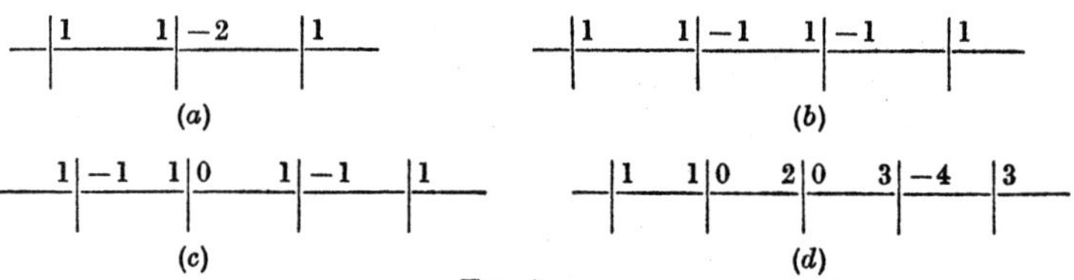

FIG. 4

It will be clear that over-relaxation is necessary when two or more adjacent residuals are of the same sign, under-relaxation when positive and negative residuals alternate freely. We notice also that if none of the points at which increments are made are adjacent to the boundary the total algebraic sum of the residuals is unchanged. Fig. 5 gives the results, similar to those of Fig. 4, when one of the altered values is next to a boundary.

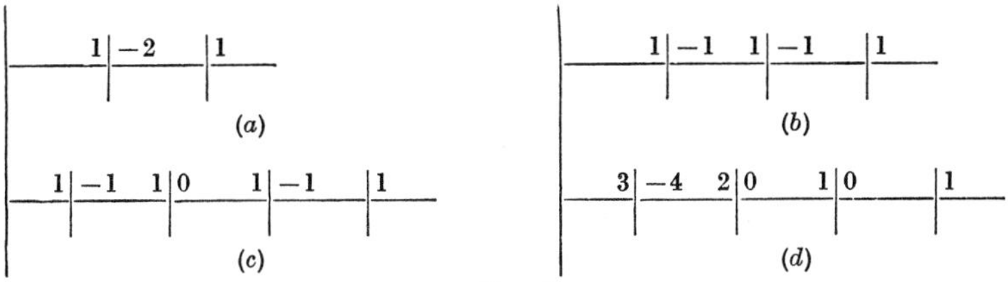

FIG. 5

In these cases the algebraic sum is decreased by positive increments. If then we start with predominantly positive residuals, the aim is to sweep them towards the boundaries, so that we can make positive increments at the penultimate points, thus reducing the algebraic sum.

22. We can best illustrate this technique by considering the solution of the equations

$$\left.\begin{aligned} y_{r+1}+y_{r-1}-2y_r+10 = 0, \qquad r = 1, 2, 3, 4, 5 \\ y_0 = 14, \quad y_6 = 20 \end{aligned}\right\}. \tag{48}$$

Taking all initial pivotal values to be zero, the residuals are calculated as shown in Fig. 6 (*a*), and we effect their liquidation by the steps shown in (*b*), (*c*), (*d*), and (*e*) of Fig. 6; in practice we should of course produce only Fig. 6 (*e*).

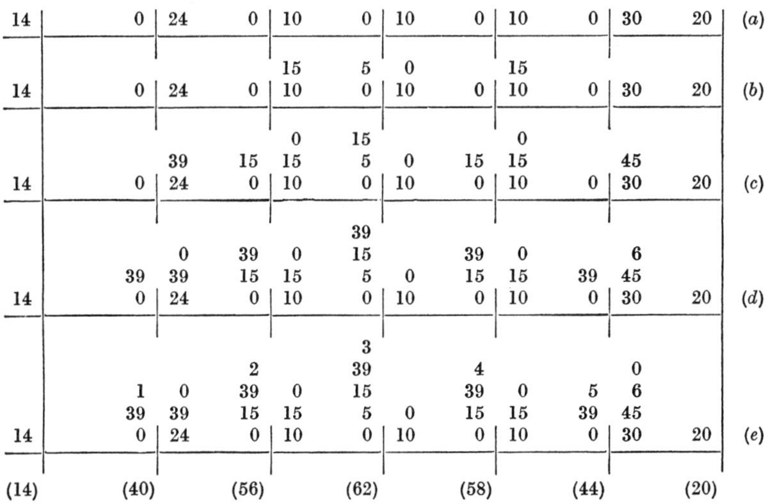

FIG. 6

When the residuals are negligible we sum the increments, as shown in Fig. 6 (*e*), and recalculate the residuals as a check on the arithmetic.

23. An alternative method might be mentioned. Starting from Fig. 6 (*a*), let us alter every value equally so that the total algebraic sum of the residuals is zero. The picture will then be as shown in Fig. 7 (*a*). Now make the algebraic sum vanish over the three central points, an operation not affecting the total residual, and obtain the situation in Fig. 7 (*b*). A single 'point relaxation' at the centre produces the result of Fig. 7 (*c*), and a self-evident 'group' produces the final result of Fig. 7 (*d*).

14	42 0	−18 24	42 0	10	42 0	10	42 0	10	42 0	−12 30	20	(*a*)
14	42 0	− 3 −18 24	15 42 0	−5 10	15 42 0	10	15 42 0	−5 10	42 0	3 −12 30	20	(*b*)
14	42 0	− 3 −18 24	15 42 0	0 −5 10	5 15 42 0	0 10	15 42 0	0 −5 10	42 0	3 −12 30	20	(*c*)
14	−2 42 0	0 − 3 −18 24	−1 15 42 0	0 −5 10	5 15 42 0	0 10	1 15 42 0	0 −5 10	2 42 0	0 3 −12 30	20	(*d*)

FIG. 7

This is of course a very simple example, but together with the remarks of § 7 should provide sufficient information to enable the reader, if he so desires, to improve his skill by practice.

24. If the boundary condition is other than 'boundary value specified' the technique needs modification. This will be discussed in detail in Chapter 4. Here we remark that it is necessary to use (45) also at that boundary point, and one external point is introduced. For the determination of the latter the boundary condition will provide yet another equation. A typical example is provided by equations (48), to be satisfied now at $r = 1, 2, 3, 4, 5$, and 6, with $y_0 = 14$ as before, and another condition given by

$$y_7 - y_5 = 40, \text{ say.} \tag{49}$$

This corresponds approximately to 'boundary derivative specified' at $r = 6$.

The six equations of the form (48) and the single equation (49) suffice for the determination of the seven unknowns $y_1,\ldots, y_7$. Equation (49) is not very satisfactory for the purpose of relaxation, and it is preferable to eliminate y_7 between (49) and (48) (for $r = 6$), obtaining, for the equation to be satisfied at $r = 6$, the result

$$2y_5 - 2y_6 + 50 = 0. \tag{50}$$

The relaxation patterns at $r = 6$ and 5 are now given respectively by the two diagrams of Fig. 8.

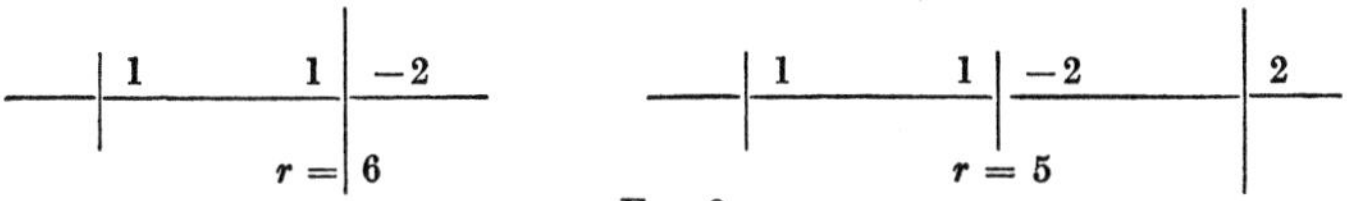

FIG. 8

The reader can follow the steps in the relaxation given in Fig. 9.

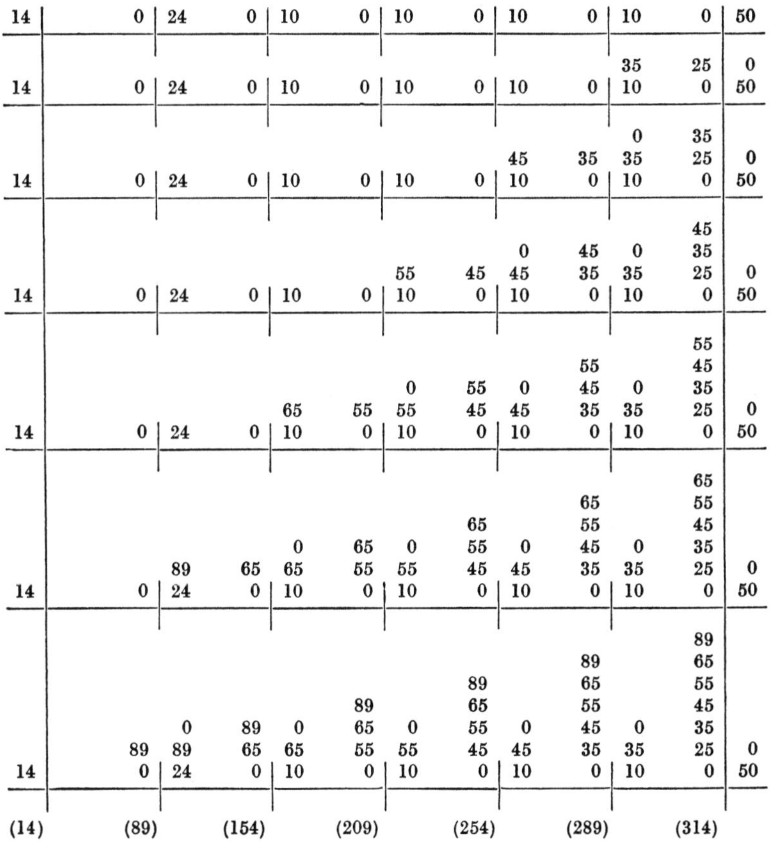

FIG. 9

A similar method may be used at a point of symmetry, often present at the mid-point of the range of integration. In this case the general equation at the mid-point x_m, containing three adjacent values, can be made to contain only two by using the exact relation $y_{m+1} = y_{m-1}$.

25. When the linear equations have the coefficients 1, -2, 1 and when the boundary values are specified, residuals can 'escape' at each end of the range. When one boundary derivative is specified, on the other hand, the residuals can escape only at one end, and much larger changes have to be made in pivotal values to liquidate residuals of similar size; in other words the equations are more ill-conditioned in the second case than in the first.

It is very important that the computer has an appreciation of the conditions under which residuals can be regarded as negligible. Suppose, for example, that in the problem corresponding to Fig. 9 the residuals at some stage were all equal to $+1$. 'Point-relaxation' would suggest to the unwary that the problem was solved, since at any point the residual is liquidated by a change of only half a unit in the pivotal value. The necessary changes for complete liquidation, however, are as shown in Fig. 10.

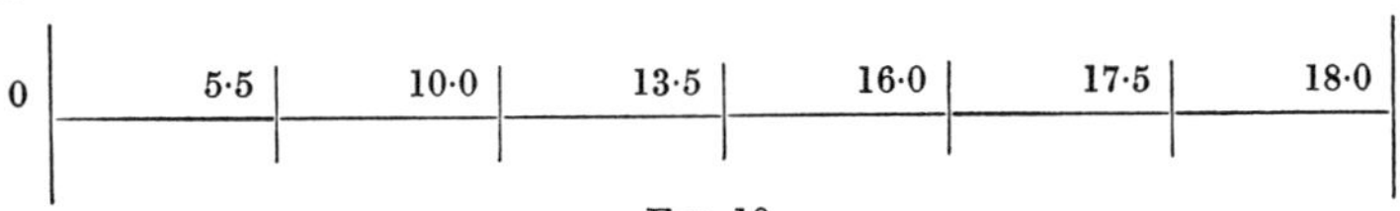

FIG. 10

If both boundary values were fixed, the effect of a unit residual at every point is as shown in Fig. 11: the precision obtainable is now considerably better.

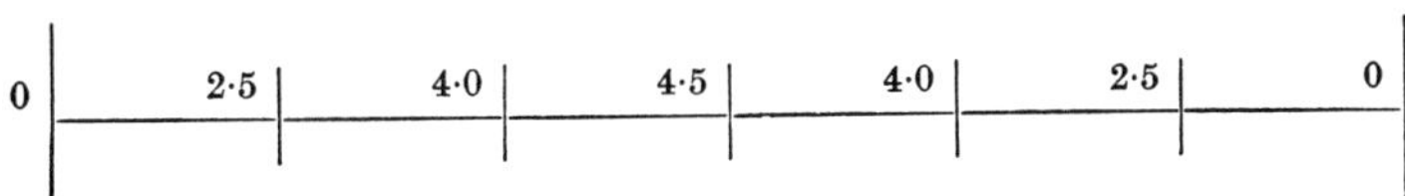

FIG. 11

Further consideration of these and similar topics concerning accuracy and precision is deferred until Chapter 9.

Relaxation: simultaneous second-order equations

26. The simultaneous solution of a pair of second-order differential equations involves the solution of a set of algebraic equations, for the unknowns y_r and z_r, typified by

$$\left.\begin{aligned} a_r y_{r+1}+b_r y_{r-1}+c_r y_r+d_r z_{r+1}+e_r z_{r-1}+f_r z_r &= g_r \\ A_r y_{r+1}+B_r y_{r-1}+C_r y_r+D_r z_{r+1}+E_r z_{r-1}+F_r z_r &= G_r \end{aligned}\right\}. \tag{51}$$

Again the boundary conditions will add enough equations to enable a unique solution to be calculable from the simultaneous algebraic equations. At a small enough interval a_r, b_r, D_r, and E_r will have values close to unity, c_r and F_r are nearly -2, and the other coefficients on the left of (51) are small.

We now have two diagrams, one for y and one for z. The residuals are given by

$$\left.\begin{aligned} R_r(y) &= a_r y_{r+1} + b_r y_{r-1} + c_r y_r + d_r z_{r+1} + e_r z_{r-1} + f_r z_r - g_r \\ R_r(z) &= A_r y_{r+1} + B_r y_{r-1} + C_r y_r + D_r z_{r+1} + E_r z_{r-1} + F_r z_r - G_r \end{aligned}\right\}, \tag{52}$$

and the effect on the residuals of changes $+1$ in y_r and $+1$ in z_r are shown in Fig. 12.

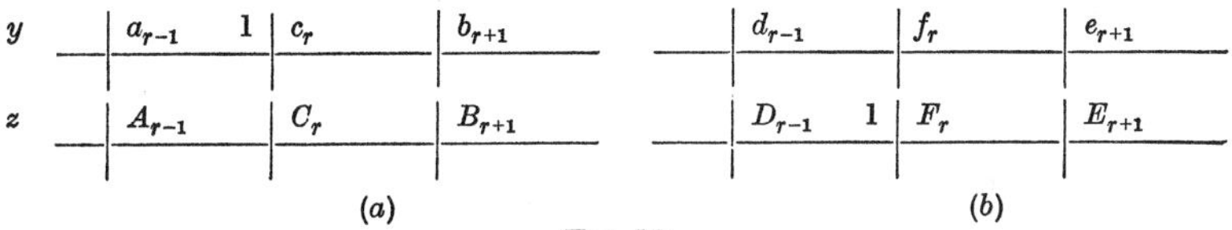

FIG. 12

The relaxation method is straightforward, and needs no illustration.

Relaxation: fourth-order equations

27. A differential equation of the fourth order has a corresponding finite-difference equation which relates five adjacent pivotal values, of the form

$$a_r y_{r+2} + b_r y_{r+1} + c_r y_r + d_r y_{r-1} + e_r y_{r-2} = f_r. \tag{53}$$

At a small interval the coefficients will not differ greatly from the binomial coefficients of the fourth order, that is equation (53) will have the approximate form

$$y_{r+2} - 4y_{r+1} + 6y_r - 4y_{r-1} + y_{r-2} = f_r. \tag{54}$$

The effect on the residuals of a unit increment in y_r is shown in Fig. 13, five residuals being affected.

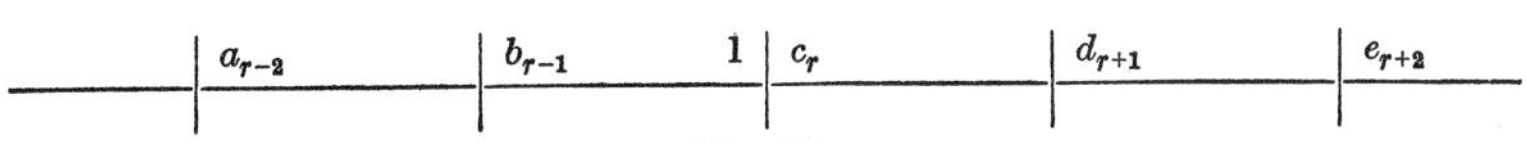

FIG. 13

Again the boundary conditions will yield enough extra equations to provide a unique solution, and the relaxation process is straightforward. Experience shows, however, that this method is rather difficult and tedious. The equations are generally ill-conditioned, and convergence is slow.

By way of illustration we take the range $r = 0$ to 6, suppose the boundary values specified at these points, and take other boundary conditions in the form

$$y_{-1} = y_1, \qquad y_5 = y_7. \tag{55}$$

Applying (54) at $r = 1, 2, 3, 4, 5$, and eliminating y_{-1} and y_7 by means of (55), we have the following five simultaneous equations for the determination of y_1, y_2, y_3, y_4, and y_5, in which the right-hand sides are calculable from the known values of f_r, y_0, and y_6.

$$\left.\begin{aligned} 7y_1-4y_2+y_3 &= S_1\\ -4y_1+6y_2-4y_3+y_4 &= S_2\\ y_1-4y_2+6y_3-4y_4+y_5 &= S_3\\ y_2-4y_3+6y_4-4y_5 &= S_4\\ y_3-4y_4+7y_5 &= S_5 \end{aligned}\right\}. \tag{56}$$

The values of y needed to liquidate unit negative residuals at every point are shown, rounded off to one decimal, in Fig. 14. The chosen boundary conditions, the range of integration, and the given residuals correspond as nearly as possible to the corresponding case of a second-order equation, and the results can be compared with those of Fig. 11. The maximum value is nearly as great, although in an unrefined relaxation process 'point' relaxation for a unit residual would suggest an initial increment of $\frac{1}{6}$ for the fourth-order equation, and $\frac{1}{2}$ for that of the second order. The ill-conditioning is therefore worse in this case by a factor of roughly three, and gets rapidly more pronounced as the number of points is increased.

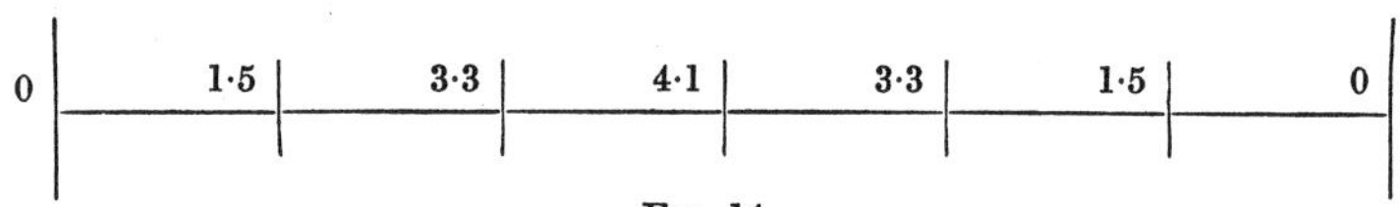

FIG. 14

We shall discuss further topics of this nature in Chapter 9.

Relaxation: equations of odd order

28. The finite-difference equations corresponding to differential equations of odd order are often inconvenient for solution by relaxation methods. For a first-order equation, for example, the finite-difference equation may have the form

$$a_r y_{r+1} + b_r y_{r-1} + c_r y_r = d_r, \tag{57}$$

in which, at a small interval, c is much less than a or b. This means that the liquidation of a residual at any point, by a change in the value of the function at that point, causes greater changes in adjacent residuals. The unrefined process is likely to be divergent, though we can always normalize the equations, by the method of § 11, thereby ensuring convergence. The lack of symmetry in the position of the associated conditions, however, means that special methods have to be devised to effect the solution of differential equations of odd order by means of boundary-value methods.

Direct methods: second-order equations

29. We turn now to the application of our direct methods to the special cases encountered in the solution of differential equations.

For a second-order equation the corresponding finite-difference equation involves only three adjacent pivotal values, and the boundary conditions can often be treated in such a way that the first and last equations contain only two unknowns.

We therefore consider a matrix of the following form, only non-zero elements being recorded.

$$\begin{bmatrix} a_{11} & a_{12} & & & & \\ a_{21} & a_{22} & a_{23} & & & \\ & a_{32} & a_{33} & a_{34} & & \\ & \cdot & \cdot & \cdot & \cdot & \cdot \\ & & & a_{n-1,n-2} & a_{n-1,n-1} & a_{n-1,n} \\ & & & & a_{n,n-1} & a_{n,n} \end{bmatrix}. \tag{58}$$

First, for the determination of the elements of the matrices L and U, with L taken to be a unit triangle, we have equations typified by the following:

$$\left.\begin{aligned} & u_{11} = a_{11}, \quad u_{12} = a_{12}, \quad u_{13} = u_{14} = \ldots = 0 \\ & l_{21} = a_{21}/u_{11} \\ & u_{22} = a_{22} - l_{21}u_{12}, \quad u_{23} = a_{23}, \quad u_{24} = u_{25} = \ldots = 0 \\ & l_{31} = 0, \quad l_{32} = a_{32}/u_{22} \end{aligned}\right\}. \tag{59}$$

The matrices L and U are found to have non-zero elements only in their main diagonals and adjacent sloping lines. In U the off-diagonal terms are identical with those in the corresponding positions of the original matrix A. The general equations are given by the simple formulae

$$\left.\begin{aligned} & l_{rr} = 1, \quad l_{r+1,r} = a_{r+1,r}/u_{rr} \\ & u_{rr} = a_{rr} - l_{r,r-1}u_{r-1,r}, \quad u_{r,r+1} = a_{r,r+1} \end{aligned}\right\}. \tag{60}$$

The solution of the matrix equations

$$\left.\begin{aligned} Lz &= b \\ Uy &= z \end{aligned}\right\} \tag{61}$$

is also straightforward. If the vectors b and z have components b_r and z_r we have

$$\left.\begin{aligned} z_1 \quad &= b_1 \\ l_{21}z_1+z_2 &= b_2 \end{aligned}\right\}, \tag{62}$$

and, in general,
$$l_{r,r-1}z_{r-1}+z_r = b_r. \tag{63}$$

Then also

$$\left.\begin{aligned} u_{n,n}y_n &= z_n \\ u_{n-1,n-1}y_{n-1}+u_{n-1,n}y_n &= z_{n-1} \end{aligned}\right\}, \tag{64}$$

and, in general,
$$u_{r,r}y_r+u_{r,r+1}y_{r+1} = z_r. \tag{65}$$

Apart from the possibility of having to keep guarding figures, the labour of this method is clearly independent of the degree of ill-conditioning of the simultaneous equations involved.

30. As a simple example consider the solution of the following set of four simultaneous equations of this type:

$$\left.\begin{aligned} -2y_1+\ y_2 \qquad\qquad &= \ \ -1 \\ y_1-2y_2+\ y_3 \qquad &= \ \ -4 \\ y_2-2y_3+\ y_4 &= \ \ -9 \\ y_3-2y_4 &= -116 \end{aligned}\right\}. \tag{66}$$

Carrying out the process of § 29 we find the following matrices L and U:

$$L = \begin{bmatrix} 1 & & & \\ -\frac{1}{2} & 1 & & \\ 0 & -\frac{2}{3} & 1 & \\ 0 & 0 & -\frac{3}{4} & 1 \end{bmatrix}, \qquad U = \begin{bmatrix} -2 & 1 & 0 & 0 \\ & -\frac{3}{2} & 1 & 0 \\ & & -\frac{4}{3} & 1 \\ & & & -\frac{5}{4} \end{bmatrix}.$$

Application of equations (62) and (63) then gives

$$z_1 = -1, \qquad z_2 = -4{\cdot}5, \qquad z_3 = -12, \qquad z_4 = -125,$$

and equations (64) and (65) yield the final results

$$y_4 = 100, \qquad y_3 = 84, \qquad y_2 = 59, \qquad y_1 = 30,$$

which may be verified by substitution in the original equations (66).

†31. Since the matrix of coefficients of equations (66) is symmetric, we could use the square-root method and compute only one triangular matrix. Since the diagonal terms of A are all negative, we write $-A = LL'$, and find L as follows:

$$L$$

$$\begin{bmatrix} \sqrt{2} & & & \\ -\dfrac{1}{\sqrt{2}} & \sqrt{\dfrac{3}{2}} & & \\ 0 & -\sqrt{\dfrac{2}{3}} & \dfrac{2}{\sqrt{3}} & \\ 0 & 0 & -\dfrac{\sqrt{3}}{2} & \dfrac{\sqrt{5}}{2} \end{bmatrix}.$$

In succession we calculate the vectors z and y from the equations $Lz = -b$, $L'y = z$, obtaining the results

$$z = \frac{1}{\sqrt{2}},\ 3\sqrt{\frac{3}{2}},\ 6\sqrt{3},\ 50\sqrt{5},$$

$$y = 30,\ 59,\ 84,\ 100.$$

Direct methods: simultaneous second-order equations

32. The finite-difference equations typified by (51), corresponding to a pair of simultaneous second-order equations, also give rise to a matrix of special type. Taking equations (51) with $r = 1$, 2, 3, 4, and 5, and assuming y_0, y_6, z_0, and z_6 to have specified values, the matrix has the following appearance.

$$\begin{array}{ccccc|ccccc} y_1 & y_2 & y_3 & y_4 & y_5 & z_1 & z_2 & z_3 & z_4 & z_5 \\ \hline c_1 & a_1 & & & & f_1 & d_1 & & & \\ b_2 & c_2 & a_2 & & & e_2 & f_2 & d_2 & & \\ & b_3 & c_3 & a_3 & & & e_3 & f_3 & d_3 & \\ & & b_4 & c_4 & a_4 & & & e_4 & f_4 & d_4 \\ & & & b_5 & c_5 & & & & e_5 & f_5 \\ C_1 & A_1 & & & & F_1 & D_1 & & & \\ B_2 & C_2 & A_2 & & & E_2 & F_2 & D_2 & & \\ & B_3 & C_3 & A_3 & & & E_3 & F_3 & D_3 & \\ & & B_4 & C_4 & A_4 & & & E_4 & F_4 & D_4 \\ & & & B_5 & C_5 & & & & E_5 & F_5 \end{array}.$$

In this form the equations corresponding to the first of (51) are represented by the first five rows of the matrix, those relating to the second of (51) occupying the last five rows. Remembering the remarks about the relative sizes of the coefficients, we see that this arrangement ensures that the diagonal terms are as large as possible.

When we construct the matrices L and U we find some simplification caused by the zeros in A, but there are no simple general expressions which are worth recording, corresponding to those of § 29 for a single second-order equation.

Direct methods: fourth-order equations

33. For a fourth-order equation, the general matrix will have the particular form shown below.

$$A$$

$$\begin{bmatrix} a_{11} & a_{12} & a_{13} & & & & & & & \\ a_{21} & a_{22} & a_{23} & a_{24} & & & & & & \\ a_{31} & a_{32} & a_{33} & a_{34} & a_{35} & & & & & \\ 0 & a_{42} & a_{43} & a_{44} & a_{45} & a_{46} & & & & \\ \cdot & \cdot & \cdot & \cdot & \cdot & \cdot & \cdot & \cdot & \cdot & \cdot \\ & & & & & a_{n-2,n-4} & a_{n-2,n-3} & a_{n-2,n-2} & a_{n-2,n-1} & a_{n-2,n} \\ & & & & & & a_{n-1,n-3} & a_{n-1,n-2} & a_{n-1,n-1} & a_{n-1,n} \\ & & & & & & & a_{n,n-2} & a_{n,n-1} & a_{n,n} \end{bmatrix}.$$

Except near the beginning and end, each row of A contains five elements, grouped round the diagonal, and again there is considerable simplification in the determination of the elements of L and U. Recalling the notation for the elements of L and U, we have the following equations for their determination:

$$\left.\begin{aligned} u_{11} &= a_{11}, \quad u_{12} = a_{12}, \quad u_{13} = a_{13}, \text{ etc.} \\ l_{21}u_{11} &= a_{21}, \quad l_{21}u_{12}+u_{22} = a_{22}, \quad l_{21}u_{13}+u_{23} = a_{23}, \text{ etc.} \\ l_{31}u_{11} &= a_{31}, \quad l_{31}u_{12}+l_{32}u_{22} = a_{32}, \quad l_{31}u_{13}+l_{32}u_{23}+u_{33} = a_{33} \end{aligned}\right\}. \tag{67}$$

The order of calculation is (i) first row of U, (ii) second row of L, (iii) second row of U, (iv) third row of L, (v) third row of U, and so on.

Pursuing this process we find, from the first of (67), that $u_{14}, u_{15},... = 0$. From the second of (67) we find that $u_{25} = u_{26} = ... = 0$, and the third of (67) gives $u_{36} = u_{37} = ... = 0$. The next equation gives $l_{41} = 0$, and the forms of L and U become clear. L has non-zero elements only in the principal diagonal and the two sloping lines beneath it, U is non-zero only in its principal diagonal and the two sloping lines above it. In comparison with the corresponding result for second-order equations, we see that both L and U have an extra non-zero sloping line.

34. Consider, as a simple example, the solution of the simultaneous equations of this type given by (56), with $S_1 = S_2 = S_3 = S_4 = S_5 = 1$.

The matrices A, L, and U are shown below:

$$\overset{A}{\begin{bmatrix} 7 & -4 & 1 & & \\ -4 & 6 & -4 & 1 & \\ 1 & -4 & 6 & -4 & 1 \\ & 1 & -4 & 6 & -4 \\ & & 1 & -4 & 7 \end{bmatrix}}, \quad \overset{L}{\begin{bmatrix} 1 & & & & \\ -\frac{4}{7} & 1 & & & \\ \frac{1}{7} & -\frac{12}{13} & 1 & & \\ & \frac{7}{26} & -\frac{8}{7} & 1 & \\ & & \frac{13}{35} & -\frac{40}{31} & 1 \end{bmatrix}}, \quad \overset{U}{\begin{bmatrix} 7 & -4 & 1 & & \\ & \frac{26}{7} & -\frac{24}{7} & 1 & \\ & & \frac{35}{13} & -\frac{40}{13} & 1 \\ & & & \frac{31}{14} & -\frac{20}{7} \\ & & & & \frac{456}{155} \end{bmatrix}}.$$

We now calculate the auxiliary variable z from the equation $Lz = S$, obtaining the results

$$z = 1, \tfrac{11}{7}, \tfrac{30}{13}, \tfrac{45}{14}, \tfrac{133}{31},$$

and the required y from $Uy = z$, finding

$$y = \tfrac{35}{24}, \tfrac{10}{3}, \tfrac{33}{8}, \tfrac{10}{3}, \tfrac{35}{24},$$

in agreement with the result of Fig. 14.

Direct methods: equations of odd order

35. Direct methods find no difficulty with equations of odd order. A first-order equation will often have a matrix already of triangular form, though occasionally it is solved by a three-term formula similar to that of second-order equations. A third-order equation may be solved by using a five-term formula as in the case of fourth-order equations, but more often we shall use a four-term formula, as described in Chapter 6, and this creates a new situation.

The original matrix may have the form shown below:

$$\begin{bmatrix} a_{11} & a_{12} & a_{13} & & & & & \\ a_{21} & a_{22} & a_{23} & a_{24} & & & & \\ & a_{32} & a_{33} & a_{34} & a_{35} & & & \\ & & a_{43} & a_{44} & a_{45} & a_{46} & & \\ \cdot & \cdot & \cdot & \cdot & \cdot & \cdot & \cdot & \cdot \\ & & & & & a_{n-2,n-3} & a_{n-2,n-2} & a_{n-2,n-1} \\ & & & & & & a_{n-1,n-2} & a_{n-1,n-1} \end{bmatrix}.$$

In the triangular resolution we shall find in this case that the lower triangle L has the same form as that corresponding to second-order equations, while U has the same shape as that corresponding to fourth-order equations. For the determination of the elements we have

$$\left.\begin{aligned} u_{11} &= a_{11}, \quad u_{12} = a_{12}, \quad u_{13} = a_{13}, \quad u_{14} = \dots = 0 \\ l_{21}u_{11} &= a_{21}, \quad l_{21}u_{12}+u_{22} = a_{22}, \quad l_{21}u_{13}+u_{23} = a_{23}, \\ & \qquad\qquad u_{24} = a_{24}, \quad u_{25} = \dots = 0 \\ l_{31} &= 0, \quad l_{32}u_{23}+u_{33} = a_{33}, \quad \text{etc.} \end{aligned}\right\}, \tag{68}$$

and we need give no further numerical illustration.

In the original matrix there are usually two elements above the main diagonal in each row, one element below. We might also have the reverse case, when the matrix L will have the form corresponding to that of fourth-order equations, U corresponding to that of second-order equations. It will be obvious by now that L will have non-zero elements below the diagonal and U above the diagonal when A has non-zero elements in the corresponding positions.

General comments

36. The direct method has one great advantage over the iterative method. The hardest part of the work is involved in the determination of L and U, and whenever several sets of equations are presented for solution for which the left-hand sides are unaltered, the right-hand sides only changing, the evaluation of the triangles need be performed once only. With an iterative method there is no such simplification. This is an important consideration, since this state of affairs is presented whenever a first approximation is improved by the incorporation of a difference correction.

On the other hand the relaxation method may be quicker when the matrix has a form 'unsatisfactory' for the direct method, as for example that of § 32, particularly when the off-diagonal terms are fairly small. Each case has to be determined on its merits, and the computer familiar with both methods will have little difficulty in deciding which is the more appropriate.

In many problems the two methods will be found to be complementary. This would happen, for example, whenever a solution has been obtained by a direct method at a certain interval, which is subsequently found to be too large for the satisfactory convergence of differences. We should then need to use a smaller interval, but the first approximation may be so good that the remaining work can be performed trivially by relaxation, and it would clearly be uneconomic to solve the new equations by direct methods. The wise computer will make himself familiar with both methods.

Additional notes and bibliography

§ 3. The difference correction is also called the 'truncation error'. We favour the former term, which suggests that it will be used to correct, while the latter suggests unavoidable error. This may be the origin of the often accepted fiction that numerical methods are 'approximate', analytical methods 'exact'. When numerical values are needed, analytical methods are no more exact than numerical methods properly applied. Both methods involve calculations with rounded numbers, so that results are almost always accurate only to a certain number of decimals or significant figures. The analytical method, for example, may involve computations with functions given in tables of fixed precision, or may involve the summation of an infinite series which is necessarily truncated at some term. If we call this 'exact' then our numerical method, which involves the solution of linear equations, is also exact. If the difference corrections are properly applied, using an interval which makes the finite-difference equations meaningful, there need be no significant extra error from this source. If the difference correction is neglected, of course, the finite-difference equations are no longer precise representations of the differential equation, and the numerical method is then rightly called 'approximate'.

Further comment on the difference correction is given in the notes on Chapter 4.

§ 4. In the present state of development, high-speed computing machines work best with ordered routines such as those given in equations (8) and (9). Relaxation, involving scanning and judgement, makes heavy demands on the programmer and wastes machine time: it is generally preferred for pencil and paper computation.

§ 11. Questions of convergence of relaxation processes have been discussed by

TEMPLE, G., 1939, The general theory of relaxation methods applied to linear systems. *Proc. Roy. Soc.* A, **169**, 476.

§ 12. The theorem $A = LU$ holds only if none of the principal minors of A vanish. In particular, there is an obvious difficulty in equations (25) if $a_{11} = 0$. The theorem still holds, however, for a matrix obtained from A by interchanging some of the rows of A. This is an unimportant restriction on the theorem. If the determinant of A vanishes, however, the last diagonal term of L or U will also vanish, and this important special case is discussed in Chapter 9.

§§ 12–14. A detailed and illustrated description of many direct methods of solving linear equations and inverting matrices has been given by

FOX, L., 1954, Practical solution of linear equations and inversion of matrices. *National Bureau of Standards Applied Mathematics Series*, **39**. Washington.

For accurate computation with desk machines close attention must be paid to the choice of position which the various numbers occupy on the computing sheet. In a matrix multiplication AB for example, in which the rule is 'row-by-column', the corresponding pairs of numbers are inconveniently placed. If we record the transpose B' beneath A, we can multiply rows of A by rows of B' to find the elements of AB, and numbers to be multiplied together lie in the same column. Similarly the recording of A' on the left of B would lead to column-by-column multiplication, corresponding elements being in the same row.

Frequent checks are also essential. In computations involving matrices the use of 'sum' checks is easy and adequate. For example, to check the elements of $C = AB$, we might form a row S of the column sums of B', and a column $\sum$ of the row sums of C. Then the rth element of $\sum$ should satisfy the relation

$$\sum_r = R_r(A) \times S,$$

where $R_r(A)$ means the rth row of A. Similar checks are used in many contexts.

The conditioning of the matrix becomes apparent in this direct method from the behaviour of the diagonal terms of the 'non-unit' triangle. If any of the principal minors have small values the corresponding diagonal terms may be small, but this is balanced by large adjacent diagonal terms and is not significant. If the complete matrix is ill-conditioned, however, so that the matrix is almost singular, the last diagonal element is small and the solution not very well defined. This is illustrated in detail in Chapter 9.

§ 31. In practical problems, of course, we should evaluate the square roots to a fixed number of decimals. The proper computational arrangement, with adequate checks, would appear as follows.

$$-A = \begin{bmatrix} 2 & -1 & & \\ -1 & 2 & -1 & \\ & -1 & 2 & -1 \\ & & -1 & 2 \end{bmatrix} \qquad \begin{matrix} -b \\ 1 \\ 4 \\ 9 \\ 116 \end{matrix} \qquad \begin{matrix} \Sigma \\ 2 \\ 4 \\ 9 \\ 117 \end{matrix}$$

$$L = \begin{bmatrix} 1{\cdot}4142 & & & \\ -0{\cdot}7071 & 1{\cdot}2247 & & \\ 0 & -0{\cdot}8165 & 1{\cdot}1547 & \\ 0 & 0 & -0{\cdot}8660 & 1{\cdot}1181 \end{bmatrix} \qquad \begin{matrix} S_2 \\ 1{\cdot}4142 \\ 0{\cdot}5176 \\ 0{\cdot}3382 \\ 0{\cdot}2521 \end{matrix} \qquad \begin{matrix} y \\ 29{\cdot}9976 \\ 58{\cdot}9951 \\ 83{\cdot}9889 \\ 99{\cdot}9880 \end{matrix}$$

z	0·7071	3·6744	10·3924	111·7966	(126·5705)
S_1	1·4142	4·0826	10·6811	112·9147	

Here $\sum$ is formed from the rows of $-A$ and the terms $-b$, the row S_1 is formed from the columns of L and the terms z, and S_2 is formed from the rows of L. The checks on elements of L and z are contained in the relation

$$R_r(L) \times S_1 = \textstyle\sum_r,$$

each being performed whenever a new element of S_1 is available. The check on the elements of y is given by

$$S_2 \times y = \text{sum of elements of } z.$$

4

Second-order Equations

Introduction

1. FOR a second-order problem of boundary-value type the differential equation is associated with two boundary conditions, one at each end of the range of integration. In the most general linear problem the differential equation is given by

$$y''+f(x)y'+g(x)y = k(x), \tag{1}$$

and the boundary conditions have the form

$$ay'+by = c, \tag{2}$$

in which a, b, and c are constants. When $a = 0$ we have the special but common case of 'boundary values specified'.

In Chapter 8 we shall consider some aspects of the solution of such problems by a step-by-step method, in which two or more trial solutions, each satisfying one of the boundary conditions, are combined to produce a solution which also satisfies the other. In this chapter we discuss methods, in which both boundary conditions are satisfied simultaneously, which involve the solution of simultaneous algebraic equations.

Finite-difference equations

†2. We have already discussed the replacement of the differential equation (1) by a set of finite-difference equations, one for each point of the range of integration. Using central-difference formulae for first and second derivatives, the finite-difference form of (1) at the pivotal point x_0 is given by

$$\left.\begin{aligned} y_1(1+\tfrac{1}{2}hf_0)+y_{-1}(1-\tfrac{1}{2}hf_0)-y_0(2-h^2g_0)-h^2k_0+Cy_0 = 0 \\ C = (-\tfrac{1}{12}\delta^4+\tfrac{1}{90}\delta^6-...)+hf_0(-\tfrac{1}{6}\mu\delta^3+\tfrac{1}{30}\mu\delta^5-...) \end{aligned}\right\}. \tag{3}$$

If the first derivative is absent from the differential equation, equations (3) take the simpler form

$$\left.\begin{aligned} y_1+y_{-1}-y_0(2-h^2g_0)-h^2k_0+Cy_0 = 0 \\ C = (-\tfrac{1}{12}\delta^4+\tfrac{1}{90}\delta^6-...) \end{aligned}\right\}, \tag{4}$$

and in this case the difference correction can be made still smaller by operating throughout with $(1+\frac{1}{12}\delta^2)$, producing the equations

$$\left.\begin{aligned} y_1(1+\tfrac{1}{12}h^2g_1)+y_{-1}(1+\tfrac{1}{12}h^2g_{-1})-y_0(2-\tfrac{5}{6}h^2g_0)- \\ -\tfrac{1}{12}h^2(k_1+k_{-1}+10k_0)+Cy_0 = 0 \\ C = \tfrac{1}{240}\delta^6-\tfrac{13}{15120}\delta^8+... \end{aligned}\right\}. \tag{5}$$

In a given problem equations of type (3), (4), or (5) have to be satisfied at every point of the range of integration. The treatment of the end points depends on the boundary conditions.

3. The required solution must be obtained by an iterative process. First, an approximate solution is obtained neglecting any difference corrections. This solution is differenced and approximations to the corrections computed and inserted in the finite-difference equations. A better approximate solution is thereby produced, and the process repeated until the full finite-difference equations are everywhere satisfied.

In the following sections we illustrate the technique by solving problems of various kinds, typical of those which arise in practice, involving both linear and non-linear equations and with various boundary conditions. In this chapter we shall concern ourselves mainly with procedure. We have already considered in Chapter 3 the most useful methods of solving simultaneous algebraic equations, and we leave until Chapter 9 a consideration of the accuracy attainable in the solution of second-order differential equations.

Simple examples: boundary values specified

4. The simplest boundary condition is that in which the boundary value is specified, and the simplest form of differential equation has no first derivative. As a first example, therefore, we consider the differential equation

$$y''-y-x = 0, \tag{6}$$

with boundary conditions $y(0) = 0$, $y(1) = 1$.

In this case equation (4) has the form

$$\left.\begin{aligned} y_1+y_{-1}-(2+h^2)y_0-h^2x_0+Cy_0 = 0 \\ C = -\tfrac{1}{12}\delta^4+\tfrac{1}{90}\delta^6-... \end{aligned}\right\}. \tag{7}$$

At an interval $h = \frac{1}{4}$ we then have the following equations, corresponding to the application of (7) at each internal pivotal point.

$$\left.\begin{aligned} y(\tfrac{1}{2})+y(0)-2{\cdot}0625y(\tfrac{1}{4})-0{\cdot}015625+Cy(\tfrac{1}{4}) = 0 \\ y(\tfrac{3}{4})+y(\tfrac{1}{4})-2{\cdot}0625y(\tfrac{1}{2})-0{\cdot}031250+Cy(\tfrac{1}{2}) = 0 \\ y(1)+y(\tfrac{1}{2})-2{\cdot}0625y(\tfrac{3}{4})-0{\cdot}046875+Cy(\tfrac{3}{4}) = 0 \end{aligned}\right\}. \tag{8}$$

No other equations are required, the boundary conditions being satisfied by giving $y(0)$ and $y(1)$ their specified values in these equations.

5. The first approximation $y^{(1)}$ is obtained by solving the simultaneous equations (8), with neglect of the difference corrections. When this approximation is differenced prior to the calculation of these corrections, we find that there are not enough pivotal values to provide all the required differences. If, however, we apply the first of (7) at the boundary point 0, with $Cy(0)$ neglected, we can use the boundary value and the adjacent internal value to produce the first external value, and successive external values are obtained by using (7) at successive points in a step-by-step process. Since the difference correction has the same form throughout, the values so obtained by recurrence at external points will form a smooth continuation of the internal values already obtained from the simultaneous equations.

Table 1 gives to five decimals the calculated values, sufficient to make available the sixth central difference at every internal pivotal point.

TABLE 1

x	$y^{(1)}$		δ^2		δ^4		δ^6	$Cy^{(1)}$
$-\frac{1}{2}$	−0·38735							
		+20712						
$-\frac{1}{4}$	− ·18023		− 2689					
		18023		+2689				
0	+ ·00000		0		+ 0			
		18023		2689		+168		
$+\frac{1}{4}$	·18023		+ 2689		168		+10	−0·00014
		20712		2857		178		
$\frac{1}{2}$	·38735		5546		346		24	−0·00029
		26258		3203		202		
$\frac{3}{4}$	0·64993		8749		548		+31	−0·00045
		35007		3751		+233		
1	1·00000		12500		+781			
		47507		+4532				
$\frac{5}{4}$	1·47507		+17032					
		+64539						
$+\frac{3}{2}$	+2·12046							

For the calculation of the difference corrections it is clear that the eighth and higher differences can be ignored, and from the second of equations (7) we find the values of $Cy^{(1)}$ shown in the table.

6. Our next task is to use these corrections to produce a better approximation to the solution of the differential equation. To avoid unnecessary arithmetic it is preferable to compute the increment z, say, to be added to the first approximation $y^{(1)}$. This increment is clearly zero at the boundary points, and at internal points therefore satisfies the equations

$$\left.\begin{aligned} z(\tfrac{1}{2})-2{\cdot}0625z(\tfrac{1}{4}) \qquad\qquad &+Cy^{(1)}(\tfrac{1}{4}) = 0 \\ z(\tfrac{3}{4})+z(\tfrac{1}{4})-2{\cdot}0625z(\tfrac{1}{2})&+Cy^{(1)}(\tfrac{1}{2}) = 0 \\ z(\tfrac{1}{2})-2{\cdot}0625z(\tfrac{3}{4}) \qquad\qquad &+Cy^{(1)}(\tfrac{3}{4}) = 0 \end{aligned}\right\}. \tag{9}$$

We find $z(\frac{1}{4})$, $z(\frac{1}{2})$, $z(\frac{3}{4}) = -0{\cdot}00032$, $-0{\cdot}00053$, $-0{\cdot}00048$ respectively, and the differences of this incremental solution are shown in Table 2.

TABLE 2

x	z		δ^2		δ^4
$+0$	$-0{\cdot}00000$				
		-32			
$\frac{1}{4}$	$\cdot 00032$		$+11$		$(+4)$
		-21		$+15$	
$\frac{1}{2}$	$\cdot 00053$		$+26$		$+2$
		$+5$		$+17$	
$\frac{3}{4}$	$\cdot 00048$		$+43$		$(+5)$
		$+48$			
$+1$	$-0{\cdot}00000$				

It would appear likely that no further correction is necessary. To verify this, however, it is necessary to evaluate z at some external points, to provide more central differences at internal points. We have, for example,

$$z(\tfrac{1}{4})+z(-\tfrac{1}{4})+Cy^{(1)}(0) = 0, \tag{10}$$

an equation from which $z(-\frac{1}{4})$ can be calculated if the value $Cy^{(1)}(0)$ is known. From Table 1 we have the fourth difference of $y^{(1)}$ at $x = 0$, and can estimate the sixth with sufficient confidence to enable us to quote the value of zero for $Cy^{(1)}(0)$. At the other end of the range, similarly, we have

$$z(1\tfrac{1}{4})+z(\tfrac{3}{4})+Cy^{(1)}(1) = 0, \tag{11}$$

and for the last term in (11) we can estimate the value

$$Cy^{(1)}(1) = -\tfrac{1}{12}(0{\cdot}00781)+\tfrac{1}{90}(0{\cdot}00040) = -0{\cdot}00065.$$

Thus we find $z(-\frac{1}{4}) = +0{\cdot}00032$, $z(1\frac{1}{4}) = +0{\cdot}00113$. From these we can calculate the remaining fourth central differences at internal points, shown in brackets in Table 2.

We should now proceed to calculate Cz and obtain further increments, which again satisfy equation (9), with $Cy^{(1)}$ replaced by Cz. In this case the increment does not exceed half a unit in the fifth decimal at any point, so that we accept the solution $y^{(1)}+z$ as our final solution. It differs from the analytical solution, $(2\sinh x/\sinh 1)-x$, by less than a unit in the fifth decimal.

†7. The estimation involved in the calculation of $Cy^{(1)}$ at the boundary points can be avoided by calculating more external values of $y^{(1)}$ by the step-by-step process. This would indeed be necessary if the estimation could not be performed with confidence, and if higher differences were significant in $Cy^{(1)}$. If several incremental solutions are necessary it is clear that the first approximation may have to be calculated at many external points, successive approximations at fewer such points, this number steadily decreasing. It depends on the order of negligible differences and hence on the chosen interval size.

We shall see later that it is desirable, as regards the solution of the simultaneous equations, to use as large an interval as possible in the interests of accuracy and economy of effort. Against this we must balance the inconvenience of calculating many external values and several incremental solutions. The interval size may also be restricted by the possible presence of infinities in the solution or one of its derivatives at an external point not too remote from a boundary.

8. We might also try the effect of using equation (5), with its smaller difference correction, given in this example by

$$\left.\begin{aligned} y_1(1-\tfrac{1}{12}h^2)+y_{-1}(1-\tfrac{1}{12}h^2)-y_0(2+\tfrac{5}{6}h^2)-& \\ -\tfrac{1}{12}h^2(x_1+x_{-1}+10x_0)+Cy_0 &= 0 \\ C = \tfrac{1}{240}\delta^6-\tfrac{13}{15120}\delta^8+\ldots & \end{aligned}\right\}. \quad (12)$$

TABLE 3

x	$y^{(1)}$		δ^2		δ^4		δ^6
$-\frac{1}{2}$	−0·38681						
		+20691					
$-\frac{1}{4}$	− ·17990		− 2701				
		17990		+2701			
0	+ ·00000		0		+ 1		
		17990		2702		+166	
$+\frac{1}{4}$	·17990		+ 2702		167		+19
		20692		2869		185	
$\frac{1}{2}$	·38682		5571		352		+15
		26263		3221		200	
$\frac{3}{4}$	0·64945		8792		552		+38
		35055		3773		+238	
1	1·00000		12565		+790		
		47620		+4563			
$\frac{5}{4}$	1·47620		+17128				
		+64748					
$+\frac{3}{2}$	+2·12368						

At the same interval $h = \frac{1}{4}$, equations (8) are now replaced by

$$\left.\begin{aligned} \tfrac{191}{192}\{y(\tfrac{1}{2})+y(0)\}-\tfrac{197}{96}y(\tfrac{1}{4})-\tfrac{1}{64} &= 0 \\ \tfrac{191}{192}\{y(\tfrac{3}{4})+y(\tfrac{1}{4})\}-\tfrac{197}{96}y(\tfrac{1}{2})-\tfrac{1}{32} &= 0 \\ \tfrac{191}{192}\{y(1)+y(\tfrac{1}{2})\}-\tfrac{197}{96}y(\tfrac{3}{4})-\tfrac{3}{64} &= 0 \end{aligned}\right\}, \tag{13}$$

in which the difference corrections are neglected and $y(0) = 0$, $y(1) = 1$. On solving these equations, and continuing the solution beyond the boundaries by applying equation (12), with Cy neglected, at the boundary and successive points, we have the results tabulated and differenced in Table 3. The difference correction is clearly negligible and the solution is correct to the number of figures retained.

9. The general differential equation (1) involves the first derivative. This can always be removed mathematically by a change of dependent variable. If we write

$$y(x) = \phi(x)Y(x),$$

and substitute in (1), we have

$$\phi Y''+(2\phi'+f\phi)Y'+(\phi''+f\phi'+g\phi)Y = k. \tag{14}$$

This equation will contain no first derivative if we choose ϕ so that

$$2\phi'+f\phi = 0, \qquad \phi = \exp\int(-\tfrac{1}{2}f)\,dx. \tag{15}$$

The ϕ so obtained can be substituted in equation (14), which then involves the function Y and its second derivative only.

Considered numerically, however, the form of ϕ may be such that the new dependent variable Y is badly behaved and not readily amenable to finite-difference techniques. Consider, for example, the equation

$$y''-x^{-1}y'+y = 0,$$

of which there is a well-behaved solution with $y(0) = 1$, $y'(0) = 0$. From equation (15) we find $\phi = x^{\frac{1}{2}}$, giving $Y = x^{-\frac{1}{2}}y$. It is therefore not possible to solve numerically the equation for Y, since the initial value is infinite.

10. These considerations show that we cannot always avoid the presence of the first derivative in the differential equation. There is no essential difference in technique but, as equation (3) shows, the difference correction is likely to be larger and, for the same accuracy, it will usually be necessary either to use a smaller interval or to apply the difference corrections more times.

As an example we consider the differential equation

$$y''+y'+4xy+1 = 0, \tag{16}$$

with the boundary conditions $y(0) = y(1) = 0$. The finite-difference equation is given by

$$\left.\begin{aligned} y_1(1+\tfrac{1}{2}h)+y_{-1}(1-\tfrac{1}{2}h)-y_0(2-4h^2x_0)+h^2+Cy_0 &= 0 \\ C = h(-\tfrac{1}{6}\mu\delta^3+\tfrac{1}{30}\mu\delta^5-\ldots)+(-\tfrac{1}{12}\delta^4+\tfrac{1}{90}\delta^6-\ldots) & \end{aligned}\right\}. \tag{17}$$

At an interval $h = \frac{1}{4}$ we then have the following three equations for solution, in which $y(0)$ and $y(1)$ have been given their specified zero boundary values.

$$\left.\begin{aligned} 1{\cdot}1250y(\tfrac{1}{2})-1{\cdot}9375y(\tfrac{1}{4}) \qquad\qquad\qquad\qquad +0{\cdot}0625+Cy(\tfrac{1}{4}) &= 0 \\ 1{\cdot}1250y(\tfrac{3}{4})+0{\cdot}8750y(\tfrac{1}{4})-1{\cdot}8750y(\tfrac{1}{2})+0{\cdot}0625+Cy(\tfrac{1}{2}) &= 0 \\ 0{\cdot}8750y(\tfrac{1}{2})-1{\cdot}8125y(\tfrac{3}{4})+0{\cdot}0625+Cy(\tfrac{3}{4}) &= 0 \end{aligned}\right\}. \tag{18}$$

The first approximation $y^{(1)}$, obtained with neglect of difference corrections, and with external values calculated by subsequent step-by-step methods, is shown together with its differences in Table 4.

TABLE 4

x	$y^{(1)}$		δ^2		δ^4		δ^6	$Cy^{(1)}$
$-\frac{1}{2}$	−0·6143							
		+3840						
$-\frac{1}{4}$	− ·2303		−1537					
		2303		+470				
0	+ ·0000		1067		−302			
		1236		168		+227		
$+\frac{1}{4}$	·1236		899		− 75		−74	+0·0002
		+ 337		93		+153		
$\frac{1}{2}$	·1573		806		+ 78		−77	−0·0012
		− 469		171		+ 76		
$\frac{3}{4}$	·1104		635		+154		−94	−0·0024
		1104		325		− 18		
1	+ ·0000		− 310		+136			
		1414		+461				
$\frac{5}{4}$	− ·1414		+ 151					
		−1263						
$+\frac{3}{2}$	−0·2677							

Differences of higher order than the sixth are clearly negligible, and we can calculate the difference corrections with confidence from the second of (17). These values are given on the right of Table 4. We now proceed to calculate a correction z which satisfies the equations

$$\left.\begin{aligned} 1{\cdot}1250z(\tfrac{1}{2})-1{\cdot}9375z(\tfrac{1}{4}) \qquad\qquad\qquad\qquad +0{\cdot}0002 &= 0 \\ 1{\cdot}1250z(\tfrac{3}{4})+0{\cdot}8750z(\tfrac{1}{4})-1{\cdot}8750z(\tfrac{1}{2})-0{\cdot}0012 &= 0 \\ 0{\cdot}8750z(\tfrac{1}{2})-1{\cdot}8125z(\tfrac{3}{4})-0{\cdot}0024 &= 0 \end{aligned}\right\}. \tag{19}$$

The calculated values and their differences are shown in Table 5, together with the approximate solution $y^{(1)}+z$.

TABLE 5

x	z		δ^2		δ^4	$y^{(1)}+z$
$+0$	$-0{\cdot}0000$					$+0{\cdot}0000$
		-18				
$\frac{1}{4}$	$\cdot0018$		$+\ 4$			$\cdot1218$
		-14		$+13$		
$\frac{1}{2}$	$\cdot0032$		$+17$		-4	$\cdot1541$
		$+\ 3$		$+\ 9$		
$\frac{3}{4}$	$\cdot0029$		$+26$			$\cdot1075$
		$+29$				
$+1$	$-0{\cdot}0000$					$+0{\cdot}0000$

It would appear that the difference corrections for z are not likely to exceed half a unit, but this might affect the rounding of the last figure of the functional values, and we cannot without further work guarantee the accuracy of this figure.

11. In this case we can transform the dependent variable by the method of § 9, by writing

$$y = Ye^{-x/2}, \tag{20}$$

and producing for Y the differential equation and boundary conditions

$$Y''+(4x-\tfrac{1}{4})Y+e^{x/2} = 0, \qquad Y(0) = Y(1) = 0. \tag{21}$$

The typical finite-difference equation is now obtained from (5), and when the three relevant equations are solved for $Y(\frac{1}{4})$, $Y(\frac{1}{2})$, and $Y(\frac{3}{4})$ and the corresponding values of y recovered from (20), we find the values

$$y(\tfrac{1}{4}) = 0{\cdot}1218, \qquad y(\tfrac{1}{2}) = 0{\cdot}1541, \qquad y(\tfrac{3}{4}) = 0{\cdot}1076.$$

The difference corrections are almost negligible to four decimals, but would have the effect of decreasing slightly the computed pivotal values.

Boundary condition involving derivative. Simple example

12. The technique is rather more complicated when one or both boundary conditions have the form of equation (2), involving a first derivative. The boundary value is not given, and its determination involves an extra equation. To fix ideas we consider a simple example, the evaluation of the solution of the differential equation

$$y''-y = 0, \tag{22}$$

with boundary conditions $y'(0) = 1$, $y(1) = 0$.

At an interval h our most accurate simple finite-difference equation for (22) is given by

$$\left.\begin{aligned} y_{r+1}(1-\tfrac{1}{12}h^2)+y_{r-1}(1-\tfrac{1}{12}h^2)-y_r(2+\tfrac{5}{6}h^2)+Cy_r &= 0 \\ C &= \tfrac{1}{240}\delta^6-\dots \end{aligned}\right\}. \tag{23}$$

With reference to Fig. 15, equation (23) must be used at the internal points 1, 2,..., $n-1$.

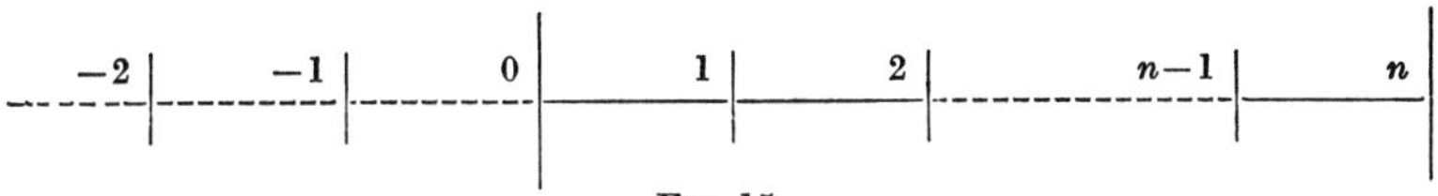

FIG. 15

At the end point n the value of y is specified so that, as in previous examples, no other equation is needed here. At the beginning of the range, however, we need an extra equation for the determination of y_0, and we proceed to consider various possibilities.

(i) The most obvious method is to use a forward-difference formula for the first derivative at point 0, in the form

$$hy_0' = (\Delta-\tfrac{1}{2}\Delta^2+\tfrac{1}{3}\Delta^3-\tfrac{1}{4}\Delta^4+\dots)y_0. \tag{24}$$

To be useful this must be expressed in its Lagrangian form, depending on the number of differences retained: thus, for example, we might write

$$\left.\begin{aligned} hy_0' &= -\tfrac{3}{2}y_0+2y_1-\tfrac{1}{2}y_2+C_1y_0 \\ C_1 &= \tfrac{1}{3}\Delta^3-\tfrac{1}{4}\Delta^4+\dots \end{aligned}\right\}. \tag{25}$$

We could use the first of (25) to provide the extra required equation, noticing that it has a difference correction $C_1 y$ attached to it, analogous to the Cy of the differential equations.

(ii) One drawback to method (i) lies in the large size of the coefficients in equation (24), a remark first made and illustrated in Chapter 2. This can partly be overcome by applying equation (23) at the boundary point 0, thus introducing an external point -1. The boundary condition must involve this extra point and is given by

$$hy_0' = (\Delta+\tfrac{1}{2}\Delta^2-\tfrac{1}{6}\Delta^3+\tfrac{1}{12}\Delta^4-\dots)y_{-1}. \tag{26}$$

Again this must be expressed partly in Lagrangian form, and we can use any one of the following formulae:

$$\left.\begin{aligned} hy_0' &= -\tfrac{1}{2}y_{-1}+\tfrac{1}{2}y_1+(-\tfrac{1}{6}\Delta^3+\tfrac{1}{12}\Delta^4-\tfrac{1}{20}\Delta^5+\dots)y_{-1} \\ hy_0' &= -\tfrac{1}{3}y_{-1}-\tfrac{1}{2}y_0+y_1-\tfrac{1}{6}y_2+(\tfrac{1}{12}\Delta^4-\tfrac{1}{20}\Delta^5+\dots)y_{-1} \\ hy_0' &= -\tfrac{1}{4}y_{-1}-\tfrac{5}{6}y_0+\tfrac{3}{2}y_1-\tfrac{1}{2}y_2+\tfrac{1}{12}y_3+(-\tfrac{1}{20}\Delta^5+\dots)y_{-1} \end{aligned}\right\} \tag{27}$$

etc.

(iii) By applying (23) at some external points also we can use the central-difference formulae

$$\left.\begin{aligned} hy'_0 &= (\mu\delta - \tfrac{1}{6}\mu\delta^3 + \tfrac{1}{30}\mu\delta^5 - \ldots)y_0 \\ hy'_0 &= \tfrac{1}{2}(y_1 - y_{-1}) + (-\tfrac{1}{6}\mu\delta^3 + \tfrac{1}{30}\mu\delta^5 - \ldots)y_0 \\ hy'_0 &= \tfrac{2}{3}(y_1 - y_{-1}) - \tfrac{1}{12}(y_2 - y_{-2}) + (\tfrac{1}{30}\mu\delta^5 - \ldots)y_0 \end{aligned}\right\} \tag{28}$$

etc.

If one external point is included, corresponding to the second of (28), we must apply the main equation (23) at the boundary point 0: if we use the second of (28), involving two external points, we apply (23) at the boundary and the first external point, and so on.

(iv) If the pivotal value does not enter into the boundary condition, which then involves only the first derivative and some function of x, it might pay to arrange the pivotal points in the form shown in Fig. 16, where the point 0 marks the end at which the derivative is specified.

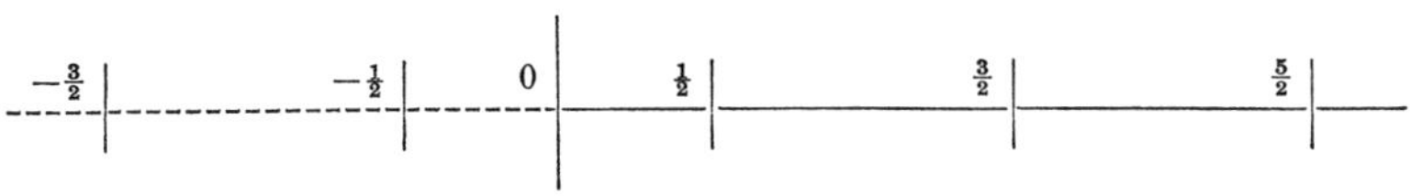

FIG. 16

We can then use, for the derivative condition at 0, one of the equations

$$\left.\begin{aligned} hy'_0 &= (\delta - \tfrac{1}{24}\delta^3 + \tfrac{3}{640}\delta^5 - \ldots)y_0 \\ hy'_0 &= y_{\frac{1}{2}} - y_{-\frac{1}{2}} + (-\tfrac{1}{24}\delta^3 + \tfrac{3}{640}\delta^5 - \ldots)y_0 \\ hy'_0 &= -\tfrac{1}{24}y_{\frac{3}{2}} + \tfrac{9}{8}y_{\frac{1}{2}} - \tfrac{9}{8}y_{-\frac{1}{2}} + \tfrac{1}{24}y_{-\frac{3}{2}} + (\tfrac{3}{640}\delta^5 - \ldots)y_0 \end{aligned}\right\}. \tag{29}$$

If we use the last of (29), then we must apply the differential equation at the point $-\frac{1}{2}$ as well as at the internal points $\frac{1}{2}$, $\frac{3}{2}$,... .

This method can still be applied if the boundary condition also involves y at 0, but this value must now be expressed by interpolation in terms of pivotal values, by an equation of the form

$$y_0 = -\tfrac{1}{16}(y_{\frac{3}{2}} - 9y_{\frac{1}{2}} - 9y_{-\frac{1}{2}} + y_{-\frac{3}{2}}) + \tfrac{3}{128}\mu\delta^4 y_0 - \ldots . \tag{30}$$

This extra interpolation, however, is an unsatisfactory feature of the method, and in any case involves a larger difference correction than the third of (28) or (29).

The choice of method will depend largely on the convenience of the calculations involved and on the form of the differential equation. If the differential equation contains a first derivative the general finite-difference equation has a fairly large difference correction, so that we may allow an equally large correction at the boundary. If the equation has no first derivative and we use the powerful formula (23), we would like similar accuracy in the boundary condition. In the first case, therefore, it may be sufficient to introduce just one external point and to use the second of (28) at the boundary; in the second the use of the third of (28) is desirable. Equations (28) are generally preferred to (27), and the latter are used only when it is desirable, for some reason, not to include more than one external point.

†13. In the simple example of equation (22) we shall use equation (23) at a general point, and include the two external points -1, -2 of Fig. 15 for the satisfaction of the boundary condition. We then have the following simultaneous equations, the first five representing the differential equation at the successive points $x = \frac{3}{4}, \frac{1}{2}, \frac{1}{4}, 0, -\frac{1}{4}$, and the sixth representing the boundary condition at $x = 0$:

$$\left.\begin{aligned}
-\tfrac{197}{96}y(\tfrac{3}{4})+\tfrac{191}{192}y(\tfrac{1}{2}) \qquad\qquad\qquad\qquad\qquad\qquad &+ Cy(\tfrac{3}{4}) = 0\\
\tfrac{191}{192}y(\tfrac{3}{4})-\tfrac{197}{96}y(\tfrac{1}{2})+\tfrac{191}{192}y(\tfrac{1}{4}) \qquad\qquad\qquad\qquad &+ Cy(\tfrac{1}{2}) = 0\\
\tfrac{191}{192}y(\tfrac{1}{2})-\tfrac{197}{96}y(\tfrac{1}{4})+\tfrac{191}{192}y(0) \qquad\qquad &+ Cy(\tfrac{1}{4}) = 0\\
\tfrac{191}{192}y(\tfrac{1}{4})-\tfrac{197}{96}y(0)+\tfrac{191}{192}y(-\tfrac{1}{4}) \quad &+ Cy(0) = 0\\
\tfrac{191}{192}y(0)-\tfrac{197}{96}y(-\tfrac{1}{4})+\tfrac{191}{192}y(-\tfrac{1}{2}) &+ Cy(-\tfrac{1}{4}) = 0\\
-\tfrac{1}{12}y(\tfrac{1}{2})+\tfrac{2}{3}y(\tfrac{1}{4}) \qquad -\tfrac{2}{3}y(-\tfrac{1}{4})+\tfrac{1}{12}y(-\tfrac{1}{2}) &+ C_1\,y(0) = \tfrac{1}{4}
\end{aligned}\right\} \qquad (31)$$

where $\qquad C = \frac{1}{240}\delta^6 - \ldots, \qquad C_1 = \frac{1}{30}\mu\delta^5 - \ldots,$

and we have used the boundary condition $y(1) = 0$.

The method of solution is to obtain first an approximation $y^{(1)}$ in which both Cy and $C_1 y$ are neglected. The latter are then calculated for $y^{(1)}$ and used to find an increment z which will have the boundary conditions $z'(0) = 0, z(1) = 0$. The process is repeated until the full finite-difference equations are everywhere satisfied.

With regard to the solution of the simultaneous equations we note that the last equation, representing the derivative condition, is not convenient for the relaxation process, and it upsets the regular pattern of the matrix as far as the direct method is concerned. For relaxation we should need to produce suitably effective group operations. The direct method is not greatly affected, the matrix U being unchanged in form, but in the last row of L only the first element is zero.

The first approximation, with several external values included, is shown in Table 6. The sixth differences contribute almost negligible amounts to the difference corrections of type Cy, but the fifth differences at $x = 0$ make a small contribution to the difference correction $C_1 y$ of the boundary condition: $C_1 y(0) = +\frac{1}{60}(199) = +3{\cdot}3$ in units of the fifth decimal. This should now be inserted in (31) and an incremental correction computed.

We notice, however, a nice computational point in this particular example which makes this calculation easy to perform. If Cy is completely negligible the differential equation is everywhere satisfied. The differential equation is homogeneous and $y(1) = 0$. If therefore our approximate solution $y^{(1)}$ gives a derivative $dy^{(1)}/dx = A$ at $x = 0$, we need merely multiply all our pivotal values by the factor $1/A$, the true boundary condition being $y'(0) = 1$, to produce an accurate solution.

TABLE 6

x	$y^{(1)}$		δ^2		δ^4		δ^6
$-\frac{3}{4}$	$-1{\cdot}80859$						
		$+42852$					
$\frac{1}{2}$	$1{\cdot}38007$		-8671				
		34181		$+2147$			
$-\frac{1}{4}$	$1{\cdot}03826$		6524		-408		
		27657		1739		$+105$	
0	$0{\cdot}76169$		4785		303		-11
		22872		1436		94	
$+\frac{1}{4}$	$\cdot 53297$		3349		209		18
		19523		1227		76	
$\frac{1}{2}$	$\cdot 33774$		2122		133		9
		17401		1094		67	
$\frac{3}{4}$	$-\ \cdot 16373$		-1028		66		$-\ 1$
		16373		1028		$+\ 66$	
1	$\cdot 00000$		0		$-\ 0$		
		16373		$+1028$			
$\frac{5}{4}$	$+\ \cdot 16373$		$+1028$				
		$+17401$					
$+\frac{3}{2}$	$+0{\cdot}33774$						

The central-difference formula gives

$$2hA = 0{\cdot}50529 - \tfrac{1}{6}(0{\cdot}03175) + \tfrac{1}{30}(0{\cdot}00199) = 0{\cdot}50006.$$

We therefore multiply all pivotal values by $0{\cdot}5/0{\cdot}50006 = 0{\cdot}99988$. The solution so obtained differs from the analytical solution by a maximum of half a unit in the fifth decimal.

14. As an alternative treatment of the boundary condition at $x = 0$ we might use the last of equations (27), in which the first neglected difference is again the fifth, but its coefficient is rather larger. We now need to introduce only the one external value $y(-\frac{1}{4})$, and the last two simultaneous equations in (31) are replaced by the single equation

$$\tfrac{1}{12}y(\tfrac{3}{4}) - \tfrac{1}{2}y(\tfrac{1}{2}) + \tfrac{3}{2}y(\tfrac{1}{4}) - \tfrac{5}{6}y(0) - \tfrac{1}{4}y(-\tfrac{1}{4}) = \tfrac{1}{4} - C_2\, y(-\tfrac{1}{4}), \tag{32}$$

where $$C_2 = -\tfrac{1}{20}\Delta^5 + \tfrac{1}{30}\Delta^6 - \dots .$$

As might be expected, the error in the first approximation is now about 50 per cent. greater than that given in Table 6, and its sign is changed.

15. Two further points might be mentioned. First, if the boundary condition involves y as well as y' there is no new problem. In the example of § 12, for instance, the replacement of the boundary condition $y'(0) = 1$ by the more general condition $y'(0)+ay(0) = 1$ would merely introduce a term $ahy(0)$ in the last of equations (31).

Second, although we suggested in § 12 that if the differential equation contained a first derivative the boundary condition could contain a difference correction involving differences of order three and higher, it is usually advisable to use a more accurate boundary formula. The reason is connected with the fact that the third-difference correction is multiplied by h in the differential equation but not in the boundary condition, so that its effect is usually greater in the latter than in the former.

An alternative to the difference correction. Iteration

16. We now extend in more general terms the theory behind the iterative process, involving successive applications of the difference correction, which we have used in our previous examples.

Consider again the differential equation (1) and suppose that an approximation $y^{(1)}$ is substituted in the differential equation, and the derivatives calculated using accurate finite-difference formulae. We find that the differential equation is not satisfied accurately, but there is a residual error $\epsilon(x)$ given by

$$\frac{d^2y^{(1)}}{dx^2}+f(x)\frac{dy^{(1)}}{dx}+g(x)y^{(1)} = k(x)+\epsilon(x). \tag{33}$$

Similarly, if the boundary conditions involve derivatives, being given at the ends of the range by the respective equations

$$a_1 y'+b_1 y = c_1, \qquad a_2 y'+b_2 y = c_2, \tag{34}$$

the substitution of $y^{(1)}$ into these equations produces residual errors γ_1 and γ_2 given by

$$a_1\frac{dy^{(1)}}{dx}+b_1 y^{(1)} = c_1+\gamma_1, \qquad a_2\frac{dy^{(1)}}{dx}+b_2 y^{(1)} = c_2+\gamma_2. \tag{35}$$

The correction z to be made to $y^{(1)}$ thus satisfies the differential equation and boundary conditions

$$\left.\begin{aligned} &\frac{d^2z}{dx^2}+f(x)\frac{dz}{dx}+g(x)z = -\epsilon(x) \\ &a_1\frac{dz}{dx}+b_1 z = -\gamma_1, \qquad a_2\frac{dz}{dx}+b_2 z = -\gamma_2 \end{aligned}\right\}. \tag{36}$$

When z is computed using approximate formulae, it will again be found to satisfy equations (36) except for error terms, probably small, and further incremental functions can be calculated until the error terms are negligible.

It is clearly not necessary in the foregoing that the approximate finite-difference equations used to produce $y^{(1)}$, say, should be truncated forms of the accurate equations used for the calculation of the error terms ϵ and γ. In many cases it is convenient to use similar equations, but the use, for example, of a sloping-difference formula for a derivative, in the production of a first approximation, does not prevent the use of a central-difference formula for its correction.

17. We might indeed go further in the interests of convenience. The differential equation and boundary conditions are easiest to handle numerically when no first derivative exists. This remark suggests a more ruthless iterative procedure for the solution of second-order equations. Suppose, for example, that we evaluate a sequence of functions y_n satisfying respectively the differential equations

$$y''_{n+1}+g(x)y_{n+1} = k(x)-f(x)y'_n \tag{37}$$

and boundary conditions

$$b_1 y_{n+1} = c_1-a_1 y'_n, \qquad b_2 y_{n+1} = c_2-a_2 y'_n. \tag{38}$$

If the process converges, the limit is clearly a solution of the differential equation (1) with boundary conditions (2). We have nowhere used a first derivative formula in the coefficients of the linear equations replacing the finite-difference equation, but only to calculate the first derivative of a function already given numerically.

In certain cases it is even possible to iterate according to the scheme

$$g(x)y_{n+1} = k(x)-y''_n-f(x)y'_n, \tag{39}$$

with boundary conditions given by (38).

The choice of iterative method depends on which terms in the differential equation happen to be dominant. The only criterion, however, for the success of the method is that it should converge, and this is true also of the more common difference-correction method.

Example

†18. For linear equations direct methods are usually satisfactory, convenient, and rapid. As an example, however, of an uncommon but convenient iterative procedure consider the solution of the differential equation

$$y''-\left(\frac{1}{x^4}-\frac{2}{x^3}\right)y+\frac{1}{x^2} = 0, \tag{40}$$

which is satisfied by the function $e^{-(1/x)}\mathrm{Ei}(1/x)-x$, where $\mathrm{Ei}(z)$ is the exponential integral, defined by

$$\mathrm{Ei}(z) = \int_{-\infty}^{z} t^{-1}e^{t}\,dt.$$

For a well-behaved solution y must clearly vanish at $x = 0$, and we shall find that no other boundary condition need be specified. We iterate according to the scheme

$$y_{n+1} = \frac{x^2}{1-2x}(1+x^2y_n''), \quad \text{with} \quad y_0 = 0, \tag{41}$$

calculating the derivatives numerically with the use of central-difference formulae. In Table 7 we give the first three sequences derived from (41), together with the true solution. The discrepancy gets larger with increasing x, as could be inferred from (41), and the method may break down for much larger values of x. In the range $x = 0$ to $0{\cdot}1$, however, it appears to be converging to a good solution, and the calculations involved are almost trivial. We note that, as with difference-correction methods, a larger range of x must be covered with approximate solutions than is required for the correct solution. Near the origin we have extrapolated the differences given in brackets, but we could clearly extend the table backwards for negative values of x, to produce results connected with $\mathrm{Ei}(-x)$, thereby filling in these differences accurately. For negative x, moreover, the factor $x^2/(1-2x)$ is even smaller than in the previous case, so that sequences of the same order will have greater accuracy than those on the positive side of x.

Simultaneous second-order equations

19. We now treat briefly the solution of two simultaneous second-order differential equations, the extension to more than two equations needing no additional fundamental consideration. The differential equations, if linear, have the general forms

$$\left.\begin{aligned} y''+f_1y'+f_2y+f_3z'+f_4z &= f_5 \\ z''+g_1z'+g_2z+g_3y'+g_4y &= g_5 \end{aligned}\right\}, \tag{42}$$

and the boundary conditions consist of equations of the form

$$\left.\begin{aligned} a_1y'+a_2y+a_3z'+a_4z &= a_5 \\ b_1z'+b_2z+b_3y'+b_4y &= b_5 \end{aligned}\right\}, \tag{43}$$

a pair of such conditions holding at each end of the range of integration. The conditions (43) are perhaps more general than is strictly necessary.

TABLE 7

x	y_1		δ^2		δ^4	y_2		δ^2		δ^4		δ^6	y_3	*Analytical solution*
+0·00	+0·000000					+0·000000							+0·000000	+0·000000
		+ 417					+ 417							
·02	417		+ 905		(10)	417		+ 912		(50)			417	417
		1322		+125			1329		+ 147					
·04	1739		1030		+ 21	1746		1059		+ 56			1746	1746
		2352		146			2388		203		+ 33			
·06	4091		1176		31	4134		1262		89		+31	4137	4138
		3528		177			3650		292		64			
·08	7619		1353		36	7784		1554		153		+26	7807	7819
		4881		213			5204		445		90			
·10	12500		1566		49	12988		1999		243		+98	+0·013118	+0·013147
		6447		262			7203		688		+188			
·12	18947		1828		60	20191		2687		+431				
		8275		322			9890		+1119					
·14	27222		2150		81	30081		+3806						
		10425		403			+13696							
·16	37647		2553		+108	+0·043777								
		12978		+511										
·18	50625		+3064											
		+16042												
+0·20	+0·066667													

For example, we can produce equations like (43) in which the coefficients a_3 and b_3 are replaced by zero. For the sake of completeness, however, we take the full forms (43) in the following discussion.

It is clearly possible to adapt the techniques of this chapter to the solution of this problem. The general finite-difference equations replacing (42) are given by

$$\left.\begin{aligned} &y_1(1+\tfrac{1}{2}hf_1)+y_{-1}(1-\tfrac{1}{2}hf_1)-y_0(2-h^2f_2)+ \\ &\qquad +\tfrac{1}{2}hf_3(z_1-z_{-1})+h^2f_4z_0+C_1y_0+C_2z_0 = h^2f_5 \\ &z_1(1+\tfrac{1}{2}hg_1)+z_{-1}(1-\tfrac{1}{2}hg_1)-z_0(2-h^2g_2)+ \\ &\qquad +\tfrac{1}{2}hg_3(y_1-y_{-1})+h^2g_4y_0+C_3z_0+C_4y_0 = h^2g_5 \end{aligned}\right\}, \tag{44}$$

where

$$\left.\begin{aligned} C_1 &= (-\tfrac{1}{12}\delta^4+\tfrac{1}{90}\delta^6-...)+hf_1(-\tfrac{1}{6}\mu\delta^3+\tfrac{1}{30}\mu\delta^5-...) \\ C_2 &= hf_3(-\tfrac{1}{6}\mu\delta^3+\tfrac{1}{30}\mu\delta^5-...) \end{aligned}\right\}, \tag{45}$$

and C_3 and C_4 are the same as C_1 and C_2 respectively, but with f replaced by g. All functions f and g are evaluated at the pivotal point x_0.

At the boundaries we can use for derivatives one or other of the sloping-difference formulae or the central-difference formula. Applying the latter, the boundary conditions are replaced, at each end of the range, by the equations

$$\left.\begin{aligned} \tfrac{1}{2}a_1(y_1-y_{-1})+ha_2y_0+\tfrac{1}{2}a_3(z_1-z_{-1})+ha_4z_0+C_5y_0+C_6z_0 &= ha_5 \\ \tfrac{1}{2}b_1(z_1-z_{-1})+hb_2z_0+\tfrac{1}{2}b_3(y_1-y_{-1})+hb_4y_0+C_7z_0+C_8y_0 &= hb_5 \end{aligned}\right\}, \tag{46}$$

where

$$\left.\begin{aligned} C_5 &= a_1(-\tfrac{1}{6}\mu\delta^3+\tfrac{1}{30}\mu\delta^5-...), \qquad C_6 = a_3(-\tfrac{1}{6}\mu\delta^3+\tfrac{1}{30}\mu\delta^5-...) \\ C_7 &= b_1(-\tfrac{1}{6}\mu\delta^3+\tfrac{1}{30}\mu\delta^5-...), \qquad C_8 = b_3(-\tfrac{1}{6}\mu\delta^3+\tfrac{1}{30}\mu\delta^5-...) \end{aligned}\right\}. \tag{47}$$

There are clearly just enough equations for the determination of y and z at every pivotal point, including one external point at each end of the range. Additional external points can then be calculated by applying (44) at successive external points, the evaluation of y and z involving at each such point the solution of the pair of simultaneous equations (44). The method of obtaining an approximation with difference corrections neglected, and incorporating them at a later stage to effect an improvement, is applied as before.

20. As a simple illustration we take the equations

$$\left.\begin{aligned} y''-z &= 0 \\ z''-y &= 0 \end{aligned}\right\}, \tag{48}$$

with the boundary conditions

$$\left.\begin{aligned} y = 1,\ y' &= 1 \text{ at } x = 0 \\ y = z,\ y' &= z' \text{ at } x = 1 \end{aligned}\right\}. \tag{49}$$

The analytical solution is $y = z = e^x$.

We take the interval $h = 0{\cdot}2$, and replace (48) by the finite-difference equations

$$\left.\begin{aligned} y_1+y_{-1}-2y_0-0{\cdot}04z_0+C_1 y_0 &= 0 \\ z_1+z_{-1}-2z_0-0{\cdot}04y_0+C_1 z_0 &= 0 \end{aligned}\right\}, \tag{50}$$

where

$$C_1 = -\tfrac{1}{12}\delta^4+\tfrac{1}{90}\delta^6-\ldots, \tag{51}$$

and the boundary conditions by

$$\left.\begin{aligned} &y_0 = 1,\quad \tfrac{1}{2}(y_1-y_{-1})+C_2 y_0 = 0{\cdot}2 \quad \text{at } x = 0 \\ &y_0 = z_0,\quad \tfrac{1}{2}(z_1-z_{-1})-\tfrac{1}{2}(y_1-y_{-1})+C_2 z_0-C_2 y_0 = 0 \quad \text{at } x = 1 \end{aligned}\right\}, \tag{52}$$

where

$$C_2 = -\tfrac{1}{6}\mu\delta^3+\tfrac{1}{30}\mu\delta^5-\ldots. \tag{53}$$

Applying the first of (50) at the points $x = 0{\cdot}0(0{\cdot}2)1{\cdot}0$ and the second at $x = 0{\cdot}2(0{\cdot}2)1{\cdot}0$, and using the four equations (52), we have fifteen equations for the determination of the pivotal values of y at $x = -0{\cdot}2(0{\cdot}2)1{\cdot}2$ and of z at $x = 0{\cdot}0(0{\cdot}2)1{\cdot}2$. The approximations obtained by neglecting the difference corrections are shown in Table 8. Only $y^{(1)}$ is given, since the values of $z^{(1)}$ are found to be identical with those of $y^{(1)}$. We give also, to an extra figure, the difference corrections of types C_1 and C_2, but since $y^{(1)}$ and $z^{(1)}$ are identical the contributions from $C_2 y$ and $C_2 z$ cancel in the second of (52). Differences of order greater than six are clearly negligible.

We now proceed to use the difference corrections to find a correction to the first approximation. The correction $y^{(2)}$ to $y^{(1)}$, again equal to that of $z^{(2)}$ to $z^{(1)}$, is shown, with its differences, in Table 9.

The difference corrections of $y^{(2)}$ are not completely negligible, and we ought to proceed to use them and obtain an additional correction. We shall not pursue this here, but it might be noted that the new difference correction $C_2 y$ is about $-1{\cdot}3$ units at $x = 0$, and $C_1 y$ lies between $-0{\cdot}1$ and $-0{\cdot}3$ units, in the fifth decimal. Comparison with the corresponding figures of Table 8 shows that we would expect further increments in y and z of magnitude roughly 1 per cent. of $y^{(2)}$, and of the same sign. Such a correction would give a final solution which is in error by about one unit in the fifth decimal.

This accuracy, however, cannot be guaranteed: the main difference correction in Table 8 is $C_2 y$ at $x = 0$, and the resulting maximum component of $y^{(2)}$ is almost ten times as great. The extra figure given in $C_2 y$,

TABLE 8

x	$y^{(1)}$		δ^2		δ^4		δ^6	$C_1\, y^{(1)}$	$C_2\, y^{(1)}$
−0·6	+0·55251								
		+12029							
·4	0·67280		+ 2691						
		14720		+ 589					
− ·2	0·82000		3280		+131				
		18000		720		+29			
+ ·0	1·00000		4000		160		+ 6	−13·3	−132·3
		22000		880		35			
·2	1·22000		4880		195		9	16·1	
		26880		1075		44			
·4	1·48880		5955		239		7	19·8	
		32835		1314		51			
·6	1·81715		7269		290		14	24·0	
		40104		1604		65			
0·8	2·21819		8873		355		13	29·4	
		48977		1959		78			
1·0	2·70796		10832		433		+19	−35·9	−359·7
		59809		2392		+97			
1·2	3·30605		13224		+530				
		73033		+2922					
1·4	4·03638		+16146						
		+89179							
+1·6	+4·92817								

TABLE 9

x	$y^{(2)}$		δ^2		δ^4
−0·2	−0·00125				
		+125			
+ ·0	+ ·00000		+14		
		139		+ 8	
·2	·00139		22		+1
		161		9	
·4	·00300		31		4
		192		13	
·6	·00492		44		+2
		236		+15	
0·8	·00728		+59		
		+295			
+1·0	+0·01023				

however, is obviously not very reliable, and hence the last figure of our final solution is doubtful. To attain five-figure accuracy in the final solution we should need to work with at least one extra figure throughout. We return to questions of this kind in Chapter 9.

21. Methods of iteration similar to those of § 17 can be applied to solve simultaneous equations. For example, we might iterate equations (48) according to the scheme

$$\left.\begin{aligned} y^{(s+1)''} &= z^{(s)} \\ z^{(s+1)''} &= y^{(s+1)} \end{aligned}\right\}, \tag{54}$$

starting with $z^{(0)} = 0$, the index now denoting the order of the iterate involved. The first of (49) provides *initial* conditions on y and y', so that the first of (54) may be solved for $y^{(1)}$, with $s = 0$, by a step-by-step method. When $y^{(1)}$ is obtained we can calculate $z^{(1)}$ from the second of (54), working backwards from $x = 1$ using the second of (49) in the form

$$z^{(1)} = y^{(1)}, \qquad z^{(1)\prime} = y^{(1)\prime} \quad \text{at } x = 1, \tag{55}$$

both $y^{(1)}$ and $y^{(1)\prime}$ being known.

If the boundary conditions specified y and z at each end of the range then the above procedure would lead to successive solution of second-order equations by boundary-value techniques.

It is not easy to decide when to use a particular method. Much depends on the structure of the differential equations and boundary conditions, and on the computational equipment available.

Non-linear equations

22. We can write the general non-linear equation of the second order in the form

$$f(y'', y', y, x) = 0, \tag{56}$$

and the boundary conditions in the form

$$g_1(y', y) = 0, \qquad g_2(y', y) = 0. \tag{57}$$

The principle of superposition no longer holds, and we cannot use the methods of §§ 4 to 15 without some modification. The iterative methods of §§ 16 and 17 can, however, be applied here, provided only that a convergent process can be found.

Iterative methods. Examples

23. As a first example we take the differential equation

$$y'' + \sin y' + 1 = 0, \tag{58}$$

with the boundary conditions $y(0) = 0$, $y(1) = 1$. We shall use the iterative scheme

$$y^{(n+1)''} = -1 - \sin y^{(n)'}, \tag{59}$$

where each $y^{(n)}$ satisfies the correct boundary conditions. Using an interval $h = 0{\cdot}2$, the first step is to solve the equation

$$y'' = -1 \tag{60}$$

by means of the simultaneous equations typified by

$$y_{r+1} + y_{r-1} - 2y_r + 0{\cdot}04 = 0, \qquad y(0) = 0, \qquad y(1) = 1. \tag{61}$$

On the left of Table 10 we give the solution $y^{(1)}$ of (61) obtained at internal points from the simultaneous equations and at external points by subsequent step-by-step processes. The values of $\sin y^{(1)\prime}$ are also given, the derivative being obtained from the approximate formula

$$2hy'_r = y_{r+1} - y_{r-1}. \tag{62}$$

In this case both $y^{(1)}$ and $y^{(1)\prime}$ have no truncation errors, but in general we would make no elaboration at this stage, since the truncation errors would usually be much smaller than the neglected term $\sin y'$.

The next step is to insert the calculated values of $\sin y^{(1)\prime}$ in (59), obtaining a second approximation $y^{(2)}$ satisfying the equations

$$y^{(2)}_{r+1}+y^{(2)}_{r-1}-2y^{(2)}_r+0\cdot04+0\cdot04\sin y^{(1)\prime}_r = 0, \tag{63}$$

and we repeat the process again, at which stage the internal values show no further change. The second and third approximations are also given in Table 10.

The approximate solution $y^{(3)}$ would be correct if the difference corrections, for the second derivative in (61) and the first derivative in (62), were really negligible. We now take account of them and produce a fourth approximation $y^{(4)}$ from the equations typified by

$$(y_{r+1}+y_{r-1}-2y_r) = -0\cdot04-0\cdot04\sin y'_r+(\tfrac{1}{12}\delta^4-\tfrac{1}{90}\delta^6+\ldots)y_r, \tag{64}$$

where the y on the left refers to the new approximation $y^{(4)}$, that on the right to the previous solution $y^{(3)}$, and where we now use the accurate central-difference formula for the calculation of $y^{(3)\prime}$. For this purpose we difference $y^{(3)}$ as shown in Table 10, finding that the term $\frac{1}{12}\delta^4 y$ on the right of (64) is not quite negligible, and that third differences make a small contribution to the first derivative. The calculated values of the right-hand sides of (64) are given in Table 10 in the column labelled S, for internal points only.

When equations (64) are solved we find that the approximation $y^{(3)}$ is reduced by little more than 1 unit in the third decimal at any internal point, so that this solution is nearly accurate. If more significant figures were needed it would be necessary to evaluate S also at some external points, extending the approximation $y^{(4)}$ into these regions by the use of (64) in a step-by-step technique, and recomputing the right-hand side of (64).

This procedure, though apparently lengthy, is quite easy to carry out in practice: in particular we have avoided any solution of non-linear algebraic equations. The method has also shown fairly rapid convergence to its final solution, but this is not always certain, particularly when a boundary condition involves a derivative.

The 'η' method

†24. An alternative is provided by a method called the 'η-process'. If y is an approximate solution of the differential equation

$$y''+f(y',y,x) = 0, \tag{65}$$

TABLE 10

x	$y^{(1)}$	$y^{(1)\prime}$	$0{\cdot}04 \sin y^{(1)\prime}$	$y^{(2)}$	$y^{(2)\prime}$	$0{\cdot}04 \sin y^{(2)\prime}$	$y^{(3)}$		δ^2		δ^4		δ^6	S
−0·6	−1·080			−1·528			−1·469							
								+549						
·4	0·680	+1·90	+0·038	0·940	+2·75	+0·015	0·920		−55					
								494		−13				
− ·2	− ·320	1·70	·040	− ·430	2·35	·028	− ·426		68		+4			
								426		− 9		+2		
+ ·0	+ ·000	1·50	·040	+ ·000	1·95	·037	+ ·000		77		6		−2	
								349		− 3		0		
·2	·280	1·30	·039	·350	1·55	·040	·349		80		6		0	−0·0795
								269		+ 3		0		
·4	·520	1·10	·036	·621	1·16	·037	·618		77		6		−5	·0759
								192		9		−5		
·6	·720	0·90	·031	·816	0·80	·029	·810		68		1		+6	·0682
								124		10		+1		
0·8	0·880	0·70	·026	0·941	0·46	·018	0·934		58		+2		−4	−0·0578
								66		12		−3		
1·0	1·000	0·50	·019	1·000	+0·15	+ ·006	1·000		46		−1		+2	
								+ 20		11		−1		
1·2	1·080	0·30	·012	1·000	−0·13	− ·005	1·020		35		−2			
								− 15		+ 9				
1·4	1·120	+0·10	+0·004	0·948	−0·37	−0·014	1·005		−26					
								− 41						
+1·6	+1·120	−0·10		+0·852			+0·964							

with boundary conditions

$$g_1(y,y') = 0 \quad \text{at } x = x_1, \qquad g_2(y,y') = 0 \quad \text{at } x = x_2, \tag{66}$$

then a better approximation is given by $y+\eta$, where, neglecting η^2,

$$\eta''+\eta'\frac{\partial f}{\partial y'}+\eta\frac{\partial f}{\partial y} = -\{y''+f(y',y,x)\}, \tag{67}$$

and

$$\left.\begin{aligned} \eta'\frac{\partial g_1}{\partial y'}+\eta\frac{\partial g_1}{\partial y} &= -g_1(y,y') \quad \text{at } x = x_1 \\ \eta'\frac{\partial g_2}{\partial y'}+\eta\frac{\partial g_2}{\partial y} &= -g_2(y,y') \quad \text{at } x = x_2 \end{aligned}\right\}. \tag{68}$$

The right-hand sides of (67) and (68) and the coefficients of η'', η', and η on the left are known, and the correction η satisfies a linear differential equation with linear boundary conditions. If y were only a poor approximation, we now take $y+\eta$ as a better approximation and repeat the process.

This technique can also be applied directly to the finite-difference equations used to replace the differential equation and boundary conditions.

25. As an example consider the differential equation and boundary conditions given by

$$y''-2y^3 = 0, \qquad y(1) = 1, \qquad y'(2)+\{y(2)\}^2 = 0, \tag{69}$$

of which the solution is $y = x^{-1}$. The first correction η to the first approximation y satisfies the equations

$$\left.\begin{aligned} \eta''-6y^2\eta = -(y''-2y^3), \qquad \eta(1) &= 0 \\ \eta'(2)+2y(2)\eta(2) = -y'(2)-\{y(2)\}^2 & \end{aligned}\right\}. \tag{70}$$

We take a first approximation $y = 1$ so that (70) reduce to

$$\eta''-6\eta = 2, \qquad \eta(1) = 0, \qquad \eta'(2)+2\eta(2) = -1. \tag{71}$$

At an interval $h = 0{\cdot}2$ we have the finite-difference equations

$$\left.\begin{aligned} \eta_1+\eta_{-1}-2{\cdot}24\eta_0-0{\cdot}08+C\eta_0 &= 0 \\ C = -\tfrac{1}{12}\delta^4+\tfrac{1}{90}\delta^6-... & \end{aligned}\right\}, \tag{72}$$

which hold at the pivotal points 1·2, 1·4, 1·6, 1·8, and 2·0, and we satisfy the boundary condition at the latter point by the equation

$$\left.\begin{aligned} \eta_1-\eta_{-1}+0{\cdot}8\eta_0+0{\cdot}4+C_1\,\eta_0 &= 0 \\ C_1 = -\tfrac{1}{3}\mu\delta^3+\tfrac{1}{15}\mu\delta^5-... & \end{aligned}\right\}. \tag{73}$$

There is at this stage little point in including the difference corrections, since their effect is likely to be less than that of the squares and higher powers of η neglected in the formulation of the equations. We therefore ignore $C\eta$ and $C_1\eta$ completely, using the approximate equations to produce values of $\eta^{(1)}$ incorporated in the $y+\eta^{(1)}$ of Table 11.

The second correction $\eta^{(2)}$ comes from replacing the η of (70) by $\eta^{(2)}$, and the y by $y+\eta^{(1)}$. For the coefficients and right-hand sides of the new equations we use the simplest form of the central-difference formulae for derivatives, retaining only the first term. The new correcting equations are found to be

$$\left.\begin{array}{ll} \eta_1+\eta_{-1}-2{\cdot}180\eta_0-0{\cdot}005 = 0 & (x = 1{\cdot}2) \\ \eta_1+\eta_{-1}-2{\cdot}146\eta_0-0{\cdot}011 = 0 & (x = 1{\cdot}4) \\ \eta_1+\eta_{-1}-2{\cdot}124\eta_0-0{\cdot}017 = 0 & (x = 1{\cdot}6) \\ \eta_1+\eta_{-1}-2{\cdot}108\eta_0-0{\cdot}023 = 0 & (x = 1{\cdot}8) \\ \eta_1+\eta_{-1}-2{\cdot}094\eta_0-0{\cdot}029 = 0 & \multirow{2}{*}{$(x = 2{\cdot}0)$} \\ \eta_1-\eta_{-1}+0{\cdot}501\eta_0+0{\cdot}056 = 0 & \end{array}\right\}, \tag{74}$$

and their solution provides the second approximation $y+\eta^{(1)}+\eta^{(2)}$. The next approximation is obtained in a similar manner.

TABLE 11

x	$y+\eta^{(1)}$	$y+\eta^{(1)}+\eta^{(2)}$	$y+\eta^{(1)}+\eta^{(2)}+\eta^{(3)}$				
+1·0	+1·000	+1·000	+1·000		+80		
				−165		−34	
1·2	0·866	0·837	0·835		46		+18
				119		16	
1·4	·779	·721	·716		30		6
				89		10	
1·6	·719	·634	·627		20		3
				69		7	
1·8	·672	·567	·558		13		+ 5
				56		2	
2·0	·626	·513	·502		11		− 1
				−45		− 3	
+2·2	+0·571	+0·467	+0·457		+ 8		

At this stage further changes are so small that it is desirable to introduce refinements. These consist in using the accurate finite-difference formulae for any derivatives involved, and for this purpose we need to extrapolate our last approximation, to produce values external to the range from which enough central differences can be obtained at all points. For this purpose we write the first of equations (69) in the form

$$y_1+y_{-1}-2y_0-2h^2y_0^3 = 0, \tag{75}$$

identify y with our last approximation, and use (75) to produce external values, the non-linear terms being already known. In Table 11 we give the 'missing' differences obtained by this procedure: differences of order greater than four are clearly negligible. We now calculate accurately the expressions on the right of equations (70), obtaining, for a further correction $\eta^{(4)}$, the equations

$$\left.\begin{aligned} \eta_1+\eta_{-1}-2\cdot167\eta_0+C\eta_0 &= +0\cdot0021 \quad (x=1\cdot2)\\ \eta_1+\eta_{-1}-2\cdot123\eta_0+C\eta_0 &= -0\cdot0001 \quad (x=1\cdot4)\\ \eta_1+\eta_{-1}-2\cdot094\eta_0+C\eta_0 &= 0\cdot0000 \quad (x=1\cdot6)\\ \eta_1+\eta_{-1}-2\cdot075\eta_0+C\eta_0 &= +0\cdot0013 \quad (x=1\cdot8)\\ \eta_1+\eta_{-1}-2\cdot060\eta_0+C\eta_0 &= -0\cdot0009 \\ \eta_1-\eta_{-1}+0\cdot402\eta_0+C_1\eta_0 &= -0\cdot0006 \end{aligned}\right\} (x=2\cdot0) \tag{76}$$

of which, with $C\eta$ and $C_1\eta$ neglected, the solution is given by

x	1·0	1·2	1·4	1·6	1·8	2·0
$\eta^{(4)}$	0·000	−0·002	−0·002	−0·002	−0·002	−0·002

The only remaining neglected quantities are the difference corrections $C\eta^{(4)}$ and $C_1\eta^{(4)}$, and these are clearly negligible. We would therefore expect $y+\eta^{(1)}+\eta^{(2)}+\eta^{(3)}+\eta^{(4)}$ to be a good solution: it agrees with the analytical solution to the last figure given.

26. We note the rapid convergence of this method, and in this respect it is interesting to compare it with the method of § 23, which would attempt a solution by iteration according to the scheme

$$\frac{d^2y^{(s+1)}}{dx^2} = 2\{y^{(s)}\}^3, \qquad y^{(0)} \equiv 1$$

with $\quad y^{(s+1)} = 1$ at $x = 1$, and $\dfrac{dy^{(s+1)}}{dx} = -\{y^{(s)}\}^2$ at $x = 2$.

Successive values at $x = 2$ obtained by this procedure are given by

$$1\cdot000, \quad -1\cdot000, \quad 0\cdot354, \quad 0\cdot636, \quad 0\cdot248,$$

and it is not even certain that the method is convergent.

Boundary condition at infinity

27. We may be presented with a problem for which one boundary condition is specified at $x = \infty$. Two possible techniques are illustrated by the following example.

Consider the differential equation

$$y''-\left(1+\frac{1}{x}\right)y+\frac{1}{x^2} = 0, \tag{77}$$

with $y(1) = 0$, $y(\infty) = 0$. We replace (77) by the finite-difference equation

$$\left.\begin{aligned} y_1+y_{-1}-\left\{2+h^2\left(1+\frac{1}{x_0}\right)\right\}y_0+\frac{h^2}{x_0^2}+Cy_0 = 0 \\ C = -\tfrac{1}{12}\delta^4+\tfrac{1}{90}\delta^6-\ldots \end{aligned}\right\}. \tag{78}$$

In the first method we apply the second boundary condition at several finite values of x, observing the variations in the solutions so obtained. Suppose, for example, we choose $x = 5$ to 'represent' infinity, and take an interval $h = 1$, we find the following first approximation, with Cy neglected:

x	1	2	3	4	5
y	0·000	0·093	0·074	0·042	0·000

Taking infinity at $x = 10$, we have the following approximation:

x	1	2	3	4	5	6	7	8	9	10
y	0·000	0·094	0·078	0·055	0·038	0·027	0·020	0·014	0·008	0·000

This procedure should be pursued until at least two adjacent values at 'infinity' are zero to the number of figures retained: the boundary condition is then accurately satisfied. The difference correction can of course be inserted at any convenient stage, for example, when the first approximation satisfies this boundary condition.

In this example the required solution tends to zero rather slowly, but it may not be necessary to pursue the method to its full conclusion. The solution may be required, for example, merely for the sake of determining the derivative at $x = 1$, and the two approximations obtained suggest that the second is almost sufficiently accurate for this purpose (though the interval is too large to permit confident estimation of the derivative).

28. The previous method works well if the function tends rapidly to zero. If the rate is slow, we may save labour by anticipating this rate and using it as a boundary condition. In the present problem, for example, we see that for large x the differential equation has the solution $g(x)e^{-x}$, where $g(x)$ varies slowly, and we can allow for this by using as the second boundary condition the equation

$$y(x) = e^h y(x+h) \tag{79}$$

at any convenient point x.

Choosing $x = 5$ for this purpose, we have the boundary condition

$$y(5) = 2{\cdot}718y(6),$$

and, using (78) with Cy neglected to represent the differential equation, we find the following solution:

x	1	2	3	4	5
y	0·000	0·094	0·077	0·053	0·033

This does not differ greatly from the result obtained from the first method using twice as many points. The new criterion for approximate satisfaction of the true boundary condition is that (79) should hold at successive points. On this occasion it is clearly not satisfied at $x = 4$ so that the boundary should be moved farther away. This approximation, however, furnishes a very good start for later approximations, especially if we use relaxation methods for the solution of the linear equations.

Mixed boundary conditions

29. In all previous examples each boundary condition has involved the function or its derivative, or both, at a single point. We are occasionally presented with mixed boundary conditions, which in the most general linear case we can take to be

$$\left.\begin{aligned} a_1 y_0 + b_1 y'_0 + c_1 y_n &= d_1 \\ a_2 y_n + b_2 y'_n + c_2 y_0 &= d_2 \end{aligned}\right\}. \tag{80}$$

This leads to no new problem. The simple central-difference representations of (80), with difference corrections neglected, together with the approximate finite-difference equivalents of the differential equation applied at all pivotal points, including the boundaries, lead to a set of algebraic equations in number sufficient to provide pivotal values at x_0-h, x_0, x_0+h,..., x_n-h, x_n, x_n+h. External values, difference corrections, and correcting solutions are calculated in the usual way.

The matrix of coefficients has the following form, in which the first and last rows correspond to the boundary conditions:

$$\begin{bmatrix} \times & \times & \times & & & & \times & \\ \times & \times & \times & & & & & \\ & \times & \times & \times & & & & \\ & & & \cdot\;\cdot\;\cdot\;\cdot & \cdot & & & \\ & & & & \times & \times & \times & \\ & & & & & \times & \times & \times \\ & \times & & & & \times & \times & \times \end{bmatrix}.$$

Before solution starts, by either direct or relaxation methods, it is best to eliminate the external values, so that all the diagonal elements dominate their respective equations and the matrix then has the following appearance:

$$\begin{bmatrix} \times & \times & & & & & & \times \\ \times & \times & \times & & & & & \\ & & & \cdot & \cdot & \cdot & \cdot & \\ & & & & & \times & \times & \times \\ \times & & & & & & \times & \times \end{bmatrix}.$$

It cannot easily be further reduced to the simple type discussed in § 29 of Chapter 3. The isolated terms cause little inconvenience in the relaxation process, but their effect on the matrices L and U in the triangular resolution process is to give U a non-zero term at the end of its first row, and L non-zero terms in all positions in its last row.

Change of interval

30. The only major arithmetical topic we discuss in this chapter concerns techniques for changing the interval between successive pivotal points. We shall need to use a smaller interval whenever the differences of the first approximate solution are divergent, in the sense defined in Chapter 2.

If the differences diverge throughout the range, it is necessary to take a smaller interval everywhere, and all the coefficients in the finite-difference equations are altered. For the direct method of solution the whole process has to be performed again with the new equations: for the relaxation method, on the other hand, though the new equations are different from the old, the first approximate solution will usually provide a good starting-point, and the relaxation process may be quite short.

The more usual condition is that the interval needs to be reduced over only part of the range, usually near one boundary point, and in this case it would be uneconomic to use a smaller interval everywhere. The natural choice is that the new interval should be half the old one, and we shall consider this case only.

The obvious method is to bridge the junction between the large and small interval by special finite-difference formulae, but this is not essential. Consider the situation shown in Fig. 17, where the points 0, 1, 2, and 3 are separated by an interval h, points 3, 4, 5,... by an interval $2h$. The point labelled 0 is one boundary point.

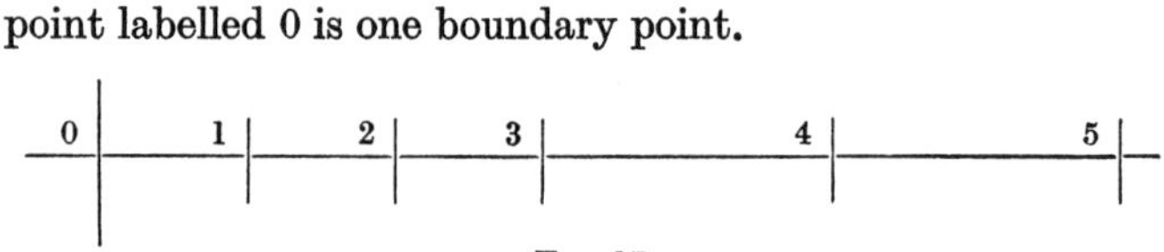

FIG. 17

For the points 2, 1, and 0 (if necessary), the usual finite-difference equation can be used at an interval h: for points 4, 5,... we use the same equation with an interval $2h$, and the only point apparently calling for special treatment is that labelled 3. However, its treatment need not be special, since we use the same type of formula for 3 as we used for 4, 5,... and there is the right number of equations for the determination of the unknowns.

31. Consider as an illustration the determination of the function $(1+x^2)^{-1}$, which has 'difficult' behaviour near the origin, from the facts that it satisfies the differential equation

$$y''+\left(\frac{4x}{1+x^2}\right)y'+\left(\frac{2}{1+x^2}\right)y = 0, \tag{81}$$

and the boundary values $y(0) = 1$, $y(2) = 0{\cdot}2$. At an interval of 0·2 the approximate finite-difference formula is given by

$$y_1\left(1+\frac{0{\cdot}4x_0}{1+x_0^2}\right)+y_{-1}\left(1-\frac{0{\cdot}4x_0}{1+x_0^2}\right)-y_0\left(2-\frac{0{\cdot}08}{1+x_0^2}\right) = 0. \tag{82}$$

The approximation obtained, using this interval throughout and with points external to the range computed in the usual way, is shown in Table 12.

TABLE 12

x	y		δ^2		δ^4		δ^6
−0·4	+0·8514						
		+1054					
− ·2	0·9568		−622				
		+ 432		−178			
+ ·0	1·0000		800		+343		
		− 368		+165		−153	
·2	0·9632		635		+190		− 95
		1003		355		248	
·4	·8629		−280		− 58		+158
		1283		297		− 90	
·6	·7346		+ 17		148		123
		1266		149		+ 33	
0·8	·6080		166		115		+ 32
		1100		+ 34		65	
1·0	·4980		200		50		− 31
		900		− 16		34	
1·2	·4080		184		16		20
		716		32		14	
1·4	·3364		152		− 2		− 4
		564		34		+ 10	
1·6	·2800		118		+ 8		
		446		− 26			
1·8	·2354		+ 92				
		− 354					
+2·0	+0·2000						

According to the rules of Chapter 2 the differences near the origin show signs of divergence and, though it might be safe to apply the difference correction at this interval, we shall take no chances and use the smaller interval $h = 0{\cdot}1$ in the range $x = 0{\cdot}0$ to $0{\cdot}8$. Four more unknowns are introduced, the approximate equations at $x = 0{\cdot}8$ and onwards being given by (82), those at $x = 0{\cdot}1(0{\cdot}1)0{\cdot}7$ by

$$y_1\left(1+\frac{0{\cdot}2x_0}{1+x_0^2}\right)+y_{-1}\left(1-\frac{0{\cdot}2x_0}{1+x_0^2}\right)-y_0\left(2-\frac{0{\cdot}02}{1+x_0^2}\right) = 0. \tag{83}$$

32. The matrix of coefficients is not quite in the best possible form for the 'triangular resolution' method of solution. The equations at $x = 0{\cdot}6$, $0{\cdot}7$, $0{\cdot}8$, and $1{\cdot}0$ are as follows:

$$\left.\begin{aligned}
a_1y(0{\cdot}5)+b_1y(0{\cdot}6)+c_1y(0{\cdot}7) &= 0\\
a_2y(0{\cdot}6)+b_2y(0{\cdot}7)+c_2y(0{\cdot}8) &= 0\\
a_3y(0{\cdot}6)+b_3y(0{\cdot}8)+c_3y(1{\cdot}0) &= 0\\
a_4y(0{\cdot}8)+b_4y(1{\cdot}0)+c_4y(1{\cdot}2) &= 0
\end{aligned}\right\}. \tag{84}$$

The first element of the third of (84) is in the 'wrong' place, and the forms of L and U are consequently not as simple as possible. If, however, we use the second of (84) to eliminate $y(0{\cdot}6)$ from the third of (84), we replace the latter by an equation of the required form

$$a_5y(0{\cdot}7)+b_5y(0{\cdot}8)+c_5y(1{\cdot}0) = 0. \tag{85}$$

The solution of the equations is then straightforward, and we produce the results shown in Table 13. In both ranges the external values were obtained by extrapolation in the usual way so that, for example, the 'external' value $y(1{\cdot}0)$ for the first range does not quite agree with that obtained from the linear equations and used in the second range. This process ensures continuity and smoothness of differences.

We can now neglect seventh and higher differences and calculate with confidence the difference corrections, given to an extra figure on the right of Table 13. The formula for Cy is given by

$$Cy_0 = (-\tfrac{1}{12}\delta^4+\tfrac{1}{90}\delta^6-\ldots)y_0+\left(\frac{4x_0h}{1+x_0^2}\right)(-\tfrac{1}{6}\mu\delta^3+\tfrac{1}{30}\mu\delta^5-\ldots)y_0. \tag{86}$$

When these quantities are introduced into the linear equations we find the following corrections z, in units of the fourth decimal, to our first approximation:

x	0·0	0·1	0·2	0·3	0·4	0·5	0·6	0·7	0·8	1·0	1·2	1·4	1·6	1·8	2·0
z	0	+2	+7	+12	+17	+22	+26	+29	+31	+30	+25	+18	+11	+5	0

TABLE 13

x	y		δ^2		δ^4		δ^6	Cy
−0·2	+0·9614							
		+287						
− ·1	0·9901		−188					
		+ 99		−12				
+ ·0	1·0000		200		+23			
		−101		+11		− 1		
·1	0·9899		189		22		−11	−2·0
		290		33		12		
·2	·9609		156		10		+ 6	1·3
		446		43		6		
·3	·9163		113		+ 4		− 5	1·2
		559		47		11		
·4	·8604		66		− 7		+11	0·4
		625		40		0		
·5	·7979		− 26		7		− 5	−0·4
		651		33		− 5		
·6	·7328		+ 7		12		+10	+0·2
		644		21		+ 5		
·7	·6684		28		7		− 5	0·0
		616		14		0		
·8	·6068		42		− 7			
		574		+ 7				
0·9	·5494		+ 49					
		−525						
+1·0	+0·4969							
+0·2	+0·9605							
		− 998						
0·4	·8607		−280					
		1278		+297				
0·6	·7329		+ 17		−150			
		1261		147		+39		
0·8	·6068		164		111		+20	+4·2
		1097		+ 36		59		
1·0	·4971		200		52		−25	4·0
		897		− 16		34		
1·2	·4074		184		− 18		14	3·3
		713		34		20		
1·4	·3361		150		+ 2		16	1·9
		563		32		+ 4		
1·6	·2798		118		6		− 6	1·2
		445		26		− 2		
1·8	·2353		92		4		+ 4	+1·1
		353		22		+ 2		
2·0	·2000		70		+ 6			
		283		− 16				
2·2	·1717		+ 54					
		− 229						
+2·4	+0·1488							

The new difference correction is negligible, so that our accepted solution is $y+z$. The size of z, in comparison with that of the difference correction, indicates that the algebraic equations are fairly ill-conditioned, and we would expect to have to work to an extra figure to guarantee the accuracy of the last figure in our solution: it has a maximum error of 2 units.

33. The relaxation process can, of course, be applied in the usual way to the set of equations, of which (84) is an extract, without first performing the elimination resulting in (85). The only unusual feature is that a change in $y(0{\cdot}7)$ affects the residuals at only two points, $x = 0{\cdot}6$ and $0{\cdot}7$, while a change in every other value not adjacent to a boundary has the usual effect on three adjacent residuals.

34. This example suggests another method for the treatment of the problem of § 27. It is clear from the differential equation that the solution contains the function e^{-x}. The diagonal coefficients in the linear equations are always greater than 2, so that the equations are well-conditioned. As the interval increases, moreover, these coefficients become still larger. This means that we can keep increasing the interval, as x gets larger, adding more points in the knowledge that their residuals can be liquidated by changes in y which decrease rapidly with x and which do not greatly affect other residuals. We need therefore specify no precise boundary to 'represent' infinity, merely stopping when residuals are negligible.

Singularities

35. This discussion regarding a change of interval is very relevant in the treatment of a differential equation whose solution has a singularity in function or some derivative. Whenever possible, and particularly when high accuracy is required, singularities should be removed by previous mathematical treatment. Occasionally, however, one or two figures can be obtained directly from the finite-difference equations.

We cannot deal with an infinity in the function, and an infinite derivative in the middle of the range is not easily managed. A not uncommon case, however, arises in which the solution has an infinite derivative at one boundary point. We assume no knowledge of the form of singularity.

The differential equation

$$y'' + \frac{1}{x} = 0, \text{ with } y(0) = y(1) = 0, \tag{87}$$

has the solution $-x \log x$, whose first derivative is infinite at $x = 0$. Although the differential equation contains no first derivative we cannot use the more accurate finite-difference formula (5) at all points, since the equation at $x = h$ contains an infinity in its constant term. Since, further, any difference correction is likely to have its greatest value near the origin, it is not worth while in this particular problem to use the better formula at other points. We therefore take the approximate finite-difference equation in the form

$$y_1 + y_{-1} - 2y_0 + h^2/x_0 = 0, \tag{88}$$

and in Table 14 we give the results obtained with the respective intervals $h = \frac{1}{2}, \frac{1}{4}, \frac{1}{8}$.

TABLE 14

x	y	x	y	x	y		δ^2		δ^4
+0	+0·000	+0	+0·000	+0	+0·000				
						+215			
$\frac{1}{2}$	·250	$\frac{1}{4}$	·271	$\frac{1}{8}$	·215		−125		
						90		+62	
+1	+0·000	$\frac{1}{2}$	·292	$\frac{1}{4}$	·305		63		−40
						+ 27		22	
		$\frac{3}{4}$	·188	$\frac{3}{8}$	·332		41		13
						− 14		9	
		+1	+0·000	$\frac{1}{2}$	·318		32		2
						46		7	
				$\frac{5}{8}$	·272		25		− 3
						71		4	
				$\frac{3}{4}$	·201		21		0
						92		+ 4	
				$\frac{7}{8}$	·109		− 17		
						−109			
				+1	+0·000				

We see that the solutions are showing steady but not very rapid improvement, the correct answer at $x = \frac{1}{2}$ being 0·347. It is of course virtually impossible to use the difference-correction method, since the latter is greatest near the origin, and there incalculable. The best we can do is to go on decreasing the interval, observing that this need be done only over part of the range. Taking an interval $\frac{1}{16}$ as far as $x = \frac{1}{2}$, while retaining $h = \frac{1}{8}$ from $x = \frac{1}{2}$ to 1, we find the value 0·330 at $x = \frac{1}{2}$, which is in error by 17 units in the third decimal. The convergence is clearly slow.

36. Better results are obtained with weaker forms of singularity. Consider, for example, the differential equation

$$y'' - \frac{2}{x^2}y = -3, \text{ with } y(0) = y(1) = 0, \tag{89}$$

which has the solution $y = -x^2 \log x$. With the approximate finite-difference equation

$$y_1 + y_{-1} - \left(2 + \frac{2h^2}{x_0^2}\right) y_0 = -3h^2, \tag{90}$$

and the intervals $h = \frac{1}{2}, \frac{1}{4}$, and $\frac{1}{8}$, we find the respective solutions shown in Table 15.

The true value at $x = \frac{1}{2}$ is 0·1733, and we see that convergence to this value is now more rapid. This is partly in virtue of the larger diagonal coefficients in the matrix corresponding to (90): a disturbance near $x = 0$ has much less effect on other values than in the previous example.

TABLE 15

x	y	x	y	x	y		δ^2		δ^4
+0	+0·0000	+0	+0·0000	+0	+0·0000				
						+ 337			
$\frac{1}{2}$	·1875	$\frac{1}{4}$	·0912	$\frac{1}{8}$	·0337		+205		
						542		−234	
+1	+0·0000	$\frac{1}{2}$	·1771	$\frac{1}{4}$	·0879		− 29		+102
						513		132	
		$\frac{3}{4}$	·1641	$\frac{3}{8}$	·1392		161		42
						352		90	
		+1	+0·0000	$\frac{1}{2}$	·1744		251		20
						+ 101		70	
				$\frac{5}{8}$	·1845		321		12
						− 220		58	
				$\frac{3}{4}$	·1625		379		+ 10
						599		− 48	
				$\frac{7}{8}$	·1026		−427		
						−1026			
				+1	+0·0000				

When the singularity gets still weaker, being present only in derivatives higher than the second, it even becomes possible to use the difference correction. We shall not, however, pursue these methods further; they are tentative, the results not easy to justify, and every case has to be treated on its merits.

37. Equation (90) suggests a less serious difficulty which the computer can overcome. Suppose that the differential equation is of the form

$$y'' + \frac{a}{x}y' + \frac{b}{x^2}y = f(x), \tag{91}$$

and that we know the required solution has no singularity. If the boundary values are given we can obtain a first approximation in the usual way to the solution of (91). If one boundary is at $x = 0$, however, we cannot now proceed to calculate values external to the range by direct use of the finite-difference equation. The boundary value is certainly zero, but there is an apparent indeterminacy in y'/x and y/x^2. We can, however, go further and assert that y is of order x^2 at least, y' of order x at least. We can then write (91), for $x = 0$, in the form

$$y'' + ay'' + \tfrac{1}{2}by'' = f(0), \tag{92}$$

using a well-known rule for indeterminate forms, and with y_0 zero we can express y_{-1} in terms of y_1 and a known quantity.

A problem of this kind is always presented when we are dealing with cylindrical coordinates. A typical equation is then given by

$$\frac{d^2y}{dr^2}+\frac{1}{r}\frac{dy}{dr}+f(r)y = g(r), \tag{93}$$

which can be replaced at the origin by the equation

$$2\frac{d^2y}{dr^2}+f(0)y = g(0). \tag{94}$$

In this case y is not necessarily zero, but y' must be zero if the solution is well behaved.

The use of Taylor's series for boundary conditions

†**38.** In most of our examples we have used finite-difference formulae to replace derivatives wherever they occurred, either in differential equations or boundary conditions. For the former this is essential: for the latter we find that the labour of our work is much greater than that of the simplest case 'boundary values specified', and that the largest contribution to the difference correction often arises through inadequate satisfaction of the boundary conditions in the first approximation. For this reason alone it may be necessary to make more than one correction.

By use of Taylor's series we can often find formulae which satisfy accurately the boundary conditions of the general type (2), so that in subsequent corrections we need consider only the difference corrections arising from the differential equation, as in the simplest case. Consider, for example, the general equation (1) with the general boundary conditions (2). The finite-difference form of (1), applied at all pivotal points, including the boundaries, gives n simultaneous equations in $n+2$ unknowns, the extra two being the first external pivotal value at each end of the range. We have been accustomed to finding two more equations by using simple approximate finite-difference forms of (2), one at each end of the range, but we can make these extra equations as exact as desired.

If one boundary point is labelled 0 and the corresponding external point -1, Taylor's series for y_{-1} is given by

$$y_{-1} = y_0-hy_0'+\frac{h^2}{2!}y_0''-\frac{h^3}{3!}y_0'''+\dots\,. \tag{95}$$

The boundary condition (2) gives y_0' in terms of y_0, the differential equation (1) gives y_0'' in terms of y_0' and y_0 and hence in terms of y_0, and by successive differentiation of (1) we can express all the higher derivatives in terms of y_0 so that, using as many terms in (95) as are significant, we can finally find

$$y_{-1} = \alpha_0 y_0 + \beta_0, \tag{96}$$

and, at the other end of the range,

$$y_{n+1} = \alpha_n y_n + \beta_n. \tag{97}$$

We can thus eliminate from the algebraic equations the external points y_{-1} and y_{n+1}, finding n equations in the required n unknowns, of which the first and last equations are of as simple a form, and as accurate, as in the favourable case of boundary-values specified. The only disadvantage of this method is the necessity for successive differentiation of the differential equation, an undertaking not always attractive.

39. The same process can be used for simultaneous equations, and this we illustrate by reworking the example of § 20. Applying the standard equations (50) at the pivotal points $x = 0\cdot2,\ 0\cdot4,\ 0\cdot6,\ 0\cdot8,\ 1\cdot0$, we introduce seven pivotal values of y and seven of z at the points $x = 0\cdot0(0\cdot2)1\cdot2$, with ten equations for their evaluation. Four more equations must be produced from the boundary conditions. The conditions $y(0) = 1$ and $y(1) = z(1)$ reduce the number of unknowns by two, so that we need two extra equations from the two remaining boundary conditions.

At the first boundary point we can write

$$z(0\cdot2) = z(0)+0\cdot2z'(0)+0\cdot02z''(0)+0\cdot001\dot{3}z'''(0)+ \\ +0\cdot0000\dot{6}z^{\mathrm{iv}}(0)+0\cdot000002\dot{6}z^{\mathrm{v}}(0)+\ldots, \tag{98}$$

and, using the differential equations (48), we can replace $z''(0)$ by $y(0)$, $z'''(0)$ by $y'(0)$, $z^{\mathrm{iv}}(0)$ by $z(0)$, $z^{\mathrm{v}}(0)$ by $z'(0)$, so that to seven-decimal accuracy (98) becomes

$$z(0\cdot2) = 1\cdot0000667z(0)+0\cdot2000027z'(0)+0\cdot0213333, \tag{99}$$

when $y(0)$ and $y'(0)$ are given their known values. A similar equation for $y(0\cdot2)$ gives

$$y(0\cdot2) = 0\cdot02z(0)+0\cdot0013333z'(0)+1\cdot2000693. \tag{100}$$

Eliminating $z'(0)$ between (99) and (100) we have one of the required equations, with six accurate decimals in its coefficients, given by

$$y(0\cdot2) = 0\cdot013333z(0)+0\cdot006667z(0\cdot2)+1\cdot199927. \tag{101}$$

At the second boundary point we can write

$$\left.\begin{aligned} y(1{\cdot}2) &= y(1)+0{\cdot}2y'(1)+0{\cdot}02z(1)+0{\cdot}001\dot{3}z'(1)+0{\cdot}0000\dot{6}y(1)+0{\cdot}000002\dot{6}y'(1) \\ z(1{\cdot}2) &= z(1)+0{\cdot}2z'(1)+0{\cdot}02y(1)+0{\cdot}001\dot{3}y'(1)+0{\cdot}0000\dot{6}z(1)+0{\cdot}000002\dot{6}z'(1) \end{aligned}\right\}, \tag{102}$$

and, using the conditions $y(1) = z(1)$, $y'(1) = z'(1)$ we are left with the simple equation

$$y(1{\cdot}2) = z(1{\cdot}2). \tag{103}$$

We then have the following eleven equations, with the difference corrections defined by equation (51):

$y(0{\cdot}2)$	$y(0{\cdot}4)$	$y(0{\cdot}6)$	$y(0{\cdot}8)$	$y(1{\cdot}0)$	$y(1{\cdot}2)$	$z(0{\cdot}0)$	$z(0{\cdot}2)$	$z(0{\cdot}4)$	$z(0{\cdot}6)$	$z(0{\cdot}8)$	
−2	1						−0·04				$= -1-Cy(0{\cdot}2)$
1	−2	1						−0·04			$= -Cy(0{\cdot}4)$
	1	−2	1						−0·04		$= -Cy(0{\cdot}6)$
		1	−2	1						−0·04	$= -Cy(0{\cdot}8)$
			1	−2·04	1						$= -Cy(1{\cdot}0)$
−0·04						1	−2	1			$= -Cz(0{\cdot}2)$
	−0·04						1	−2	1		$= -Cz(0{\cdot}4)$
		−0·04						1	−2	1	$= -Cz(0{\cdot}6)$
			−0·04	1					1	−2	$= -Cz(0{\cdot}8)$
				−2·04	1					1	$= -Cz(1{\cdot}0)$
1						$-0{\cdot}01\dot{3}$	$-0{\cdot}00\dot{6}$				$= 1{\cdot}199927$

(104)

It is easy to see that $y = z$ at every pivotal point, and we solve the equations, with the difference corrections neglected, finding the values

$$y(0{\cdot}2) = 1{\cdot}22140, \quad y(0{\cdot}4) = 1{\cdot}49166, \quad y(0{\cdot}6) = 1{\cdot}82159,$$
$$y(0{\cdot}8) = 2{\cdot}22438, \quad y(1{\cdot}0) = 2{\cdot}71614.$$

The maximum difference between this first approximation and the true solution e^x is only about 20 per cent. of that of Table 8, giving the first approximation by the previous method, and one application of the difference correction produces five-decimal accuracy.

The application of the method to more complicated second-order differential equations, simultaneous or otherwise, needs no further illustration.

Additional notes and bibliography

§ 2. Equations (3) provide the simplest finite-difference equivalent of the differential equation. The difference correction could be made smaller by using more complicated formulae. At points remote from the boundaries, for example, we could exclude the third, fourth, and perhaps higher-difference terms from Cy, expressing these differences in terms of pivotal values at internal points. Near the boundaries this would be impracticable, since external pivotal values would be involved and the resulting algebraic equations would be fewer in number than the unknowns. In this case we would have to use some unsymmetric formula involving only internal points. For a range of five internal and two boundary points, for example, we could use formulae like

$$\left.\begin{aligned} hy_1' &= \tfrac{1}{720}(-120y_0-924y_1+1800y_2-1200y_3+600y_4-180y_5+24y_6)\\ hy_2' &= \tfrac{1}{720}\ (+24y_0-288y_1-\ 420y_2+\ 960y_3-360y_4+\ 96y_5-12y_6)\\ hy_3' &= \tfrac{1}{720}\ (-12y_0+108y_1-\ 540y_2\qquad\qquad +540y_4-108y_5+12y_6)\end{aligned}\right\},$$

and

$$\left.\begin{aligned} h^2y_1'' &= \tfrac{1}{360}(+274y_0-294y_1-\ 510y_2+\ 940y_3-570y_4+186y_5-26y_6)\\ h^2y_2'' &= \tfrac{1}{360}(-\ 26y_0+456y_1-\ 840y_2+\ 400y_3+\ 30y_4-\ 24y_5+\ 4y_6)\\ h^2y_3'' &= \tfrac{1}{360}(+\ \ 4y_0-\ 54y_1+\ 540y_2-\ 980y_3+540y_4-\ 54y_5+\ 4y_6)\end{aligned}\right\},$$

the remaining equations following from considerations of symmetry. A collection of such formulae has been given by

BICKLEY, W. G., 1941, Formulae for numerical differentiation. *Math. Gaz.* **25**, 19–27.

These forms have several disadvantages. First, they lead to a set of algebraic equations whose matrix is no longer specially simple. Second, the error does not vary smoothly from point to point. Third, the formulae cannot easily be memorized. Fourth, the solution obtained from such formulae must still be tested, by differencing, and if necessary corrected by use of a difference-correction method. For pencil and paper computation, therefore, this method cannot be recommended: even for high-speed automatic computation it is convenient only in special circumstances.

In the case of equation (5) the reduction in the size of the difference correction has been effected with much smaller sacrifice. Even here, however, the coefficients of the resulting matrix are more complicated than those involved in equation (4), and the transformation is not always advantageous.

§ 7. The effect of an infinity gets progressively weaker, with the order of derivative in which it occurs.

§ 13. In this example we could have ignored the derivative condition in producing the first approximation, taking an arbitrary value for $y^{(1)}(0)$, calculating $dy^{(1)}/dx$ at $x = 0$, and using an appropriate multiplying factor. With $y(1) = 0$ the same method could be used for any boundary condition of the form $ay'+by = 0$ at $x = 0$, provided the differential equation is homogeneous and the difference correction of the first approximation is negligible after multiplication by the appropriate factor.

§ 18. This example is interesting, since the differential equation has an essential singularity at $x = 0$. The series solution is given by

$$y = x^2+2!x^3+3!x^4+...,$$

which has zero radius of convergence. The series is asymptotic, but it can be shown that the error committed by truncating the series at any point is less than the next term. We therefore cannot, without further consideration, use this series to give more than a certain precision in y. At $x = 0{\cdot}1$, for example, the smallest term is about 0·000036, so that the fifth decimal in the computed value at this point would be uncertain.

Our iterative process can also be performed analytically. We find the results

$$y_0 = 0, \qquad y_1 = \frac{x^2}{1-2x}, \qquad y_2 = \frac{x^2}{1-2x}\left\{1+\frac{2x^2}{(1-2x)^3}\right\},$$

$$y_3 = \frac{x^2}{1-2x}\left\{1+\frac{2x^2}{(1-2x)^3}+\frac{24x^4+32x^5}{(1-2x)^6}\right\},$$

$$y_4 = \frac{x^2}{1-2x}\left\{1+\frac{2x^2}{(1-2x)^3}+\frac{24x^4+32x^5}{(1-2x)^6}+\frac{720x^6+2496x^7+1088x^8}{(1-2x)^9}\right\}, \text{ etc.}$$

The terms of the series in brackets behave like those of an asymptotic series, first decreasing but subsequently increasing, so that our iterative process would give better results only up to a certain stage, successive iterates then diverging. If the denominators in the expressions for y_r are expanded by the binomial theorem, we find that y_r agrees with the previous asymptotic series as far as the term in x^{2r+1}. The remainder, however, behaves in such a way that, at $x = 0{\cdot}1$, y_3 is a better approximation than y_4, but has an error slightly greater than the 'best' value obtained from the asymptotic series by stopping at the term $9!(0{\cdot}1)^{10}$.

In the range of negative x, for which the asymptotic series has alternating signs, we find that our iterative method gives a better result than that obtainable from the asymptotic series.

It is also interesting to note that the differential equation can be solved, by our standard method of using algebraic equations, and with two correct boundary values at $x = 0$ and $x = 0{\cdot}1$, say, with full accuracy. See, for example,

Fox, L., and Miller, J. C. P., 1951, Table-making for large arguments. The exponential integral. *Mathematical Tables and Other Aids to Computation*, **5**, 163–7.

§ 24. This method was suggested for the correction of first approximations in initial-value problems by

Clenshaw, C. W., and Olver, F. W. J., 1951, Solution of differential equations by recurrence relations. *Mathematical Tables and Other Aids to Computation*, **5**, 34–39.

§ 38. This method of satisfying the boundary conditions appears to be quite new. It is, of course, an obvious adaptation of that used to give starting values for an initial-value problem.

5

First-order Equations

Introduction

1. First-order equations are generally amenable only to step-by-step methods. Only one extra condition is needed to provide a unique solution, there is only one boundary point, and in the sense of our definition we can have only an initial-value problem, never one of boundary-value type.

It is nevertheless instructive and interesting to examine various finite-difference formulae and certain transformations of the given problem, to see whether first-order equations can be solved by techniques suited to true boundary-value problems. We shall find that techniques are possible which in some cases are easy to apply and may involve less labour than those of step-by-step methods.

Finite-difference equations and boundary conditions

2. The linear problem involves the solution of the differential equation

$$y'+f(x)y = g(x), \tag{1}$$

with y specified at some point in the range of integration. In previous chapters we have replaced the first derivative by a central-difference expression, which in this case leads to the replacement of (1) by the typical algebraic equation

$$\left.\begin{aligned} y_{r+1}+2hf_r y_r-y_{r-1}+Cy_r &= 2hg_r \\ C &= -\tfrac{1}{3}\mu\delta^3+\tfrac{1}{15}\mu\delta^5-\ldots \end{aligned}\right\}. \tag{2}$$

If we now consider a range of points x_0 to x_n, and apply (2) at all internal points, we have $n-1$ equations in $n+1$ unknowns, and it would appear that *two* extra conditions are in general needed to provide a unique solution, just as for a second-order equation.

The specification of y_0 is one condition, and another could be obtained from the fact that y'_0 and all successive derivatives can be determined from (1) and other equations obtained by successive differentiation of (1), so that Taylor's series can be used to calculate y_1. The set of equations (2), with Cy neglected, could then be used to calculate $y(x_r)$ for successive r, and this is a possible step-by-step process.

3. Occasionally, however, we might get the second condition in the form of another value of y known at the second boundary $x = x_n$, obtained possibly from physical considerations or from mathematical arguments such as the use of asymptotic series. We want then to consider whether it is possible to use a boundary-value technique which incorporates this extra information. The first problem concerns the 'boundary values', which cannot be independent, as they are for second-order equations, but must be 'compatible' with each other.

Compatibility of boundary values. Simple examples

†4. To illustrate the problems which can arise, we consider first a simple differential equation for which we can obtain an analytical solution to the corresponding difference equations.

The problem defined by the equations

$$y'-y = 0, \qquad y(0) = 1, \tag{3}$$

has the solution e^x, and $y(1) = e = 2{\cdot}718...$. The difference equation corresponding to (2), with Cy neglected, now has the form

$$y_{r+1}-y_{r-1}-2hy_r = 0, \tag{4}$$

with $y_0 = 1$. The general solution of equation (4) is given by

$$y = Pe^{r\theta}+(-1)^r Qe^{-r\theta}, \tag{5}$$

where $\sinh\theta = h$ and P and Q are arbitrary constants. Now let the second given condition be $y_n = A$, where y_n is the value at $x = 1$, so that $nh = 1$. The solution of (4) under these conditions is given by

$$y_r = \frac{\{Ae^{r\theta}-(-1)^n e^{(r-n)\theta}\}+(-1)^r\{e^{(n-r)\theta}-Ae^{-r\theta}\}}{e^{n\theta}-(-1)^n e^{-n\theta}}. \tag{6}$$

The solution evidently consists of two parts, the first bracketed expression in the numerator being quite well behaved, the second oscillating due to the presence of the factor $(-1)^r$. It will be noted that for arbitrary values of A this oscillation does not necessarily get smaller with h.

This type of oscillation does not occur in the solution of the differential equation so that, to relate the solution (6) to that of the differential equation, we should like to choose A so that the oscillation disappears. In this case we have

$$A = e^{n\theta} = e^{n\sinh^{-1}h}. \tag{7}$$

For $h = 0{\cdot}25$, we find $A = 2{\cdot}691$; for $h = 0{\cdot}1$, $A = 2{\cdot}714$, and as $h \to 0$ we find $A \to e$, the expected result.

5. In general we are unlikely to be able to solve analytically the difference equation resulting from the differential equation. We can, however, carry out a numerical analogue of this analysis. Taking the set of equations (4), for example, at an interval $h = \frac{1}{4}$, and letting A take separately the values 2·0, 2·5, and 3·0, we find the values tabulated and differenced in Table 1 (the factor of 900 is introduced to avoid decimals).

In each case there is wild oscillation in the differences, but if we insist on smoothness, defining this in the best way open to us, to mean that the fourth difference of y should vanish, we can interpolate inversely to produce an A for which $\delta^4 y(\frac{1}{2}) = 0$: we find $A = 2{\cdot}692$.

6. We note particularly that, at a finite interval, the value of A which gives a smooth solution to the approximate difference equation is not the true value of the solution of the differential equation. The use of the latter as boundary value would in fact produce an oscillatory approximate solution. Nevertheless, if we are lucky enough to know this boundary value, it will be the true solution of the differential equation, and the question then arises as to whether we can improve and correct the oscillatory solution thereby produced. If the interval is small enough, of course, the first approximation will be correct, Cy will be negligible, and there is no oscillation. For reasons of economy, however, we want to use as large an interval as possible, calculate the difference correction and improve the approximate solution by replacing the calculated Cy in equations (2). We are encouraged to try this technique, even though the approximation and its differences oscillate, because Cy involves the *average* value of successive differences.

†7. Returning to the differential equation (3), with its difference equation (4), and taking also $y(1) = e = 2{\cdot}718$ to three decimals, we solve the simultaneous equations at interval $h = \frac{1}{4}$, with Cy neglected, and produce a first approximation; values at points outside the range are calculated by the standard extrapolation process described in the previous chapter, and used for the determination of Cy near the end points of the range. Table 2 gives this first approximation, its differences, and the calculated values of Cy. The contribution to the latter from fifth differences is at most two units, so that it is reasonable to neglect seventh differences.

When the calculated Cy are inserted in equations (4), a second approximation is obtained whose errors are about one-half of those of the first approximation. The third and fourth approximations (for whose

TABLE 1

x	$900y$		δ^2		δ^4	$900y$		δ^2		δ^4	$900y$		δ^2		δ^4
+0	+ 900					+ 900					+ 900				
		−300					+100					+500			
$\frac{1}{4}$	600		+900			1000		+300			1400		−300		
		+600		−1500			400		−400			200		+700	
$\frac{1}{2}$	1200		−600		+2700	1400		−100		+750	1600		+400		−1200
		0		+1200			300		+350			600		−500	
$\frac{3}{4}$	1200		+600			1700		+250			2200		−100		
		+600					+550					+500			
+1	+1800					+2250					+2700				

TABLE 2

x	10^3y		δ^2		δ^4		Cy
$-\frac{1}{2}$	$+598$						
		$+207$					
$-\frac{1}{4}$	805		-12				
		195		$+122$			
$+0$	1000		$+110$		-190		
		305		-68		$+348$	
$\frac{1}{4}$	1305		42		$+158$		$-1{\cdot}1$
		347		$+90$		-272	
$\frac{1}{2}$	1652		132		-114		$-12{\cdot}9$
		479		-24		$+215$	
$\frac{3}{4}$	2131		108		$+101$		$-7{\cdot}1$
		587		$+77$		-162	
1	2718		185		-61		
		772		$+16$			
$\frac{5}{4}$	3490		$+201$				
		$+973$					
$+\frac{3}{2}$	$+4463$						

determination we need more external values than are given in Table 2) get steadily, though not monotonically, closer to the analytical solution, and the process is evidently converging.

Table 3 gives the successive approximate solutions, together with the true solution e^x.

TABLE 3

Approximations to 10^3y

x	(1)	(2)	(3)	(4)	10^3e^x
$+0$	$+1000$	$+1000$	$+1000$	$+1000$	$+1000$
$\frac{1}{4}$	1305	1292	1279	1287	1284
$\frac{1}{2}$	1652	1647	1650	1647	1649
$\frac{3}{4}$	2131	2128	2111	2120	2117
$+1$	$+2718$	$+2718$	$+2718$	$+2718$	$+2718$

The convergence is fairly slow, and we might try to accelerate it by using a device of Aitken. From the sequence $y_{r-1}, y_r, y_{r+1},\ldots$ we produce a second sequence z_r defined by the equation

$$z_r = \frac{y_{r-1}y_{r+1}-y_r^2}{y_{r+1}+y_{r-1}-2y_r} = y_{r+1}-\frac{(y_r-y_{r+1})^2}{y_{r+1}+y_{r-1}-2y_r}, \tag{8}$$

the suffixes here referring to successive terms of the sequence at a particular pivotal point. Under certain conditions z_r converges to the same limit as y_r, and at a faster rate. Applying this algorithm to the approximations (2), (3), and (4) of Table 3, at each of the points $x = \frac{1}{4}, \frac{1}{2}$, and $\frac{3}{4}$, we produce the respective values 1284·0, 1648·5, and 2116·9, no value being in error by more than half a unit.

8. In taking account of fifth differences in Cy we may have done more work than is really necessary: the fifth differences may not be significant in the final solution, though they are in the first approximation. It is therefore worth while to ignore fifth differences until this point is settled. If after the first approximation we calculate the Cy using only third differences, we find that the next approximation has a maximum error of 6 units, and an average of only 3 units. The convergence to the final solution is much faster, the third approximation having a maximum error of 2 units. Had the fifth differences been really significant they could of course be introduced at any subsequent stage, but in the final solution they are in fact negligible.

A practical example

9. Though problems for which the technique of the previous example can be used are not of frequent occurrence they are not unknown, and in some cases the method given here is less laborious than any step-by-step method of integration. The following example, arising in a practical problem, also illustrates a technique for solving non-linear equations by relaxation.

In work involving the tabulation of the Airy integral in certain regions of the complex plane, it was necessary to solve two simultaneous first-order equations given by

$$\left.\begin{aligned} \frac{dp}{dx} &= p^2-q^2+A(x) \\ \frac{dq}{dx} &= 2pq+B(x) \end{aligned}\right\}, \tag{9}$$

in which $A(x)$ and $B(x)$ are known functions of the independent variable. These equations had to be solved for a certain range of x: at one end of the range values of p and q were available from the use of an ascending series, and at the other by means of asymptotic series. The conditions are therefore similar to those of the last example.

The finite-difference equations replacing (9) are given by

$$\left.\begin{aligned} p_1-p_{-1}-2h\{p_0^2-q_0^2+A(x_0)\}+Cp_0 &= 0 \\ q_1-q_{-1}-2h\{2p_0q_0+B(x_0)\} \quad +Cq_0 &= 0 \end{aligned}\right\}, \tag{10}$$

where the difference-correction operator is given in equation (2). Now if a guess is made for values of p and q at all points in the range we can substitute these values in the left of equations (10), with Cp and Cq neglected as usual in the first approximation, and calculate the residuals R and S say. We now wish to make such changes in p and q at the various pivotal points that the residuals steadily reduce in magnitude. If these changes are not large, we can express the effects on the residuals R and S, caused by changes in p_0 and q_0 only, by the two equations

$$\left.\begin{aligned}\delta R_0 &= -4h(p_0\,\delta p_0 - q_0\,\delta q_0)\\ \delta S_0 &= -4h(p_0\,\delta q_0 + q_0\,\delta p_0)\end{aligned}\right\}. \tag{11}$$

We can thus make both residuals R_0 and S_0 almost vanish at a particular point by altering the values of p_0 and q_0 in accordance with the equations

$$\left.\begin{aligned}p_0\,\delta p_0 - q_0\,\delta q_0 &= R_0/4h\\ q_0\,\delta p_0 + p_0\,\delta q_0 &= S_0/4h\end{aligned}\right\}, \tag{12}$$

whose solution is given by

$$\delta p_0 = \frac{S_0 q_0 + R_0 p_0}{4h(p_0^2+q_0^2)}, \qquad \delta q_0 = \frac{S_0 p_0 - R_0 q_0}{4h(p_0^2+q_0^2)}. \tag{13}$$

By these increments the residuals R_0 and S_0 are not reduced completely to zero, since terms like $(\delta p_0)^2$, $(\delta q_0)^2$ were neglected in the analysis. The effect on residuals at neighbouring points is known exactly, and is given by

$$\left.\begin{aligned}\delta R_1 &= -\delta p_0, & \delta R_{-1} &= \delta p_0\\ \delta S_1 &= -\delta q_0, & \delta S_{-1} &= \delta q_0\end{aligned}\right\}. \tag{14}$$

The iterative technique of solution is obvious. Applications of equations (13) at successive pivotal points, together with calculation of residuals made more frequent by the non-linearity of the equations, gave rapid convergence to the desired solution. The subsequent introduction of the difference corrections caused little difficulty, the corrections being fairly small, and the oscillations of the first approximations disappeared.

In this work three further points were of considerable interest. First, with this boundary-value method it was possible to use with convenience a bigger interval than with a step-by-step method. Second, many cases were computed for values of a parameter which showed regular change; after one or two cases it was possible to make such good guesses, for a first approximation, that convergence to the final answer was extremely rapid. Finally, in the few cases for which errors in previous computations had produced incompatible boundary values, this was revealed by the behaviour of the differences, which refused to settle down to smooth behaviour.

It is, of course, somewhat fortuitous that this relaxation method works so well in this case. The non-linear algebraic equations evidently form a very stable system, analogous to the well-conditioned linear equations encountered in Chapter 3.

A smaller difference correction

10. The oscillation of approximate solutions is an unsatisfactory feature of the method of §§ 7 and 8, and when they become too violent the correcting procedure may not converge, or at best converge only slowly. Now this feature is absent in the true solution, so that we would like to use a method which would produce, as its first approximation, a result more accurate than that, for example, of Table 2.

One possibility is to reduce the interval so that the difference corrections, neglected in the first approximation, are really negligible or at least small: this method was used effectively in the last example. A second method, more generally attractive, is to use from the beginning formulae which involve smaller correction terms. We cannot in this connexion use the term $\frac{1}{3}\mu\delta^3$ of equation (2) in its Lagrangian form, since this would immediately bring in points outside the range of integration. We can, however, produce a formula which introduces no more points and has a much smaller difference correction.

For the differential equation $y' = y$, for example, we write

$$(\mu\delta - \tfrac{1}{6}\mu\delta^3 + \tfrac{1}{30}\mu\delta^5 - \ldots)y_0 = hy_0, \tag{15}$$

operate on both sides with $1+\frac{1}{6}\delta^2$, and produce the equation

$$(\mu\delta + \tfrac{1}{180}\mu\delta^5 - \ldots)y_0 = hy_0 + \tfrac{1}{6}h(y_1 + y_{-1} - 2y_0). \tag{16}$$

Expressing $\mu\delta y_0$ in terms of pivotal values, we obtain finally the equation

$$\left.\begin{aligned} y_1(1-\tfrac{1}{3}h) - y_{-1}(1+\tfrac{1}{3}h) - \tfrac{4}{3}hy_0 + Cy_0 &= 0 \\ C &= \tfrac{1}{90}\mu\delta^5 - \ldots \end{aligned}\right\}, \tag{17}$$

the Lagrangian terms in (17) still involving only three adjacent points.

There is now a much better chance of producing an accurate first approximation without the necessity for decreasing the interval. In the case of the exponential function, for example, we find that the use of equations typified by (17), with $h = \frac{1}{4}$ and C_y neglected, gives three correct decimals at every pivotal point.

Alternative methods

A knowledge of two boundary values for a first-order equation must in general be regarded as a luxury, and we now examine alternative methods of turning a first-order equation, with its single specified boundary value, into a problem tractable by methods applied to boundary-value problems.

Method 1

11. The obvious method is to use, at the second boundary, a formula representing the differential equation at this point which involves no pivotal values outside the field of integration. Returning to the example of § 4, we have the equations typified by

$$y_{r+1}-y_{r-1}-2hy_r+Cy_r = 0, \tag{18}$$

which are to be satisfied at every internal point. The equation $y_0 = 1$ is given by the boundary condition, and we now impose at the other end of the range an equation of the form

$$ay_n+by_{n-1}+cy_{n-2}+\dots = 0, \tag{19}$$

which is to replace the differential equation $y' = y$ at the last point x_n.

We can obviously determine equation (19) by using a backward-difference formula for the first derivative at x_n, given by

$$hy'_n = (\nabla+\tfrac{1}{2}\nabla^2+\tfrac{1}{3}\nabla^3+\dots)y_n, \tag{20}$$

so that

$$(\nabla+\tfrac{1}{2}\nabla^2+\tfrac{1}{3}\nabla^3+\dots)y_n-hy_n = 0 \tag{21}$$

is the equation to be used, instead of (18), at the end point x_n. The form (19) is obtained by expressing some of the differences in terms of pivotal values; for example we can write (21) in one or other of the forms

$$\left.\begin{aligned} (1-h)y_n-y_{n-1}+(\tfrac{1}{2}\nabla^2+\tfrac{1}{3}\nabla^3+\dots)y_n &= 0 \\ (\tfrac{3}{2}-h)y_n-2y_{n-1}+\tfrac{1}{2}y_{n-2}+(\tfrac{1}{3}\nabla^3+\dots)y_n &= 0 \\ (\tfrac{11}{6}-h)y_n-3y_{n-1}+\tfrac{3}{2}y_{n-2}-\tfrac{1}{3}y_{n-3}+(\tfrac{1}{4}\nabla^4+\dots)y_n &= 0 \end{aligned}\right\}. \tag{22}$$

In each case we have some Lagrangian terms and a difference correction.

For our particular problem, with $h = \frac{1}{4}$, we then have the following set of equations at internal points:

$$\left.\begin{aligned} y(\tfrac{1}{2})-y(0)-\tfrac{1}{2}y(\tfrac{1}{4})+Cy(\tfrac{1}{4}) &= 0 \\ y(\tfrac{3}{4})-y(\tfrac{1}{4})-\tfrac{1}{2}y(\tfrac{1}{2})+Cy(\tfrac{1}{2}) &= 0 \\ y(1)-y(\tfrac{1}{2})-\tfrac{1}{2}y(\tfrac{3}{4})+Cy(\tfrac{3}{4}) &= 0 \end{aligned}\right\}, \tag{23}$$

where Cy is defined in equation (2). At the end points we have $y(0) = 1$, and one of the following:

$$\left.\begin{aligned} \tfrac{3}{4}y(1)-y(\tfrac{3}{4})+(\tfrac{1}{2}\nabla^2+\tfrac{1}{3}\nabla^3+\dots)y(1) &= 0 \\ \tfrac{5}{4}y(1)-2y(\tfrac{3}{4})+\tfrac{1}{2}y(\tfrac{1}{2})+(\tfrac{1}{3}\nabla^3+\dots)y(1) &= 0 \\ \tfrac{19}{12}y(1)-3y(\tfrac{3}{4})+\tfrac{3}{2}y(\tfrac{1}{2})-\tfrac{1}{3}y(\tfrac{1}{4})+(\tfrac{1}{4}\nabla^4+\dots)y(1) &= 0 \end{aligned}\right\}. \tag{24}$$

12. Before producing any numerical results we can show analytically that the oscillation which affected previous methods is also present here. The particular solution of the difference equations, with the correct initial condition and the use of the first of (24) at the other end of the range, and with all difference corrections neglected, is obtainable from equation (5) in the form

$$y = \frac{e^{r\theta}\{3e^{-n\theta}+4e^{-(n-1)\theta}\}+(-1)^r e^{-r\theta}\{4e^{(n-1)\theta}-3e^{n\theta}\}}{2\{4\cosh(n-1)\theta-3\sinh n\theta\}}, \tag{25}$$

with a non-vanishing oscillatory term.

13. The numerical results, corresponding to each of the three boundary conditions (24), with all difference corrections neglected and with two external points included, are given in Table 4.

TABLE 4

x	10^3y		δ^2		10^3y		δ^2		10^3y		δ^2		δ^4
$-\frac{1}{4}$	+577				+759				+775				
		+423				+241				+225			
+0	1000		−346		1000		+18		1000		+50		
		77		+730		259		+94		275		+38	
$\frac{1}{4}$	1077		+384		1259		112		1275		88		−33
		461		−537		371		−39		363		5	
$\frac{1}{2}$	1538		−153		1630		73		1638		93		+37
		308		+461		444		+76		456		42	
$\frac{3}{4}$	1846		+308		2074		149		2094		135		−17
		616		−309		593		−1		591		+25	
1	2462		−1		2667		+148		2685		+160		
		+615				+741				+751			
$+\frac{5}{4}$	+3077				+3408				+3436				

We see, by comparison with the analytical solution given in Table 3, that the solutions increase in accuracy and smoothness with the accuracy of the formula used at the point $x = 1$. It might be expected that the best of these formulae would be that having the same order of error as those at other points, in this case involving the third and higher differences. Experience indicates, however, that it is nearly always advantageous to use, at a boundary, the most accurate formula available.

The number of significant differences in equations (24) is not of course known until the final accurate solution is obtained, and we have to make a compromise between retaining too few differences, giving a poor first approximation, and retaining too many differences and wasting labour. In the present example it would appear that not enough differences were retained, but, as we noticed in § 8, the large fourth difference may not be present in the final solution, and at this stage it is advisable to correct this approximate solution by incorporating the main difference correction into the equations at internal points, but ignoring for the moment any apparent correction in the 'boundary condition'. For this purpose values external to the boundary, needed to produce central differences at internal pivotal points, can be obtained by use of (18), with Cy neglected, at the boundaries and successive external points.

Carrying out this process and ignoring fifth differences we find that the corrections to be made in the third of Table 4 to $y(\frac{1}{4})$, $y(\frac{1}{2})$, $y(\frac{3}{4})$, and $y(1)$ are respectively 9, 12, 23, and 35, and the maximum discrepancy between this new solution and e^x is 2 units in the last figure. When this solution is differenced, moreover, we find that the single fourth difference available is 10, so that we were justified in regarding with suspicion the large figure 37 in the last column of Table 4. Once the significance of the various differences in the boundary formula has been established, all significant terms in the formula can be incorporated in future approximations, exactly as with the main difference correction.

The danger of introducing the boundary-difference correction at an earlier stage is not difficult to understand. This correction is a series of backward differences, and the full effect of the oscillations is felt because there is no alleviating averaging process. Even in the 'best' approximation given in Table 4, the use of the boundary-difference correction given by the third of (24), retaining only $\frac{1}{4}\nabla^4 y(1)$, together with the other difference corrections applied to produce the second approximation, gives a result with a maximum error of as much as 7 units.

Method 2

14. In this respect the use of the backward-difference formula (20), with its large coefficients, is not very satisfactory. A more attractive formula is that given by

$$y_{r+1}-y_r = \tfrac{1}{2}h(y'_{r+1}+y'_r)-\tfrac{1}{12}\delta^3 y_{r+\frac{1}{2}}+\tfrac{1}{120}\delta^5 y_{r+\frac{1}{2}}-\ldots, \tag{26}$$

leading to the equation, replacing any of (24) for use at the end point, given by

$$\left.\begin{aligned} y(1) &= \tfrac{9}{7}y(\tfrac{3}{4})+\tfrac{8}{7}C_1 y(\tfrac{7}{8}) \\ C_1 &= -\tfrac{1}{12}\delta^3+\tfrac{1}{120}\delta^5-\ldots \end{aligned}\right\}. \tag{27}$$

If we use this equation, together with the set (23) and with all difference corrections neglected, we produce a first approximation identical with the second of Table 4. This is not surprising, since at the end point the same order of differences was neglected in each case, and indeed if $y(\frac{1}{2})$ is eliminated from the third of (23) and the second of (24) we obtain the first of (27), with all difference corrections neglected.

This approximation is not so good as that of the previous method in which $\nabla^3 y$ was retained at the boundary, but it can be improved. Following the technique of previous methods, neglecting at this stage differences of order five and higher in Cy and ignoring for the moment the

$C_1 y$ of (27), we find that a second approximation is obtained whose error is a maximum of 7 units, compared with 51 in the first approximation. Further corrections, moreover, using the second of (27) as the difference correction at the end point, can be made with fair confidence, $C_1 y$ being affected by oscillations or rounding errors in the differences to a less extent than the difference corrections of (24).

Method 3

15. Using either Method 1 or Method 2 for the satisfaction of the differential equation at the second boundary, we can incorporate the method of § 10 to give a smaller difference correction at internal pivotal points. This is not so powerful as in the case when two boundary values were known, since we now have to use at the second boundary either equation (24), with its large coefficients, or equation (27) which has a larger difference correction. Nevertheless, for the exponential function the first approximation obtained by the use of (17) and (27) has a maximum error of 6 units in the fourth figure, which compares favourably with the results of Table 4, and which is capable of easy improvement, the oscillations being small.

Solution of the linear equations

16. Except in § 9 we have not disclosed the method by which we solved the simultaneous algebraic equations arising from the finite-difference equations. When the equations are linear there is no difficulty in using straightforward elimination methods, or the triangulation process of Chapter 3, but in their original form they are not easily soluble by relaxation. This will be true in general. The three-term finite-difference form of the general linear first-order equation (1) is given by equation (2). Writing this equation without its difference correction in the form

$$y_1 - y_{-1} + 2hf_0 y_0 - 2hg_0 = R_0, \tag{28}$$

we see that the effect on the residual R_0 of making a change δy_0 is given by

$$\delta R_0 = 2hf_0 \delta y_0,$$

while the effects on adjacent residuals R_1 and R_{-1} are $-\delta y_0$ and δy_0 respectively.

Now it will usually happen that the interval h at which a satisfactory solution to the differential equation can be found will be such that hf is fairly small. When $2hf$ is less than unity, the liquidation of a particular residual by a change in the value at that point will cause bigger increments to adjacent residuals, and the chance of convergence of the unrefined relaxation process is therefore not great. We have either to construct group-relaxation operators or to recast the equations into a more tractable form.

Normalization

17. We know that in theory the device of normalization will provide a set of equations for which the relaxation method will converge. In the present context normalization is easy, so that it is worth while to try the device in practice.

Returning to the example of the exponential function, we shall treat the case in which we know a second boundary value in addition to the one originally prescribed. Using the weaker method, with difference correction starting with $\frac{1}{3}\mu\delta^3 y_0$, we have for three internal points the following three equations, in which $y(0)$ and $y(1)$ are known quantities, and difference corrections are neglected:

$$\left.\begin{aligned} y(\tfrac{1}{2})-y(0)-2hy(\tfrac{1}{4}) &= 0 \\ y(\tfrac{3}{4})-y(\tfrac{1}{4})-2hy(\tfrac{1}{2}) &= 0 \\ y(1)-y(\tfrac{1}{2})-2hy(\tfrac{3}{4}) &= 0 \end{aligned}\right\}. \tag{29}$$

The normalized equations are given by

$$\left.\begin{array}{lllll} 2hy(0)+(1+4h^2)y(\tfrac{1}{4}) & & -y(\tfrac{3}{4}) & & = 0 \\ -y(0) & +(2+4h^2)y(\tfrac{1}{2}) & & -y(1) & = 0 \\ \quad -y(\tfrac{1}{4}) & & +(1+4h^2)y(\tfrac{3}{4}) & -2hy(1) & = 0 \end{array}\right\}. \tag{30}$$

These equations are in a form in which relaxation is quite practicable. They also have another striking feature, that in the new equation for the point $x = \frac{1}{2}$ values at the adjacent points $\frac{1}{4}$ and $\frac{3}{4}$ no longer appear. In fact the values at points 0, $\frac{1}{2}$, and 1, which we might call the 'even' points, and which are separated by an interval $2h$, now have a solution, the 'even solution', which is quite independent of the 'odd' values at points $\frac{1}{4}$, $\frac{3}{4}$.

Writing the second of (30) in the form

$$y(0)+y(1)-2y(\tfrac{1}{2})-4h^2y(\tfrac{1}{2}) = 0, \tag{31}$$

we see that this is the approximate finite-difference equation derived from the second-order differential equation $y''-y = 0$, obtainable by differentiation from the original first-order equation. In this finite-difference equation the interval is $2h$.

We can therefore obtain the 'even' solution from the derived second-order equation and with the two given boundary values, without consideration of intermediate 'odd' values.

What then has happened to the odd values? If we had taken more points in the range we should find that the equations at odd points do not involve even values except at the two points adjacent to the boundaries. Except at these points, moreover, the equations are of the form of the second of (30), and therefore also are approximations to the derived second-order equation.

For the penultimate points, we can, for example, write the first of (30) in the form

$$\{y(-\tfrac{1}{4})+y(\tfrac{3}{4})-2y(\tfrac{1}{4})-4h^2y(\tfrac{1}{4})\}+\{y(\tfrac{1}{4})-y(-\tfrac{1}{4})-2hy(0)\} = 0, \quad (32)$$

the first part of which is the form corresponding to (31), representing the second-order derived equation applied at $x = \frac{1}{4}$, and the second part is the form (29), corresponding to the original first-order equation applied at the boundary point $x = 0$. Similar analysis can be applied to the third of (30), and we can regard these two equations as being obtained by satisfying the derived second-order equation, eliminating external values by using the original first-order equation as boundary conditions: this is quite analogous to the methods of § 12 of Chapter 4, for satisfying boundary conditions involving a derivative with a second-order differential equation.

18. We conclude that

(i) the odd and even solutions both satisfy the second-order equation obtained from the original equation by differentiation;

(ii) for the even solution, the given boundary values provide the boundary conditions;

(iii) for the odd solution, the boundary conditions are provided by the original first-order differential equation, applied at the boundary points.

This throws further light on the oscillation encountered in our approximate solutions. The even and odd solutions, separately satisfying a second-order differential equation with two boundary conditions, are both smooth, but do not fit well together unless the two sets of boundary conditions are compatible: in the final accurate solution compatibility exists, but in the approximate finite-difference equations there is an element of incompatibility.

Transformation to second-order differential equation

19. These arguments suggest still another method for first-order equations. The even and odd solutions can clearly be made to fit together if we solve the derived second-order equation, with, as boundary conditions, the given value at one end and the original first-order equation at the other. This method has the advantage of always giving smooth approximations, and a knowledge of a second boundary value is unnecessary. The second-order equation has two solutions, but the use of the second boundary condition ensures that we obtain the one required.

In general, then, for the first-order equation

$$y'+f(x)y = g(x), \quad \text{with } y(0) = A, \text{ known,}$$

we solve either

$$y''+fy'+f'y = g', \tag{33}$$

or

$$y''+(f'-f^2)y = g'-fg, \tag{34}$$

obtained from (33) by elimination of the first derivative, with the boundary conditions

$$\left.\begin{aligned} y(0) &= A \\ y'(1)+f(1)y(1) &= g(1) \end{aligned}\right\}, \tag{35}$$

by the methods already described in Chapter 4.

20. We note in particular that a second-order equation can always be produced in which the first derivative is absent, so that we can use finite-difference equations involving only three pivotal values and for which the difference correction is much smaller than that of other methods of this chapter. If a second boundary value is available, we can of course use it instead of the second of (35), and in no case will oscillation occur in approximate solutions.

In Chapter 4 we saw that the satisfaction of a boundary condition which included a derivative usually involved, as the algebraic equation corresponding to this boundary point, an equation which included several of the unknowns and a small difference correction, or one involving few unknowns and a large difference correction. In the case of the first-order equation, however, solved in the form of a second-order equation, one boundary condition is the first-order equation, and this is satisfied not only at the boundary, but at all points. The finite-difference form of the boundary condition, given by

$$(\mu\delta-\tfrac{1}{6}\mu\delta^3+\tfrac{1}{30}\mu\delta^5-\ldots)y_0 = -hf_0 y_0+hg_0, \tag{36}$$

can therefore be operated on by $1+\frac{1}{6}\delta^2$ to produce the result

$$\left.\begin{aligned} y_1(1+\tfrac{1}{3}hf_1)-y_{-1}(1-\tfrac{1}{3}hf_{-1})+\tfrac{4}{3}hy_0f_0+Cy_0 &= \tfrac{1}{3}h(4g_0+g_1+g_{-1}) \\ C &= \tfrac{1}{90}\mu\delta^5-\ldots \end{aligned}\right\}, \tag{37}$$

an equation involving only three adjacent pivotal points and a small difference correction.

†**21.** We thus have a very powerful method for solving a first-order equation with a specified boundary value. First we produce a second-order equation which contains no first derivative, and for which the finite-difference equation can be written in a form which has a small difference correction. The general equation (1), for example, is turned into the form (34), for which a very accurate finite-difference equation is given by

$$\left.\begin{aligned} y_1\{1+\tfrac{1}{12}h^2(f'_1-f_1^2)\}+y_{-1}\{1+\tfrac{1}{12}h^2(f'_{-1}-f_{-1}^2)\}-y_0\{2-\tfrac{5}{6}h^2(f'_0-f_0^2)\}+Cy_0 \\ = \tfrac{1}{12}h^2\{10(g'_0-f_0g_0)+(g'_1-f_1g_1)+(g'_{-1}-f_{-1}g_{-1})\} \\ C = +\tfrac{1}{240}\delta^6-\ldots \end{aligned}\right\}. \tag{38}$$

At the first boundary point we have the given boundary value, and at the other end we have the condition expressed in the form (37). There are just enough equations for the determination of the unknowns.

Alternatively we could use (37) at the first boundary point, as well as the given value, and this would enable us to proceed in a step-by-step manner.

22. This method is not restricted to linear equations. The non-linear equation

$$y' = f(x,y) \tag{39}$$

can be replaced by the second-order equation

$$y'' = \frac{\partial f}{\partial x}+f\frac{\partial f}{\partial y}, \tag{40}$$

and the finite-difference equations corresponding to (39) and (40), though non-linear, can be written as before in forms involving the corrections of the same order as those of (37) and (38) respectively.

23. As an example of the power of this method, we consider again the basic example of this chapter, for which the relevant equations of the new method are as follows:

$$\left.\begin{aligned} -\tfrac{394}{192}y(\tfrac{1}{4})+\tfrac{191}{192}y(\tfrac{1}{2}) &= -\tfrac{191}{192} \\ +\tfrac{191}{192}y(\tfrac{1}{4})-\tfrac{394}{192}y(\tfrac{1}{2})+\tfrac{191}{192}y(\tfrac{3}{4}) &= 0 \\ +\tfrac{191}{192}y(\tfrac{1}{2})-\tfrac{394}{192}y(\tfrac{3}{4})+\tfrac{191}{192}y(1) &= 0 \\ +\tfrac{191}{192}y(\tfrac{3}{4})-\tfrac{394}{192}y(1)+\tfrac{191}{192}y(\tfrac{5}{4}) &= 0 \\ -\tfrac{13}{12}y(\tfrac{3}{4})-\tfrac{4}{12}y(1)+\tfrac{11}{12}y(\tfrac{5}{4}) &= 0 \end{aligned}\right\}. \tag{41}$$

These equations have the following solution, which has a maximum error of 2 units in the fourth decimal. Further, the error is a smooth function and can easily be removed by subsequent application of difference corrections.

x	0	$\frac{1}{4}$	$\frac{1}{2}$	$\frac{3}{4}$	1
y	1·00000	1·28405	1·64878	2·11709	2·71842

Choice of method. Illustrative example

24. This chapter is rather special in the sense that its methods, except the one which treats the derived equation instead of the first-order equation, are of somewhat rare application. We therefore propose to give one final example which shows how the intelligent computer can correlate his methods with his requirements, to produce accurate answers with a minimum of labour.

The sine and cosine integrals are defined by the equations

$$\left.\begin{aligned} \mathrm{Si}(x) &= \int_0^x \frac{\sin t}{t}\,dt \\ \mathrm{Ci}(x) &= \int_\infty^x \frac{\cos t}{t}\,dt \end{aligned}\right\}, \tag{42}$$

and have asymptotic series given by

$$\left.\begin{aligned} \mathrm{Si}(x)-\tfrac{1}{2}\pi &= -P\cos x - Q\sin x \\ \mathrm{Ci}(x) &= \quad P\sin x - Q\cos x \end{aligned}\right\}, \tag{43}$$

where

$$\left.\begin{aligned} P &= \frac{1}{x}-\frac{2!}{x^3}+\frac{4!}{x^5}-\dots \\ Q &= \frac{1!}{x^2}-\frac{3!}{x^4}+\frac{5!}{x^6}-\dots \end{aligned}\right\}. \tag{44}$$

Tabulation of P and Q for very large values of x would provide a useful adjunct to the standard tables giving $\mathrm{Si}(x)$ and $\mathrm{Ci}(x)$ for values of x increasing from 0 to about 10 or 15. The project is best undertaken with the use of x^{-1} as independent variable: we can then 'reach infinity' in our tables of P and Q, provided the latter are well behaved in this region. We consider in this connexion the range $z = x^{-1} = 0{\cdot}0(0{\cdot}01)0{\cdot}10$, corresponding to the range $x = 10$ to ∞. From (44) we see that P and Q are both zero at the origin of z, P is an odd function, Q an even function, and values for small z can be calculated to many decimals from the asymptotic series. For $z \geqslant 0{\cdot}1$, we can similarly calculate as many values of P and Q as we desire, making use of published tables. We have therefore all necessary information to treat our problem as one of either initial-value or boundary-value type.

25. Taking $z = x^{-1}$ as independent variable, we find the differential equations

$$\left.\begin{aligned} z^2P' &= Q \\ z^2Q' &= z-P \end{aligned}\right\}. \tag{45}$$

These can be solved approximately, but very rapidly, by using the finite-difference equations of type (2), written here in the form

$$\left.\begin{aligned} Q_0 &= \frac{z^2}{2h}(P_1-P_{-1}-\tfrac{1}{3}\mu\delta^3P_0+\tfrac{1}{15}\mu\delta^5P_0-\ldots) \\ P_0 &= z_0-\frac{z^2}{2h}(Q_1-Q_{-1}-\tfrac{1}{3}\mu\delta^3Q_0+\tfrac{1}{15}\mu\delta^5Q_0-\ldots) \end{aligned}\right\}. \tag{46}$$

Neglecting the difference corrections, we find that iteration or relaxation is extremely rapid because $z^2/2h$ is fairly small over the whole range, and produces results which are accurate to five decimals. If five decimals are adequate, this is probably the quickest way of getting them. Greater accuracy can be achieved by using the difference correction to produce a second approximation, but if ten decimals are used the differences show the type of oscillation discussed early in this chapter, and the process converges only slowly.

The equations can be changed to a form involving a smaller difference correction by well-known techniques. Writing (46) in the forms

$$\left.\begin{aligned} (\mu\delta-\tfrac{1}{6}\mu\delta^3+\tfrac{1}{30}\mu\delta^5-\ldots)P_0 &= \frac{hQ_0}{z_0^2} \\ (\mu\delta-\tfrac{1}{6}\mu\delta^3+\tfrac{1}{30}\mu\delta^5-\ldots)Q_0 &= \frac{h}{z_0}-\frac{hP_0}{z_0^2} \end{aligned}\right\}, \tag{47}$$

and operating on both sides with $(1+\tfrac{1}{6}\delta^2)$, we find the equations

$$\left.\begin{aligned} \tfrac{1}{2}(P_1-P_{-1})+\tfrac{1}{180}\mu\delta^5P_0-\ldots &= h\left\{\frac{2Q_0}{3z_0^2}+\frac{1}{6}\left(\frac{Q_1}{z_1^2}+\frac{Q_{-1}}{z_{-1}^2}\right)\right\} \\ \tfrac{1}{2}(Q_1-Q_{-1})+\tfrac{1}{180}\mu\delta^5Q_0-\ldots &= -h\left\{\frac{2P_0}{3z_0^2}+\frac{1}{6}\left(\frac{P_1}{z_1^2}+\frac{P_{-1}}{z_{-1}^2}\right)\right\}+ \\ &\qquad +h\left(\frac{2}{3z_0}+\frac{1}{6z_1}+\frac{1}{6z_{-1}}\right) \end{aligned}\right\}. \tag{48}$$

These equations have awkward coefficients and lead to tedious arithmetic. We can produce easier equations by the following artifice. We have, from (45),

$$\frac{d}{dz}(z^2P) = z^2P' + 2zP = Q + 2zP.$$

Then
$$(\mu\delta - \tfrac{1}{6}\mu\delta^3 + \tfrac{1}{30}\mu\delta^5 - \ldots)(z^2P)_0 = h(Q+2zP)_0$$

and, operating throughout with $1+\frac{1}{6}\delta^2$, we find

$$(\mu\delta + \tfrac{1}{180}\mu\delta^5 - \ldots)(z^2P)_0 = \tfrac{1}{6}h\{4(Q_0+2z_0P_0)+(Q_1+2z_1P_1+Q_{-1}+2z_{-1}P_{-1})\}$$

or

$$4Q_0+Q_1+Q_{-1}+8z_0P_0+2z_1P_1+2z_{-1}P_{-1} = \frac{3}{h}(z_1^2P_1 - z_{-1}^2P_{-1}+C_P). \tag{49}$$

Similar treatment of the second of (45) leads to the equation

$$-(4P_0+P_1+P_{-1})+8z_0Q_0+2z_1Q_1+2z_{-1}Q_{-1}+z_1+z_{-1}+4z_0$$
$$= \frac{3}{h}(z_1^2Q_1 - z_{-1}^2Q_{-1}+C_Q). \tag{50}$$

In (49) and (50) the difference corrections are given by

$$C_P = (\tfrac{1}{90}\mu\delta^5 - \ldots)(z^2P)_0, \qquad C_Q = (\tfrac{1}{90}\mu\delta^5 - \ldots)(z^2Q)_0. \tag{51}$$

The solution by relaxation of equations (49) and (50) is easy and rapid, the coefficients being everywhere exact and the equations well conditioned. Neglect of the difference corrections yields solutions with a maximum error of 17 units in the ninth decimal, and the oscillation in these solutions does not prevent the application of the difference corrections, which at the first attempt reduces the error to a maximum of 1 unit in the ninth decimal. A second application produces ten accurate decimals everywhere.

†**26.** Consider now the use of step-by-step techniques. With two initial values of P and Q calculated from the asymptotic series, we can use equations (46) to calculate successively other pivotal values of P and Q. We find that large oscillatory errors are introduced, and this is true also of the similar use of equations (49) and (50). A more reliable method is provided by the equations

$$\left.\begin{aligned} P_1 - P_0 &= \tfrac{1}{2}h\left\{\frac{Q_1}{z_1^2}+\frac{Q_0}{z_0^2}\right\}+CP_{\frac{1}{2}} \\ Q_1 - Q_0 &= \tfrac{1}{2}h\left\{\frac{1}{z_1}+\frac{1}{z_0}-\frac{P_1}{z_1^2}-\frac{P_0}{z_0^2}\right\}+CQ_{\frac{1}{2}} \\ C &= -\tfrac{1}{12}\delta^3+\tfrac{1}{120}\delta^5-\ldots \end{aligned}\right\}, \tag{52}$$

giving

$$\left.\begin{aligned} P_1\left(1+\frac{h^2}{4z_1^4}\right) &= P_0\left(1-\frac{h^2}{4z_0^2 z_1^2}\right)+\tfrac{1}{2}hQ_0\left(\frac{1}{z_0^2}+\frac{1}{z_1^2}\right)+ \\ &\qquad +\frac{h^2}{4z_1^2}\left(\frac{1}{z_0}+\frac{1}{z_1}\right)+CP_{\frac{1}{2}}+\frac{h}{2z_1^2}CQ_{\frac{1}{2}} \\ Q_1\left(1+\frac{h^2}{4z_1^4}\right) &= -\tfrac{1}{2}hP_0\left(\frac{1}{z_0^2}+\frac{1}{z_1^2}\right)+Q_0\left(1-\frac{h^2}{4z_0^2 z_1^2}\right)+ \\ &\qquad +\tfrac{1}{2}h\left(\frac{1}{z_0}+\frac{1}{z_1}\right)-\frac{h}{2z_1^2}CP_{\frac{1}{2}}+CQ_{\frac{1}{2}} \end{aligned}\right\}. \tag{53}$$

We can use these equations to recur outwards from the origin, any difficulty at that point being removed by use of the asymptotic series to eliminate apparent infinities. Indeed when $z_0 = 0$ we find

$$P_1\left(1+\frac{h^2}{4z_1^4}\right) = \tfrac{1}{2}h+\frac{h^2}{4z_1^3}, \qquad Q_1\left(1+\frac{h^2}{4z_1^4}\right) = -\frac{h^2}{4z_1^2}+\frac{h}{2z_1}. \tag{54}$$

The results of using equations (53) for recurrence, however, show oscillations. The first approximation is correct to about five decimals, but further correction is slow and uncertain, just as with the use of equations (46). A similar phenomenon is noticed if equations (53) are inverted to produce P_0 and Q_0 in terms of P_1 and Q_1, recurrence being directed now towards the origin. Throughout this work, moreover, the coefficients are awkward and the arithmetic tedious.

The solution of the simultaneous equations (49) and (50), by relaxation or direct methods, is undoubtedly the most satisfactory method of solving the first-order differential equations if high accuracy is required.

Additional notes and bibliography

Much of the material of this chapter has been discussed in more detail by

Fox, L., 1954, A note on the numerical integration of first-order differential equations. *Quart. J. Mech. and Appl. Math.* **7**, 367–78.

§ 4, etc. The use of a three-term difference equation to represent a first-order differential equation is likely to cause oscillation even when used in a step-by-step manner, starting with a given specified value and the next adjacent value calculated from Taylor's series. For example the general solution (5) of the difference equation (4) contains two arbitrary constants, while that of the differential equation has only one. The difference equation has produced an unwanted solution, generally associated with the factor $(-1)^r$, causing oscillation. If the unwanted solution grows faster than the solution required, this oscillation is not only inconvenient, but also dangerous, giving inaccurate and even meaningless results.

§ 7. This device for accelerating the convergence of a sequence was described in the following paper.

AITKEN, A. C., 1937, Studies in practical mathematics, II. The evaluation of the latent roots and latent vectors of a matrix. *Proc. Roy. Soc. Edin.* **57**, 269–304.

It can often be applied, however, in other contexts.

§ 21. The choice of method, boundary-value as against initial-value, for solving the derived equation, depends to some extent on the form of the solution. The solution of equation (1) is given by

$$y = P(x)+AC(x),$$

where $P(x)$ is a particular integral, $C(x)$ the complementary function, and A the arbitrary constant whose value is determined from the given initial value of y. Both these functions, as well as a third introduced by the transformation to a second-order equation, will in general contribute to the numerical results. We ignore the spurious solution, suppose $C(x)$ to be a rapidly increasing function of x, that $P(x)$ varies more slowly, and that the conditions of the problem require A to vanish, so that no multiple of the complementary function appears in the analytical solution. When we perform the step-by-step method, the pivotal values are rounded to a certain number of figures, with the result that a small multiple of $C(x)$ is introduced and the error grows with increasing x. In the boundary-value method, however, the restrictions imposed by the boundary conditions at each end of the range may prohibit the appearance of this error. In the paper quoted, Fox finds that in the problem defined by the equations

$$y'-12y+11e^{x} = 0, \qquad y(0) = 1,$$

the step-by-step solution of the derived second-order equation has errors, at interval $h = 0{\cdot}2$, of as much as 540 units in the fourth decimal of the true solution e^{x}, while the boundary-value method has a maximum error of 1 unit in this figure.

In this case the spurious solution introduced by differentiation is a multiple of e^{-12x}, which will hardly affect the numerical results by either method. If the spurious solution increased rapidly, however, the step-by-step method would have little chance of success, whereas the boundary-value method would again restrain its affect.

In some cases, however, discussed in the notes on Chapter 9, the boundary-value technique gives poor precision, and a step-by-step method, using two conditions at the first point of the range, is preferable.

§ 26. The oscillation produced by the use of (53) in a step-by-step process is not caused by the introduction of unwanted solutions, since the difference equations themselves have here the same order as the differential equations. The reason is connected with the fact that the general solutions of the differential equations (45) are given by

$$\left.\begin{aligned} P &= A\sin\frac{1}{z}+B\cos\frac{1}{z}+(z-2!z^{3}+4!z^{5}-\dots) \\ Q &= -A\cos\frac{1}{z}+B\sin\frac{1}{z}+(z^{2}-3!z^{4}+5!z^{6}-\dots) \end{aligned}\right\}.$$

The required solution has $A = B = 0$, but any numerical solution, particularly an approximate solution with a large difference correction, may introduce small multiples of the unwanted functions $\sin z^{-1}$ and $\cos z^{-1}$, which, in contrast to the required solution, are not well behaved but oscillate wildly at an interval of say 0·01 in z.

The best hope of obtaining an accurate solution at this interval therefore depends on the use of formulae which have small difference corrections. Of methods so far developed which use the first-order differential equation, formulae (48), or (49) and (50), have the smallest difference corrections, (49) and (50) being the more convenient. These difference equations are of order greater than that of the differential equations, but any spurious solution introduced behaves no worse than $\sin z^{-1}$ or $\cos z^{-1}$.

Again, however, these are best solved as simultaneous equations, since the boundary conditions ensure that A and B are zero at the two ends of the range and their associated functions have little chance to become established anywhere. With the step-by-step method, on the other hand, there is an inevitable introduction of these functions through the occurrence of rounding errors.

6

Equations of Higher Orders

Fourth-order linear equations

1. THE most general linear equation of the fourth order has the form

$$y^{\text{iv}}+f_1 y'''+f_2 y''+f_3 y'+f_4 y = f_5, \tag{1}$$

where $f_1,..., f_5$ are functions of x only. Four boundary conditions are necessary to provide a unique solution, and there are three main possibilities regarding their positions:

(i) four conditions given at one point of the range;

(ii) three conditions at one end, one condition at the other end of the range;

(iii) two conditions at each end of the range.

The first of these possibilities provides essentially an initial-value problem and will not be discussed here. We should expect the last to provide us with a true boundary-value problem. Possibility (ii), or any other arbitrary combination, such as the specification of some condition at four separate points, or of two conditions at one point and one at each of two other points, is more difficult to assess, but we shall see in § 13 that any such combination can be solved by boundary-value techniques.

The most general boundary condition has the form

$$g_0 y'''+g_1 y''+g_2 y'+g_3 y = g_4, \tag{2}$$

where $g_0,..., g_4$ are constants, and we shall proceed to show by simple arguments why two such conditions at each end of the range enable us to set up, in a natural form, a set of simultaneous equations for the determination of pivotal values.

General boundary conditions

2. Consider, for example, the distribution of pivotal points shown in Fig. 18, in which points -1, -2 are outside the region of integration and 0 is one boundary point.

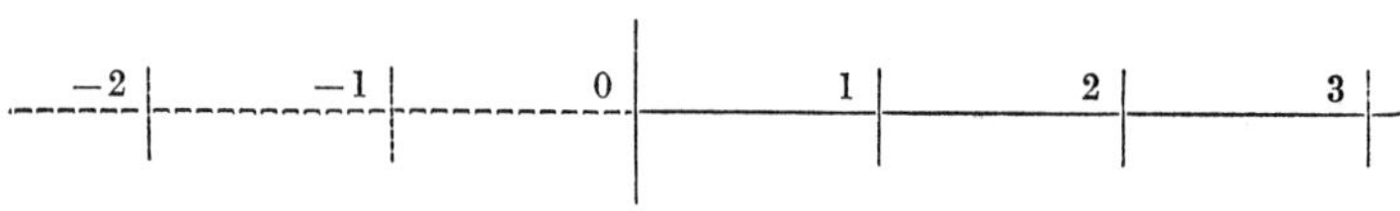

FIG. 18

With the most general differential equation (1) and boundary conditions

(2), we shall need to establish enough simultaneous algebraic equations to effect the determination of pivotal values at all points 0, 1, 2,..., inside and at the ends of the range.

Table 7 of Chapter 2 gives the following expressions for the various derivatives:

$$\left.\begin{aligned} h^4y_0^{\text{iv}} &= y_2-4y_1+6y_0-4y_{-1}+y_{-2}-\tfrac{1}{6}\delta^6y_0+\dots \\ h^3y_0''' &= \tfrac{1}{2}(y_2-2y_1+2y_{-1}-y_{-2})-\tfrac{1}{4}\mu\delta^5y_0+\dots \\ h^2y_0'' &= y_1-2y_0+y_{-1}-\tfrac{1}{12}(y_2-4y_1+6y_0-4y_{-1}+y_{-2})+\tfrac{1}{90}\delta^6y_0-\dots \\ hy_0' &= \tfrac{1}{2}(y_1-y_{-1})-\tfrac{1}{12}(y_2-2y_1+2y_{-1}-y_{-2})+\tfrac{1}{30}\mu\delta^5y_0-\dots \end{aligned}\right\}. \tag{3}$$

From these relations we see that the approximate finite-difference equation corresponding to (1) will contain five adjacent pivotal values. In particular the application of this equation at the boundary point 0 will involve the two 'unwanted' values y_{-1} and y_{-2}. We therefore need two extra equations at this point to obtain a unique solution, and these are provided by two conditions corresponding to (2), which when applied at 0 will each involve both y_{-1} and y_{-2}, but no more external values.

3. In setting up the most general equations we note that the second term in the finite-difference equations for both first and second derivatives can now be written in terms of pivotal values, so that the difference correction is a function of differences of order five and higher. The finite-difference forms of the most general boundary conditions have corrections of the same order.

Substituting into (1) the expressions (3) for derivatives, we obtain the finite-difference equation

$$\left.\begin{aligned} & y_2(1+\tfrac{1}{2}hf_1-\tfrac{1}{12}h^2f_2-\tfrac{1}{12}h^3f_3)+y_1(-4-hf_1+\tfrac{4}{3}h^2f_2+\tfrac{2}{3}h^3f_3)+ \\ &\quad +y_0(6-\tfrac{5}{2}h^2f_2+h^4f_4)+y_{-1}(-4+hf_1+\tfrac{4}{3}h^2f_2-\tfrac{2}{3}h^3f_3)+ \\ &\qquad +y_{-2}(1-\tfrac{1}{2}hf_1-\tfrac{1}{12}h^2f_2+\tfrac{1}{12}h^3f_3)+Cy_0 = h^4f_5 \\ & C = (-\tfrac{1}{4}hf_1+\tfrac{1}{30}h^3f_3)\mu\delta^5+(-\tfrac{1}{6}+\tfrac{1}{90}h^2f_2)\delta^6+ \\ &\qquad +(\tfrac{7}{120}hf_1-\tfrac{1}{140}h^3f_3)\mu\delta^7+(\tfrac{7}{240}-\tfrac{1}{560}h^2f_2)\delta^8+\dots \end{aligned}\right\}, \tag{4}$$

and the functions f are evaluated at the pivotal point x_0.

The boundary condition (2) is similarly expressible in the form

$$\left.\begin{aligned} & y_2(\tfrac{1}{2}g_0-\tfrac{1}{12}hg_1-\tfrac{1}{12}h^2g_2)+y_1(-g_0+\tfrac{4}{3}hg_1+\tfrac{2}{3}h^2g_2)+ \\ &\quad +y_0(-\tfrac{5}{2}hg_1+h^3g_3)+y_{-1}(g_0+\tfrac{4}{3}hg_1-\tfrac{2}{3}h^2g_2)+ \\ &\qquad +y_{-2}(-\tfrac{1}{2}g_0-\tfrac{1}{12}hg_1+\tfrac{1}{12}h^2g_2)+C_1y_0 = h^3g_4 \\ & C_1 = (-\tfrac{1}{4}g_0+\tfrac{1}{30}h^2g_2)\mu\delta^5+\tfrac{1}{90}hg_1\delta^6+ \\ &\qquad +(\tfrac{7}{120}g_0-\tfrac{1}{140}h^2g_2)\mu\delta^7-\tfrac{1}{560}hg_1\delta^8+\dots \end{aligned}\right\}, \tag{5}$$

and the functions g are evaluated at the pivotal point x_0.

Special boundary conditions

4. Linear fourth-order differential equations occur often in the theory of elasticity applied to the study of stresses in straight beams. The solution y of the differential equation then represents the deflexion caused by transverse loading, and the boundary conditions specify the applied constraints at the two ends. If the end is 'clamped', then y and y' are both required to vanish. Conditions of 'simple support' imply that y and y'' are both zero, and if the bending moment and shear are required to vanish then $y'' = y''' = 0$.

These special conditions involve no new problem, but merely simplify the calculation. Consider for example the case in which $y = y' = 0$. The equations corresponding to (5) now reduce to $y_0 = 0$ and

$$-\tfrac{1}{12}y_2+\tfrac{2}{3}y_1-\tfrac{2}{3}y_{-1}+\tfrac{1}{12}y_{-2}+\tfrac{1}{30}\mu\delta^5 y_0-\ldots = 0, \tag{6}$$

and the finite-difference form (4) for the differential equation at this boundary point gives another relation between y_2, y_1, y_{-1}, and y_{-2}. Again therefore we have just enough equations for the determination of all the unknowns, since one of them, y_0, is already known. Similar analysis for other combinations of boundary conditions is straightforward.

5. In some cases, however, notably those for which y and y' or y and y'' are specified at an end point, we have alternative methods, as in the case of second-order equations. Since y_0 is known, we need no other equation at this point, and can ignore the finite-difference form of the differential equation there. We must, however, apply the latter at the first internal point 1, and this involves *one* external point -1. The second boundary condition must therefore be used to provide another equation involving this value, but no more *external* values. If 'y' specified' is that condition, we can use the equation

$$hy_0' = \tfrac{1}{2}(y_1-y_{-1})+\text{difference correction}, \tag{7}$$

but this is unsatisfactory since this difference correction is much larger, starting with third differences, than that of the finite-difference form of the differential equation.

We must therefore use one of the forward difference formulae, expressing y_{-1} in terms of y_0 (known), $y_1, y_2,\ldots$. For example, if we retain forward differences as far as order four, the obviously desirable minimum, we have, as the extra equation for the determination of y_{-1}, the equation

$$\left.\begin{aligned}\tfrac{1}{12}y_3-\tfrac{1}{2}y_2+\tfrac{3}{2}y_1-\tfrac{5}{6}y_0-\tfrac{1}{4}y_{-1}-hy_0'+C_2 y_{-1} &= 0\\ C_2 = -\tfrac{1}{20}\Delta^5+\tfrac{1}{30}\Delta^6-\tfrac{1}{42}\Delta^7+\tfrac{1}{56}\Delta^8-\ldots &\end{aligned}\right\}. \tag{8}$$

Similarly, if the second boundary condition specifies y'', we can write, for the extra condition,

$$h^2y_0'' = y_1-2y_0+y_{-1}+\text{difference correction}, \tag{9}$$

but again the latter is rather large, and a forward or backward difference formula is preferable, of the type

$$h^2y_0'' = (\Delta^2-\tfrac{1}{12}\Delta^4+\tfrac{1}{12}\Delta^5-...)y_{-1}. \tag{10}$$

These methods had only rare application in the solution of second-order equations, and for equations of fourth order there is the extra disadvantage that they can be used only when the boundary conditions do not involve the third derivative, and are not therefore general. It is probable that the method of § 3 is always the most convenient for fourth-order equations.

6. The reader will notice other possibilities, only to reject them immediately. For example, the boundary conditions can both be expressed in terms of forward difference formulae, involving only the points 0, 1, 2,..., and no external points. We need then apply the finite-difference form of the differential equation only at points 2, 3,..., and the number of simultaneous equations for solution is thereby reduced. The difference corrections, however, contain large coefficients, with consequent inaccuracies in solutions.

In comparison with second-order equations, we notice also that there is nothing corresponding to 'first derivative absent', and that at least one of the two boundary conditions always involves a difference correction.

Method of solution. Simple example

7. We need not linger over the general method of solution. As in our previous boundary-value techniques we first ignore all difference corrections and solve the simultaneous equations. More external values can then be calculated as required by applying the differential equation, in its finite-difference form, at successive external points. The first approximate solution is then differenced, all difference corrections are calculated by central-difference formulae, and a correction is obtained at all relevant pivotal points. The process is repeated until there is no further change.

As a simple example which will illustrate this technique, consider the differential equation

$$y^{\text{iv}}-y = 0, \tag{11}$$

with the boundary conditions

$$\left.\begin{array}{ll} y(0) = 1, & y'(0) = 1 \\ y''(1) = y(1), & y'''(1) = y'(1) \end{array}\right\}. \tag{12}$$

Choosing the interval $h = 0{\cdot}2$ we have the distribution of points shown in Fig. 19. There are nine unknowns, since the value at $x = 0$ is specified by a boundary condition, and nine linear equations for their evaluation.

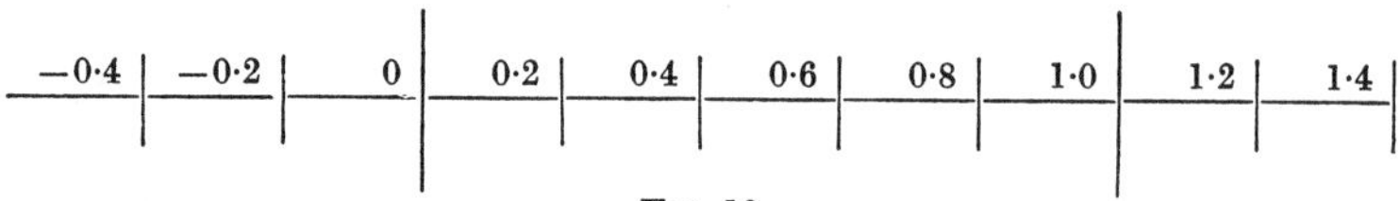

FIG. 19

At each of the points $0{\cdot}0(0{\cdot}2)1{\cdot}0$ we have the finite-difference form of the differential equation, given here by

$$\left.\begin{array}{c} y_2 - 4y_1 + 5{\cdot}9984y_0 - 4y_{-1} + y_{-2} + Cy_0 = 0 \\ C = -\tfrac{1}{6}\delta^6 + \tfrac{7}{240}\delta^8 - \dots \end{array}\right\}. \tag{13}$$

The boundary conditions at $x = 0$ provide the equations

$$y(0) = 1, \tag{14}$$

and

$$\left.\begin{array}{c} -\tfrac{1}{12}y(0{\cdot}4) + \tfrac{2}{3}y(0{\cdot}2) - \tfrac{2}{3}y(-0{\cdot}2) + \tfrac{1}{12}y(-0{\cdot}4) + C_1 y(0) = 0{\cdot}2 \\ C_1 = \tfrac{1}{30}\mu\delta^5 - \tfrac{1}{140}\mu\delta^7 + \dots \end{array}\right\}, \tag{15}$$

and those at $x = 1$ give the equations

$$\left.\begin{array}{c} -\tfrac{1}{12}y(1{\cdot}4) + \tfrac{4}{3}y(1{\cdot}2) - 2{\cdot}54y(1{\cdot}0) + \\ + \tfrac{4}{3}y(0{\cdot}8) - \tfrac{1}{12}y(0{\cdot}6) + C_2 y(1{\cdot}0) = 0 \\ C_2 = \tfrac{1}{90}\delta^6 - \tfrac{1}{560}\delta^8 + \dots \end{array}\right\}, \tag{16}$$

and

$$\left.\begin{array}{c} 0{\cdot}50\dot{3}y(1{\cdot}4) - 1{\cdot}02\dot{6}y(1{\cdot}2) + 1{\cdot}02\dot{6}y(0{\cdot}8) - \\ -0{\cdot}50\dot{3}y(0{\cdot}6) + C_3 y(1{\cdot}0) = 0 \\ C_3 = -\left(\dfrac{1}{4} + \dfrac{0{\cdot}04}{30}\right)\mu\delta^5 + \left(\dfrac{7}{120} + \dfrac{0{\cdot}04}{140}\right)\mu\delta^7 - \dots \end{array}\right\}. \tag{17}$$

The first approximation $y^{(1)}$, obtained by solving the linear equations with all difference corrections neglected, is shown in Table 1. We note that these values will provide sixth central differences at every internal point, but not at the boundaries. For the latter we need to extrapolate one value, by applying the first of (13), with Cy neglected, at the point $-0{\cdot}2$. A further extrapolation will provide the eighth central difference at the boundary. In Table 1 extrapolated values are shown in brackets, and we see from the differences that fifth, sixth, and seventh differences

TABLE 1

x	$y^{(1)}$		δ^2		δ^4		δ^6		δ^8	10^6C	z
−0·8	(+0·455099)										
		(+ 96905)									
·6	(0·552004)		(+ 22801)								
		(119706)		(+ 4850)							
·4	0·671710		(27651)		(+1075)						
		147357		(5925)		(+ 236)					
− ·2	0·819067		33576		(1311)		(+ 54)				
		180933		7236		(290)		(+ 7)			
+ ·0	1·000000		40812		1601		(61)		(+22)	−10	0
		221745		8837		351		(+29)			
·2	1·221745		49649		1952		90		(−37)	15	−0·000340
		271394		10789		441		− 8			
·4	1·493139		60438		2393		82		+58	14	·001304
		331832		13182		523		+50			
·6	1·824971		73620		2916		132		−48	22	·002830
		405452		16098		655		2			
0·8	2·230423		89718		3571		134		(+43)	22	·004844
		495170		19669		789		(45)			
1·0	2·725593		109387		4360		(179)		(−11)	−30	−0·007256
		604557		24029		(968)		(+34)			
1·2	3·330150		133416		(5328)		(+213)				
		737973		(29357)		(+1181)					
1·4	4·068123		(162773)		(+6509)						
		(900746)		(+35866)							
1·6	(4·968869)		(+198639)								
		(+1099385)									
+1·8	(+6·068254)										

$10^6C_1\,y(0) = +\ 11$
$10^6C_2\,y(1) = +\ \ 2$
$10^6C_3\,y(1) = -218$

are significant, and ought to be used in the calculation of the difference corrections, but eighth differences are oscillating about zero, and should be ignored. The difference corrections Cy for $y = 0{\cdot}0(0{\cdot}2)1{\cdot}0$ are now calculable, as well as the corrections $C_1 y$ at $x = 0$ and $C_2 y$ and $C_3 y$ at $x = 1$. All these results are given in Table 1.

Inserting the calculated difference corrections in the appropriate linear equations we calculate corrections z to $y^{(1)}$, as shown on the right of Table 1. When z is differenced, we find that fifth differences are negligible, so that there is no further correction, and $y^{(1)}+z$ should give an accurate solution. The error of this solution, compared with the analytical solution e^x, is as follows, expressed in units of the sixth decimal.

x	0·0	0·2	0·4	0·6	0·8	1·0
Error	0	2	10	22	36	55

The error is somewhat large, particularly near $x = 1$, and is caused by the ill-conditioned nature of the linear equations. This will be discussed further in Chapter 9.

8. We need give no further example of a fourth-order equation, since that of § 7 lacks generality only in the fact that all the coefficients in the differential equation were constants, and the finite-difference form of the differential equation had therefore the same form at every pivotal point. The most general equation has only the additional complication that the coefficients of the linear equations change from point to point.

The example illustrates that the direct solution of a fourth-order equation is a much more formidable undertaking than that of a second-order equation, and we therefore look for alternative possibilities.

Other methods of solution

9. Possible alternatives involve the solution of a pair of second-order equations, either simultaneously or in succession. The first alternative is always possible, the second depends on the form of the equation and its boundary conditions.

If we make the substitution $y'' = z$, we can replace the general differential equation (1) by the simultaneous second-order equations

$$\left.\begin{aligned} y'' &= z \\ z''+f_1 z'+f_2 z+f_3 y'+f_4 y &= f_5 \end{aligned}\right\}, \tag{18}$$

for which there are two boundary conditions at both ends, each of the form

$$g_0 z'+g_1 z+g_2 y'+g_3 y = g_4. \tag{19}$$

This problem we have treated already in Chapter 4.

This method will have greatest value when the differential equation and boundary conditions contain no derivatives of odd order, for then equations (18) and (19) reduce respectively to the forms

$$\left.\begin{aligned} y'' &= z \\ z''+f_2 z+f_4 y &= f_5 \end{aligned}\right\}, \tag{20}$$

and

$$g_1 z+g_3 y = g_4. \tag{21}$$

We can then use finite-difference equations, as explained in Chapter 4, whose difference corrections are small, respectively $\frac{1}{240}\delta^6 y$ and $\frac{1}{240}\delta^6 z$, and for which there is no difference correction connected with the boundary conditions.

10. Occasionally we can turn the problem into the solution of successive second-order equations. Consider, for example, the differential equation

$$\frac{d^2}{dx^2}\left(f_1\frac{d^2y}{dx^2}\right) = f_2, \tag{22}$$

with y and y'' specified at each end of the range. Writing $f_1 y'' = z$, equation (22) becomes

$$\frac{d^2z}{dx^2} = f_2, \tag{23}$$

and can be solved for z since the value of z is known at each end point. We can then solve for y, also known at each end point, from the equation

$$y'' = z/f_1. \tag{24}$$

This method can also be used more generally, whenever the original equation can be expressed in the form

$$\left(\frac{d^2}{dx^2}+f_1\frac{d}{dx}+f_2\right)\left(\frac{d^2y}{dx^2}+f_3\frac{dy}{dx}+f_4 y\right) = f_5, \tag{25}$$

where $y''+f_3 y'+f_4 y$ is known at each end of the range, the other boundary conditions involving only y and y'.

11. The solution of second-order equations is so much easier than that of fourth-order equations that it might well be advantageous to use this method whenever we have an equation expressible in the form (25), particularly if f_1 and f_3 are zero, so that there is no first derivative in the second-order equations, even if the boundary conditions do not have the most convenient form.

Consider, for example, the equation

$$\left(\frac{d^2}{dx^2}+f_2\right)\left(\frac{d^2y}{dx^2}+f_4 y\right) = f_5, \tag{26}$$

subject to the boundary conditions

$$y = y' = 0 \text{ at each end of the range.} \tag{27}$$

Suppose we make the substitution

$$y''+f_4 y = z, \tag{28}$$

so that

$$z''+f_2 z = f_5. \tag{29}$$

There are then two possible methods of procedure. In the first, we solve (29) with the conditions

$$z = 0 \text{ at } x = 0 \text{ and } 1, \tag{30}$$

and the equation

$$z''+f_2 z = 0 \tag{31}$$

with the two separate sets of conditions

$$\left.\begin{array}{ll} z = 0 \text{ at } x = 0, & z = 1 \text{ at } x = 1 \\ z = 1 \text{ at } x = 0, & z = 0 \text{ at } x = 1 \end{array}\right\}. \tag{32}$$

If these three solutions are denoted by $z^{(1)}$, $z^{(2)}$, and $z^{(3)}$, the general solution of (29) is given by

$$z = z^{(1)}+\alpha z^{(2)}+\beta z^{(3)}. \tag{33}$$

We now solve for $y^{(r)}$ the equations

$$y^{(r)''}+f_4 y^{(r)} = z^{(r)}, \qquad r = 1,\ 2, \text{ and } 3, \tag{34}$$

using the boundary conditions $y = 0$ at $x = 0$ and 1. The final solution is then

$$y = y^{(1)}+\alpha y^{(2)}+\beta y^{(3)}, \tag{35}$$

and we determine α and β from the conditions that y' should vanish at each end of the range. This is easily performed by calculating the derivatives of $y^{(1)}$, $y^{(2)}$, and $y^{(3)}$ by numerical differentiation.

This method appears formidable, but it is not much worse than the solution of two second-order equations. Referring to the direct methods for solving the simultaneous finite-difference equations, illustrated in § 29 of Chapter 3, we see that there are only two independent matrices, one corresponding to equations (29) and (31), the other to equations (34), each of them with three right-hand sides. An advantage of this method is that we can use accurate but simple finite-difference equations at a reasonably large interval.

The second possibility is merely a special case of § 9. Though z enters into none of the boundary conditions there are just enough equations for all the unknowns. If there are n internal points the unknowns comprise the n internal values of both y and z, the two boundary values of z and the values of y and z introduced at two external points, one at each end of the range. For the evaluation of these $2n+6$ unknowns we have the finite-difference equations corresponding to both (28) and (29) applied at the $2(n+2)$ pivotal points (including both boundary points), and two equations for the derivative conditions on y at the boundaries.

12. The solution of two simultaneous second-order equations does not necessarily give the same accuracy or precision as that of the corresponding fourth-order equation. The functions y and z of equations like (18) are clearly determinable with a precision dependent on the boundary conditions but, if the constants in equation (19) have approximately the same size, the calculated y and z will have approximately the same precision. When we solve the fourth-order equation, however, z will usually have fewer accurate significant figures than y, and its percentage error will be correspondingly greater. If the second derivative is the most important aspect of the solution, as is often the case, treatment of the simultaneous second-order equations is nearly always preferable. In other cases too the latter method may give best precision in both y and z. This we shall discuss further in Chapter 9. Here we note that the solution of the example of § 7 was solved in Chapter 4 as two simultaneous equations, and better precision was obtained.

Other boundary conditions

13. So far we have considered the case for which two boundary conditions are specified at each end of the range. It is interesting, however, to examine other possibilities in connexion with the boundary conditions. We have seen that if there are n pivotal points in the range, including the boundary points, then the application of the difference equations at all these points introduces the values at four extra points, two outside each end of the range. There are therefore n equations connecting $n+4$ unknowns, and we need four extra conditions to give a unique solution. These conditions may take any form, provided their finite-difference equivalents involve some or all of the $n+4$ unknowns. Thus, for example, we could have the value of the function specified at four distinct points, or y, y', and y'' given at one point, y at another, or any other conceivable combination. We could even treat in this way the case of y, y', y'', and y''' given at the same point, a true initial-value problem. Here, however, the simultaneous equations are of a kind in which the matrix of coefficients is triangular, so that the unknowns are calculated in succession, each new equation defining a new pivotal value.

We can therefore treat as a boundary-value problem any combination of boundary conditions except those defining a true initial-value problem, in which y, y', y'', and y''' are all known at the same point. This is an interesting conclusion, since the initial-value technique can only be used directly on the latter problem. Any other combination of boundary conditions involves at least two applications of the initial-value method.

Third-order equations

14. The general linear third-order equation is given by

$$y'''+f_1y''+f_2y'+f_3y = f_4, \tag{36}$$

and there are three boundary conditions of the form

$$g_0y''+g_1y'+g_2y = g_3, \tag{37}$$

where f_1, f_2, f_3, and f_4 are functions of x and g_0, g_1, g_2, and g_3 are constants. If the three conditions are specified at one point we have an initial-value problem; a boundary-value problem is presented when one boundary condition is given at one end of the range and two at the other, or when the three conditions are given at three different points.

Oscillation

15. The solution of third-order equations has difficulties similar to those encountered in applying boundary-value techniques to first-order equations, treated in Chapter 5. We saw there that the use of central differences led to a three-term difference equation whose solution involved two arbitrary constants, one more than was calculable from the boundary conditions: this led to oscillation in approximate solutions.

The same phenomenon is manifest in the solution of third-order equations. A glance at equations (3) shows that the finite-difference equations involve five adjacent values, and their application at every point in the range introduces four external pivotal points, two at each end. For their determination we have only three boundary conditions, and must therefore reduce by one the number of external points by using a sloping-difference formula for the satisfaction of the differential equation at one boundary point.

Example

16. We can make this clear by a simple example. Consider the differential equation and boundary conditions given by

$$y'''-y''+y'-y = 0; \quad y(0) = 1,\ y'(0) = 1,\ y(1) = 2{\cdot}71828..., \tag{38}$$

for which the analytical solution is $y = e^x$.

Using equations (3), we have the general finite-difference formula

$$y_2(\tfrac{1}{2}+\tfrac{1}{12}h-\tfrac{1}{12}h^2)+y_1(-1-\tfrac{4}{3}h+\tfrac{2}{3}h^2)+y_0(\tfrac{5}{2}h-h^3)+$$
$$+y_{-1}(1-\tfrac{4}{3}h-\tfrac{2}{3}h^2)+y_{-2}(-\tfrac{1}{2}+\tfrac{1}{12}h+\tfrac{1}{12}h^2)+Cy_0 = 0, \tag{39}$$

in which the difference correction involves differences of order five and higher. Consider now the range of Fig. 20, where 0 and 5 are the end

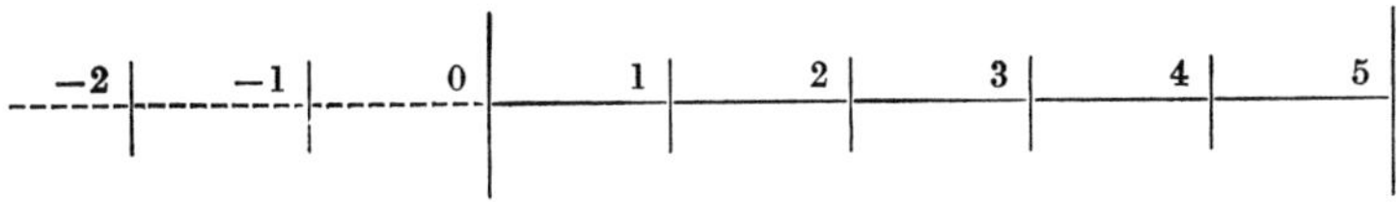

FIG. 20

points, so that $h = 0{\cdot}2$. If we apply (39) at each of the points 0, 1, 2, 3 we have four equations for the six pivotal values y_{-2}, y_{-1}, y_1, y_2, y_3, y_4 and we have already used two boundary conditions, these specifying y_0 and y_5. We have one other equation, representing the derivative condition at 0, which gives

$$-\tfrac{1}{12}y_2+\tfrac{2}{3}y_1-\tfrac{2}{3}y_{-1}+\tfrac{1}{12}y_{-2} = h, \tag{40}$$

with an error involving differences of order five and higher at 0.

We now have one more unknown than the number of equations, and have not yet applied the differential equation at point 4. Use of (39) is of no value, since it introduces yet another unknown external to the range, and we must express the differential equation only in terms of points 5, 4,... . The best form is one involving backward differences at point 5. We already have from Chapter 2 the equations

$$\left.\begin{aligned} hy_4' &= (\nabla-\tfrac{1}{2}\nabla^2-\tfrac{1}{6}\nabla^3-\tfrac{1}{12}\nabla^4-\tfrac{1}{20}\nabla^5-...)y_5 \\ h^2y_4'' &= (\nabla^2-\tfrac{1}{12}\nabla^4-\tfrac{1}{12}\nabla^5-...)y_5 \end{aligned}\right\}, \tag{41}$$

and by similar methods we can derive the other required equation, given by

$$h^3y_4''' = (\nabla^3+\tfrac{1}{2}\nabla^4+\tfrac{1}{4}\nabla^5+...)y_5. \tag{42}$$

Retaining differences up to order four, we can then express the differential equation, applied at point 4, by the algebraic equation

$$\begin{aligned} y_5(\tfrac{3}{2}-\tfrac{11}{12}h+\tfrac{1}{4}h^2)+y_4(-5+\tfrac{5}{3}h+\tfrac{5}{6}h^2-h^3)+y_3(6-\tfrac{1}{2}h-\tfrac{3}{2}h^2)+ \\ +y_2(-3-\tfrac{1}{3}h+\tfrac{1}{2}h^2)+y_1(\tfrac{1}{2}+\tfrac{1}{12}h-\tfrac{1}{12}h^2) = 0, \end{aligned} \tag{43}$$

and we now have just enough equations to solve our problem. The solution of these equations and the corresponding table of differences is shown in Table 2. The oscillation in the fifth difference is apparent. In this example it does not appear to be serious, and indeed the error in the fourth decimal is not more than five units. The exponential function is not, however, very representative of the types of functions which satisfy third-order equations, and the oscillation is often more pronounced. On the other hand we are helped by the fact that the difference correction starts with differences of a higher order than those neglected in the corresponding problem involving first-order equations. For work of moderate accuracy, therefore, this method seems quite practicable.

TABLE 2

x	y		δ^2		δ^4	
−0·4	+0·67055					
		+14835				
− ·2	0·81890		+3275			
		18110		+ 774		
+ ·0	1·00000		4049		+ 70	
		22159		844		+190
·2	1·22159		4893		260	
		27052		1104		− 76
·4	1·49211		5997		184	
		33049		1288		+143
·6	1·82260		7285		+327	
		40334		+1615		
0·8	2·22594		+8900			
		+49234				
+1·0	+2·71828					

Discussion of oscillation

†**17.** We digress for a moment to consider the general behaviour of finite-difference equations, to see why this oscillation occurs and whether it can be removed. Analysis is straightforward only in the case of linear equations with constant coefficients. The finite-difference equation for the complementary function then has the form

$$a_0 y_r + a_1 y_{r-1} + a_2 y_{r-2} + \ldots + a_n y_{r-n} = 0, \tag{44}$$

and the solution is given by

$$y_r = A_0 p_0^r + A_1 p_1^r + \ldots, \tag{45}$$

where p_0, p_1,... are the roots of the polynomial equation

$$a_0 p^n + a_1 p^{n-1} + \ldots + a_n = 0. \tag{46}$$

Undesirable oscillation will occur if any root is negative or has a fairly large negative real part.

Consider now the form of (44) derived from differential equations of various orders, using throughout central-difference formulae for derivatives at a pivotal point. For first to fourth orders we take respectively the forms

$$\left.\begin{aligned} y' + q_1 y &= 0 \\ y'' + q_1 y' + q_2 y &= 0 \\ y''' + q_1 y'' + q_2 y' + q_3 y &= 0 \\ y^{\text{iv}} + q_1 y''' + q_2 y'' + q_3 y' + q_4 y &= 0 \end{aligned}\right\}. \tag{47}$$

If only the first term is retained in each derivative formula, the approximate finite-difference equations are given by

$$\left.\begin{aligned} \tfrac{1}{2}y_{r+1}+hq_1 y_r-\tfrac{1}{2}y_{r-1} &= 0 \\ y_{r+1}(1+\tfrac{1}{2}hq_1)-y_r(2-h^2q_2)+y_{r-1}(1-\tfrac{1}{2}hq_1) &= 0 \\ \tfrac{1}{2}y_{r+2}-y_{r+1}(1-hq_1-\tfrac{1}{2}h^2q_2)-y_r(2hq_1-h^3q_3)+ & \\ +y_{r-1}(1+hq_1-\tfrac{1}{2}h^2q_2)-\tfrac{1}{2}y_{r-2} &= 0 \\ y_{r+2}(1+\tfrac{1}{2}hq_1)-y_{r+1}(4+hq_1-h^2q_2-\tfrac{1}{2}h^3q_3)+y_r(6-2h^2q_2+h^4q_4)- & \\ -y_{r-1}(4-hq_1-h^2q_2+\tfrac{1}{2}h^3q_3)+y_{r-2}(1-\tfrac{1}{2}hq_1) &= 0 \end{aligned}\right\}. \tag{48}$$

Considering the corresponding polynomials of type (46) we find

(i) for a first-order equation, one root is always negative, and there is always oscillation;

(ii) for a second-order equation, no root is negative unless $|hq_1| > 2$ or $h^2q_2 > 2$. For intervals as large as this the finite-difference equations are in nearly all cases meaningless. Oscillation is therefore extremely unlikely;

(iii) for a third-order equation, there is one negative root and oscillation is certain;

(iv) for a fourth-order equation, at an interval for which the finite-difference equations have meaning, it is fairly clear that there is no negative real root, and therefore no oscillation.

Removal of oscillation

18. This phenomenon is clearly connected with the fact that for first- and third-order equations, where oscillation occurs, the solution (45) contains more arbitrary constants than can be obtained from the given boundary conditions: for second- and fourth-order equations there is just the right number of arbitrary constants and no oscillation. This too is connected with the use of central differences for even-order equations, but of mean central differences for odd-order equations. The possibility is suggested of using central differences δ and δ^3 for odd-order equations also.

This suggestion implies that we should do better to use 'mid-way' points as the places to satisfy the differential equation, operating with μ on functional values, and on even rather than on odd derivatives, at these points. Thus we may replace the differential equation given by the first of (47) by the finite-difference equation

$$hy'_{\frac{1}{2}} = (\delta-\tfrac{1}{24}\delta^3+\tfrac{3}{640}\delta^5-\ldots)y_{\frac{1}{2}} = -h(q_1 y)_{\frac{1}{2}}. \tag{49}$$

Since $y_{\frac{1}{2}}$ does not appear explicitly in the table of pivotal values we operate throughout with μ, directly on one side and in the form $(1+\frac{1}{4}\delta^2)^{\frac{1}{2}}$ on the other, producing the well-known equation

$$(\delta+\tfrac{1}{12}\delta^3-\tfrac{1}{120}\delta^5+\ldots)y_{\frac{1}{2}} = -h\mu(q_1 y)_{\frac{1}{2}}, \tag{50}$$

expressible when q_1 is constant in the alternative form

$$y_{r+1}(1+\tfrac{1}{2}hq_1)-y_r(1-\tfrac{1}{2}hq_1)+(\tfrac{1}{12}\delta^3-\tfrac{1}{120}\delta^5+\ldots)y_{r+\frac{1}{2}} = 0. \tag{51}$$

With the difference correction neglected this is a two-term difference equation, there is one root p and one constant A in (45), and p is positive provided that $|hq_1| < 2$. Again this will be the case if the finite-difference equations are to have much meaning. We notice that the single constant A is now calculable from the single boundary condition, and everything is satisfactory.

19. Applying the same analysis to the third-order equation given by the third of (47) we use, in addition to (49), the equations

$$\left.\begin{aligned} h^3y'''_{\frac{1}{2}} &= (\delta^3-\tfrac{1}{8}\delta^5+\tfrac{37}{1920}\delta^7-\ldots)y_{\frac{1}{2}} \\ h^2y''_{\frac{1}{2}} &= (\delta^2-\tfrac{1}{12}\delta^4+\tfrac{1}{90}\delta^6-\ldots)y_{\frac{1}{2}} \end{aligned}\right\}, \tag{52}$$

insert them in the differential equation, and operate throughout with μ, using the latter in the form $(1+\frac{1}{4}\delta^2)^{\frac{1}{2}}$ when operating on odd-order terms. For the third of (47), with constant coefficients, we obtain

$$\begin{aligned} &(\delta^3-\tfrac{1}{240}\delta^7+\ldots)y_{\frac{1}{2}}+h(\mu\delta^2-\tfrac{1}{12}\mu\delta^4+\tfrac{1}{90}\mu\delta^6-\ldots)(q_1 y)_{\frac{1}{2}}+ \\ &\qquad +h^2(\delta+\tfrac{1}{12}\delta^3-\tfrac{1}{120}\delta^5+\ldots)(q_2 y)_{\frac{1}{2}}+h^3\mu(q_3 y)_{\frac{1}{2}} = 0, \end{aligned} \tag{53}$$

expressible in the Lagrangian form

$$\begin{aligned} &y_{r+2}(1+\tfrac{1}{2}hq_1+\tfrac{1}{12}h^2q_2)-y_{r+1}(3+\tfrac{1}{2}hq_1-\tfrac{3}{4}h^2q_2-\tfrac{1}{2}h^3q_3)+ \\ &\quad +y_r(3-\tfrac{1}{2}hq_1-\tfrac{3}{4}h^2q_2+\tfrac{1}{2}h^3q_3)-y_{r-1}(1-\tfrac{1}{2}hq_1+\tfrac{1}{12}h^2q_2)+Cy_{r+\frac{1}{2}} = 0, \end{aligned} \tag{54}$$

where
$$C = -\tfrac{1}{12}hq_1\,\mu\delta^4-\tfrac{1}{120}h^2q_2\,\delta^5+\tfrac{1}{90}hq_1\,\mu\delta^6+\ldots. \tag{55}$$

Equation (54) is now a four-term difference equation, and again it is clear that for an interval at which the finite-difference equations have any meaning the corresponding analytical solution has no oscillation, the roots p of equation (45), in this case a cubic, being all positive. There are now three arbitrary constants A in that equation, agreeing with the number of boundary conditions.

†20. Let us consider now the solution of a third-order equation by use of these methods. We may note that an equation in which the q_r are functions of x in the third of (47) still leads to an equation of general form (53), though the q_r in (53) are replaced by combinations of the original q_r and some of the derivatives; the relevant equations are given in 'Additional notes and bibliography' at the end of this chapter.

The remaining point of interest is the method by which the boundary conditions are satisfied. Consider, as a particular example, the arrangement of Fig. 21, in which 0 and 5 are boundary points, and the boundary conditions specify the values of y at 0 and 5, and the first derivative y' at 5.

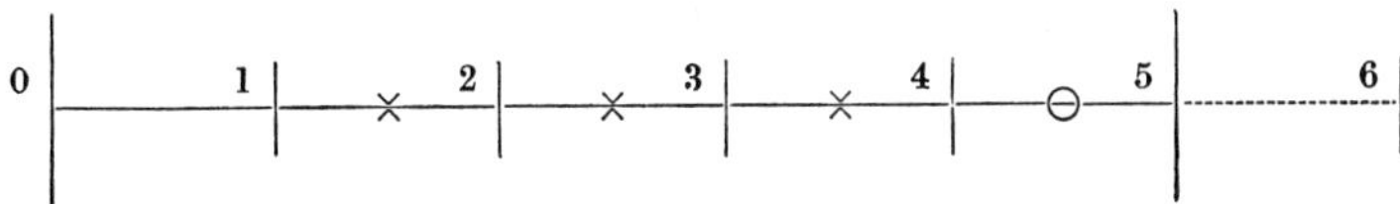

FIG. 21

The use of an equation like (54) corresponds to the satisfaction of the differential equation at the mid points of the intervals. If we use for this purpose the points marked with a cross in Fig. 21, we can use the given boundary values and produce three equations for the other four unknowns, the pivotal values at 1, 2, 3, and 4. The other required equation comes from the satisfaction of the derivative condition at 5. If we introduce one external point 6 we can use for this purpose a formula based on the backward differences of y_6, and the simultaneous application of the basic formula of type (54) at the point marked with a circle in Fig. 21 provides just enough equations for the determination of the pivotal values at 1, 2, 3, 4, and 6. If we introduce more external points, and satisfy (54) at more mid points, we can use a central-difference formula for the derivative condition, retaining as many differences as necessary. This problem, indeed, is closely analogous to the treatment of a derivative condition, discussed in Chapter 4, in the solution of second-order equations.

The treatment of other boundary conditions is almost obvious, and the somewhat unsymmetric nature of the basic finite-difference equation is explained.

Example

21. In comparison with the method of § 16 we see that the difference correction here starts with one lower order of differences, but this term is multiplied by the factor h and its effect therefore is generally decreased. In any event the removal of the oscillation is a worthwhile compensating factor. As a demonstration of the method we treat anew the example of § 16. The basic equation analogous to (54), with the difference correction neglected, is now given by

$$y_{r+2}(1-\tfrac{1}{2}h+\tfrac{1}{12}h^2)-y_{r+1}(3-\tfrac{1}{2}h-\tfrac{3}{4}h^2+\tfrac{1}{2}h^3)+$$
$$+y_r(3+\tfrac{1}{2}h-\tfrac{3}{4}h^2-\tfrac{1}{2}h^3)-y_{r-1}(1+\tfrac{1}{2}h+\tfrac{1}{12}h^2) = 0. \quad (56)$$

We use as before the interval $h = 0{\cdot}2$, so that there are four internal pivotal points. The derivative condition is at point 0, and we introduce the two external points -1 and -2, using equation (40) for the satisfaction of this condition. There are thus six unknowns, and for their determination we use (40) and the equations corresponding to (56) applied at the mid points of the intervals (4, 3), (3, 2), (2, 1), (1, 0), and (0, −1).

The solution is shown in Table 3. It is rather more accurate than that of Table 2, the maximum error at internal points being twelve units in the fifth decimal, but more important is the fact that there is no oscillation, and difference corrections can be applied with confidence to produce a better solution.

TABLE 3

x	y		δ^2		δ^4		δ^6
−0·4	+0·67017						
		+14852					
− ·2	0·81869		+3279				
		18131		+ 729			
+ ·0	1·00000		4008		+161		
		22139		890		+38	
·2	1·22139		4898		199		+ 5
		27037		1089		43	
·4	1·49176		5987		242		+10
		33024		1331		+53	
·6	1·82200		7318		+295		
		40342		+1626			
0·8	2·22542		+8944				
		+49286					
+1·0	+2·71828						

Transformation to second-order equation

22. A third-order equation can of course be solved by deriving a fourth-order equation by differentiation, and using the given equation as a fourth boundary condition: this is analogous to the treatment of first-order equations described in § 19 of Chapter 5. This procedure, however, is rarely advantageous, the resulting equations being no easier to solve, and the difference corrections cannot all be made small as in the case of first-order equations. A more attractive idea is to solve an associated second-order equation, a not uncommon possibility which we proceed to discuss.

The obvious dependent variable is $p = y'$, and if we make this substitution we can write the third-order equation (36) in the form

$$p''+f_1p'+f_2p+f_3\int p\,dx = f_4. \tag{57}$$

To facilitate the solution of this equation we must express two of the three given boundary conditions in terms of p, and the third is involved in the limits of the integral in (57).

The simplest example of this kind would have $f_3 = 0$, and boundary conditions specifying y and y' at one end point, y' at the other. If the specific conditions are $y = y_0$, $y' = y'_0$ at $x = x_0$, $y' = y'_n$ at $x = x_n$, we would replace (57) by the equation

$$p''+f_1 p'+f_2 p = f_4, \tag{58}$$

with boundary conditions

$$p_0 = y'_0, \qquad p_n = y'_n, \tag{59}$$

sufficient for the solution of (58), and y is then given by

$$y = y_0+\int\limits_0^x p\,dx. \tag{60}$$

If f_3 is not zero we have to solve (57) with boundary conditions (59), and the term $f_3\int p\,dx$ becomes $f_3\Big(y_0+\int\limits_0^x p\,dx\Big)$.

Example

†23. As a practical example of this method we take the equation

$$y'''+yy''+0{\cdot}1(y'^2-1) = 0, \tag{61}$$

with boundary conditions

$$y = y' = 0 \text{ at } x = 0, \qquad y' = 1 \text{ at } x = \infty. \tag{62}$$

The substitution $p = y'$ yields the differential equation

$$p''+yp'+0{\cdot}1(p^2-1) = 0, \tag{63}$$

where
$$y = \int\limits_0^x p\,dx. \tag{64}$$

We have then already used the first boundary condition, and the other two give
$$p = 0 \text{ at } x = 0, \qquad p = 1 \text{ at } x = \infty. \tag{65}$$

We shall solve (63) by an iterative process, expressible in the form

$$p^{(n)''}+y^{(n-1)}p^{(n)'}+0{\cdot}1\{(p^{(n)})^2-1\} = 0, \tag{66}$$

the superscript denoting the order of iterate. The finite-difference form of (66) is given by

$$\left.\begin{aligned} p_1(1+\tfrac{1}{2}hy_0)+p_{-1}(1-\tfrac{1}{2}hy_0)-2p_0+0{\cdot}1h^2(p_0^2-1)+Cp_0 = 0 \\ C = hy_0(-\tfrac{1}{6}\mu\delta^3+\tfrac{1}{30}\mu\delta^5-\ldots)+(-\tfrac{1}{12}\delta^4+\tfrac{1}{90}\delta^6-\ldots) \end{aligned}\right\}, \tag{67}$$

where the terms in y belong to a previous step and are therefore known. We shall assume initially that $x = 5$ is a sufficiently large value of x at which to apply the second of (65).

The procedure is obvious. We first make some guess at pivotal values of p, calculate y from (64) and insert it in (66), keeping it unchanged while a second approximation to p is computed from the set of simultaneous equations given by the first of (67). There is no point, in the early stages, in taking account of Cp, and the integration for y can be carried out by the simple trapezoidal rule. The simple non-linear term in (67) causes little difficulty, and it is easy, for example, to solve the equations by relaxation, by ignoring the term in p^2 in the relaxation process but incorporating it in the calculation of residuals.

24. The initial assumption is that p is linear in x, and the results of the first few iterations are shown diagrammatically in Fig. 22. The interval of tabulation is $h = 0{\cdot}5$.

0	0·100	0·200	0·300	0·400	0·500	0·600	0·700	0·800	0·900	1·000
0	0·120	0·263	0·422	0·584	0·731	0·848	0·928	0·973	0·993	1·000
0	0·149	0·320	0·502	0·675	0·818	0·916	0·970	0·992	0·999	1·000
0	0·163	0·347	0·538	0·714	0·851	0·937	0·980	0·995	0·999	1·000
0	0·169	0·359	0·554	0·731	0·865	0·946	0·984	0·997	1·000	1·000

FIG. 22

At this stage it will pay to incorporate the difference corrections and to use an accurate integration formula in the calculation of y. The intelligent computer will have noticed in the course of the work that the equations are fairly ill-conditioned: the initial residuals causing the ultimate changes in Fig. 22 are nowhere greater than 4 units in the third decimal place, and it would be expected that the new residuals would be still smaller, of a size perhaps comparable with the difference corrections or with the effect of a more accurate integration formula.

We therefore now difference our last approximation, calculating accurately y, the difference corrections, and the new residuals R defined as the left-hand sides of the first of (67). For the production of central differences near the origin we can use our standard step-by-step process, the presence of the integral causing no difficulty: in the latter, of course, we still at this stage use the trapezoidal rule. These results are shown in Table 4, with y, Cp, and R given to an extra figure.

TABLE 4

x	p		δ^2		δ^4		δ^6	y	$\frac{1}{2}hy$	10^4Cp	10^4R
−1·0	−0·266										
		+122									
−0·5	− ·144		+22								
		144		+ 3							
+0·0	+ ·000		25		− 7			+0·0000	+0·0000		
		169		− 4		− 5					
0·5	·169		21		12		+10	·0412	·0103	+11	+16
		190		16		+ 5					
1·0	·359		+ 5		7		0	·1726	·0432	9	8
		195		23		5					
1·5	·554		−18		− 2		+12	·4011	·1003	+12	32
		177		25		17					
2·0	·731		43		+15		−12	0·7236	·1809	− 2	+14
		134		−10		+ 5					
2·5	·865		53		20		−17	1·1248	·2812	19	− 7
		81		+10		−12					
3·0	·946		43		+ 8		+ 1	1·5797	·3949	28	14
		38		18		11					
3·5	·984		25		− 3		6	2·0636	·5159	28	23
		13		15		− 5					
4·0	0·997		10		8		+ 9	2·5595	·6399	16	15
		+ 3		7		+ 4					
4·5	1·000		− 3		4		− 3	3·0590	·7647	− 8	−15
		0		+ 3		+ 1					
5·0	1·000		0		− 3			+3·5590	+0·8897		
		0		0							
+5·5	+1·000		0								

25. It was clearly advisable to bring in the refinements at this point, since in all previous cases the residuals were all positive. Further correction to p is likely to be small, and we can best effect it by applying a linearization process. If the correction is called q, this clearly satisfies the equation

$$q_1(1+\tfrac{1}{2}hy_0)+q_{-1}(1-\tfrac{1}{2}hy_0)-2q_0(1-0{\cdot}1h^2p_0)+$$
$$+\tfrac{1}{2}h(p_1-p_{-1})\eta_0+Cq_0+R_0 = 0. \quad (68)$$

Here η is the change in y caused by the change q in p, and is given by

$$\eta = \int\limits_0^x q\,dx. \quad (69)$$

We first ignore Cq_0 and the term in η, obtaining the corrections q shown in Fig. 23.

0	0·0024	0·0031	0·0029	−0·0003	−0·0038	−0·0051	−0·0044	−0·0025	−0·0011	0

FIG. 23

The neglected terms will have no great effect on the third decimal in the solution, and the fourth decimal cannot be accurate since we have worked throughout with only four figures, and the equations are somewhat ill-conditioned. The final result has a maximum error of one unit in the third decimal.

26. This example shows some interesting computational features. First, the rather tentative relaxational approach has advantages over the more direct methods. As the successive results of Fig. 22 are obtained, the computer notices that at each stage the change in y causes positive residuals: he will therefore deliberately over-relax, leaving negative residuals everywhere to counteract this tendency and conserve his energy (this was not done in this illustrative example). Second, at the last stage in Fig. 23 we are justified in leaving slightly negative residuals near the origin, since the neglected term (69), though small, is certainly positive in this region. Third, there is no point in altering the 'position of infinity' until near the end of the iterative process: though in the first and even the second of Fig. 22 it is not clear whether we have gone far enough in the x-direction, it is quite clear at the end that $x = 5$ is a satisfactory 'infinity', certainly for a three-decimal solution. Finally, with an indirect method of this kind it would be quite easy, at the stage reached in Fig. 23, to add extra figures in subsequent calculations, enabling the fourth decimal, if required, to be given with confidence.

The relaxation process itself is rather interesting. The quantities $\frac{1}{2}hy$ in Table 4 show that a change in p at any point affects the residual at the point on the right to much less effect than that on the left. Residuals are therefore easily 'pushed' towards the origin and the process is quite simple to perform.

Other transformations

27. There are of course other possible methods of solving third-order equations. Many of these depend on mathematical transformations which might change the order of the equation, or put it in a more tractable form. Another possibility, which we proceed to discuss, is of general application, and involves the solution of two simultaneous equations, one of first order and one of second order.

Consider, for example, the general linear equation (36) with its associated general boundary conditions (37). The substitution $y' = p$ provides the simultaneous equations

$$\left.\begin{aligned} p''+f_1p'+f_2p+f_3y &= f_4 \\ y'-p &= 0 \end{aligned}\right\}, \tag{70}$$

and the three associated boundary conditions are given by equations of the form

$$g_0 p' + g_1 p + g_2 y = g_3. \tag{71}$$

By expressing the first of (70) in its central-difference form, and using for the second our well-known formula

$$y_1 - y_0 = \tfrac{1}{2}h(p_1 + p_0) - \tfrac{1}{12}\delta^3 y_{\frac{1}{2}} + \tfrac{1}{120}\delta^5 y_{\frac{1}{2}} - \ldots, \tag{72}$$

we have clearly just enough algebraic equations for the determination of all the unknowns, and can solve them simultaneously to obtain both y and p.

The problem given by equation (38), for example, is solved by means of the equations

$$\left.\begin{aligned} p'' - p' + p - y &= 0 \\ y' - p &= 0 \end{aligned}\right\}, \tag{73}$$

with the boundary conditions

$$y(0) = 1, \qquad y(1) = 2{\cdot}71828, \qquad p(0) = 1. \tag{74}$$

The finite-difference equations, neglecting difference corrections, are given by

$$\left.\begin{aligned} p_{r+1}(1 - \tfrac{1}{2}h) + p_{r-1}(1 + \tfrac{1}{2}h) - p_r(2 - h^2) - h^2 y_r &= 0 \\ y_r - y_{r-1} - \tfrac{1}{2}h(p_r + p_{r-1}) &= 0 \end{aligned}\right\}. \tag{75}$$

At an interval $h = 0{\cdot}2$ there are nine unknowns, and if we apply both of (75) at the points $x = 0{\cdot}2$, $0{\cdot}4$, $0{\cdot}6$, and $0{\cdot}8$, and the second of (75) also at $x = 1{\cdot}0$, we have nine equations which involve no external points, and which can therefore provide a solution. Difference corrections can be computed in the usual way and used to effect an improvement.

We could use this method also for non-linear equations. For the example of § 23 we have the two equations

$$\left.\begin{aligned} p'' + p'y + 0{\cdot}1(p^2 - 1) &= 0 \\ y' - p &= 0 \end{aligned}\right\}, \tag{76}$$

with the boundary conditions

$$y(0) = 0, \qquad p(0) = 0, \qquad p(\infty) = 1. \tag{77}$$

The unrefined finite-difference equations are now given by

$$\left.\begin{aligned} y_r - y_{r-1} - \tfrac{1}{2}h(p_r + p_{r-1}) &= 0 \\ p_{r+1}(1 + \tfrac{1}{2}hy_r) + p_{r-1}(1 - \tfrac{1}{2}hy_r) - 2p_r + 0{\cdot}1h^2(p_r^2 - 1) &= 0 \end{aligned}\right\}. \tag{78}$$

These can be solved by relaxation, but the easier method is probably to solve them in succession, by an iterative process, which reduces to that already used in § 23. The choice of method depends solely on the structure of the equation and computational convenience.

28. The substitution $y'' = q$ gives an alternative pair of simultaneous equations, (70) being replaced by

$$\left.\begin{aligned} q'+f_1q+f_2y'+f_3y &= f_4 \\ y''-q &= 0 \end{aligned}\right\}, \tag{79}$$

and the three associated boundary conditions belong to the general form

$$g_0q+g_1y'+g_2y = g_3. \tag{80}$$

This transformation may have advantages, for example, if $g_1 = 0$ in all three boundary conditions.

Equations of order higher than the fourth. Non-linear equations

29. We shall not pursue further our account of methods of solving equations of relatively high order. They are not of frequent occurrence, and the available methods involve no new principle. The computer will note, however, that the general finite-difference equations are more and more ill-conditioned as the order increases, and it becomes more difficult to obtain good precision in the solution. We have already indicated, and shall return to the point in a later chapter, that the solution of fourth-order equations has best precision when the problem is treated as a set of two simultaneous equations of second order, and the same principle should be applied when possible to equations of higher order.

Similarly there is nothing to add regarding the treatment of non-linear equations. Iterative methods, and the use of linear equations to correct first approximations, apply equally well here as in the corresponding second-order problems of Chapter 4 or first-order problems of Chapter 5.

Matrix solution of equations of fourth and third orders

30. We have not expounded in detail the methods of solving the finite-difference equations arising from the differential equations of this chapter. The relaxation method needs no further description, but it is of interest to examine the form of the matrices involved from the point of view of solution by 'triangular resolution', following the brief account in Chapter 3.

For a fourth-order equation, the finite-difference equation connects five adjacent values, so that, applying this at the pivotal points 0, 1, 2, 3, and 4, part of any such matrix will have the following composition, a cross denoting a non-zero coefficient:

$$\begin{array}{ccccccccc} (-2) & (-1) & (0) & (1) & (2) & (3) & (4) & (5) & (6) \end{array}$$
$$\left[\begin{array}{ccccccccc} \times & \times & \times & \times & \times & & & & \\ & \times & \times & \times & \times & \times & & & \\ & & \times & \times & \times & \times & \times & & \\ & & & \times & \times & \times & \times & \times & \\ & & & & \times & \times & \times & \times & \times \end{array}\right]. \qquad (81)$$

Suppose the points 0 and 4 are boundaries and two boundary conditions are applied at each point. Any boundary condition other than 'y specified' will involve five adjacent points, as seen from (3). The most general boundary conditions will give two further equations at each end of the range; at the beginning each equation will give further relations between -2, -1, 0, 1, 2, and at the end between 2, 3, 4, 5, and 6. If we use these further relations to eliminate external values, that is to express y_{-2} and y_{-1} in terms of y_0, y_1, and y_2, and similarly at the other end, the effect on the *form* of the matrix is merely to omit from (81) the columns labelled -2, -1, 5, and 6, and the resulting matrix has the shape foreshadowed in Chapter 3, and for which we have already demonstrated the triangular resolution.

If one of the boundary conditions at the first point is 'y_0 specified', then the other condition, together with the first row of (81), will still enable us to express y_{-2} and y_{-1} in terms of y_0, y_1, and y_2, and, with y_0 given, the effect is to omit the first three columns and the first row from (81), giving the same shape as before provided there are two conditions at point 4.

31. Consider now the case, mentioned in § 1, when three boundary conditions are given at the first point 0 and one at the last point 4. If all the boundary conditions are general, the effect is to add to matrix (81) three extra rows identical in form with the first row and one row identical with the last. At the beginning of the range we can use two of the first four similar rows to express y_{-2} and y_{-1} in terms of y_0, y_1, and y_2, and substitute for them in the other rows; at the end of the range we can use the second of the two similar rows to eliminate y_6, and the resulting matrix has the following appearance:

$$\begin{array}{cccccc} (0) & (1) & (2) & (3) & (4) & (5) \end{array}$$
$$\left[\begin{array}{cccccc} \times & \times & \times & & & \\ \times & \times & \times & & & \\ \times & \times & \times & \times & & \\ \times & \times & \times & \times & \times & \\ & \times & \times & \times & \times & \times \\ & & \times & \times & \times & \times \end{array}\right]. \qquad (82)$$

We could further use the first row of (82) to eliminate y_0, equivalent as regards the form of the final matrix to omitting from (82) the first row and column, and the final appearance is as follows:

$$\begin{array}{c} \begin{array}{ccccc} (1) & (2) & (3) & (4) & (5) \end{array} \\ \begin{bmatrix} \times & \times & & & \\ \times & \times & \times & & \\ \times & \times & \times & \times & \\ \times & \times & \times & \times & \times \\ & \times & \times & \times & \times \end{bmatrix} \end{array}. \tag{83}$$

In general, of course, the range will be longer than that shown, and successive rows of (83) will have 2, 3, 4, 5,..., 5, 4 non-zero adjacent elements.

The triangular matrices L and U will have the following appearance:

$$\begin{array}{cc} L & U \\ \begin{bmatrix} \times & & & & \\ \times & \times & & & \\ \times & \times & \times & & \\ \times & \times & \times & \times & \\ & \times & \times & \times & \times \end{bmatrix}, & \begin{bmatrix} \times & \times & & & \\ & \times & \times & & \\ & & \times & \times & \\ & & & \times & \times \\ & & & & \times \end{bmatrix} \end{array}. \tag{84}$$

L will have in general four non-zero elements in each row, that is three besides the diagonal, and U will have two non-zero elements in each row, that is one besides the diagonal. This corresponds, as we shall see in Chapter 8, to the fact that the use of L is equivalent to forward integration using three boundary conditions, that of U to integrating backwards using one boundary condition. If the three boundary conditions were given at the right-hand end the situation would be reversed and L would have two non-zero elements and U four in each row.

With four boundary conditions at one end, of course, the matrix is already triangular, corresponding to integration in one direction only. With more arbitrary combinations of boundary conditions the situation is more obscure, but the original matrix will always have elements grouped round the diagonal, and the non-zero elements of L and U will not extend outside this group. Suppose, for example, that the range extended from -4 to $+8$, and values were specified at points -4, 0, 2, and 8. The matrices A, L, and U would then have the following appearance:

$$A$$

$$\begin{array}{ccccccccc} (-3) & (-2) & (-1) & (1) & (3) & (4) & (5) & (6) & (7) \end{array}$$

$$\begin{bmatrix} \times & \times & \times & & & & & & \\ \times & \times & \times & \times & & & & & \\ & \times & \times & \times & & & & & \\ & & \times & \times & \times & & & & \\ & & & \times & \times & \times & & & \\ & & & \times & \times & \times & \times & & \\ & & & & \times & \times & \times & \times & \\ & & & & \times & \times & \times & \times & \times \\ & & & & & \times & \times & \times & \times \end{bmatrix},$$

$$L \qquad\qquad\qquad\qquad U$$

$$\begin{bmatrix} \times & & & & & & & & \\ \times & \times & & & & & & & \\ & \times & \times & & & & & & \\ & & \times & \times & & & & & \\ & & & \times & \times & & & & \\ & & & \times & \times & \times & & & \\ & & & & \times & \times & \times & & \\ & & & & \times & \times & \times & \times & \\ & & & & & \times & \times & \times & \times \end{bmatrix}, \quad \begin{bmatrix} \times & \times & \times & & & & & & \\ & \times & \times & \times & & & & & \\ & & \times & \times & & & & & \\ & & & \times & \times & & & & \\ & & & & \times & \times & & & \\ & & & & & \times & \times & & \\ & & & & & & \times & \times & \\ & & & & & & & \times & \times \\ & & & & & & & & \times \end{bmatrix}. \tag{85}$$

32. If third-order equations are solved by using a five-term recurrence formula, we have seen in § 16 that central-difference formulae cannot be used throughout, and some sloping-difference formulae are needed. If the function is specified at the points 0 and 6 and the derivative at 0, the matrix, obtained by the methods of § 16, has the following appearance:

$$\begin{array}{ccccccc} (-2) & (-1) & (1) & (2) & (3) & (4) & (5) \end{array}$$

$$\begin{bmatrix} \times & \times & \times & \times & & & \\ \times & \times & \times & \times & & & \\ & \times & \times & \times & \times & & \\ & & \times & \times & \times & \times & \\ & & \times & \times & \times & \times & \times \\ & & & \times & \times & \times & \times \\ & & & \times & \times & \times & \times \end{bmatrix}. \tag{86}$$

Here the first row represents the derivative condition at 0, and the last row the differential equation evaluated at point 5 and expressed as backward differences of y_6. With a larger range the only difference is the presence of more rows with five non-zero coefficients. We might treat this matrix without change, or use the first two rows to express y_{-2} and y_{-1} in terms of y_1 and y_2, so that the final matrix looks like (86) with the first two rows and columns deleted.

With two boundary conditions at the first boundary point 0, and one at a second boundary point 4, all of general form, we might express both the differential equation at the point 3 and boundary condition at point 4 in terms of backward differences of y_4 and, using central differences elsewhere, our matrix takes the form

$$\begin{array}{ccccccc} (-2) & (-1) & (0) & (1) & (2) & (3) & (4) \end{array}$$
$$\begin{bmatrix} \times & \times & \times & \times & \times & & \\ \times & \times & \times & \times & \times & & \\ \times & \times & \times & \times & \times & & \\ & \times & \times & \times & \times & \times & \\ & & \times & \times & \times & \times & \times \\ & & \times & \times & \times & \times & \times \\ & & \times & \times & \times & \times & \times \end{bmatrix}, \tag{87}$$

and we can treat this either without change or with obvious preliminary operations.

33. Of more interest is the use of four-term formulae for third-order equations. We consider the most general case with two general boundary conditions at the point 0, one at point 4, and use central differences throughout. Satisfaction of the differential equation at the points $-\frac{1}{2}, \frac{1}{2}, \frac{3}{2}, \frac{5}{2}, \frac{7}{2}, \frac{9}{2}$ gives the following arrangement:

$$\begin{array}{ccccccccc} (-2) & (-1) & (0) & (1) & (2) & (3) & (4) & (5) & (6) \end{array}$$
$$\begin{bmatrix} \times & \times & \times & \times & & & & & \\ & \times & \times & \times & \times & & & & \\ & & \times & \times & \times & \times & & & \\ & & & \times & \times & \times & \times & & \\ & & & & \times & \times & \times & \times & \\ & & & & & \times & \times & \times & \times \end{bmatrix}. \tag{88}$$

To this we have to add the equations for the boundary conditions, most generally giving two relations between y_{-2}, y_{-1}, y_0, y_1, and y_2, and one between y_2, y_3, y_4, y_5, and y_6. If we eliminate the external values we are left with the matrix

$$\begin{array}{ccccc} (0) & (1) & (2) & (3) & (4) \end{array}$$
$$\begin{bmatrix} \times & \times & \times & & \\ \times & \times & \times & & \\ \times & \times & \times & \times & \\ & \times & \times & \times & \times \\ & & \times & \times & \times \end{bmatrix}, \tag{89}$$

and by subtracting a multiple of the second equation from the first we can reduce this to the form

$$\begin{array}{ccccc} (0) & (1) & (2) & (3) & (4) \end{array}$$
$$\begin{bmatrix} \times & \times & & & \\ \times & \times & \times & & \\ \times & \times & \times & \times & \\ & \times & \times & \times & \times \\ & & \times & \times & \times \end{bmatrix}. \tag{90}$$

Similarly, if there is one general boundary condition at the point 0 and two at point 4, we can express the matrix in the form

$$\begin{array}{ccccc} (0) & (1) & (2) & (3) & (4) \end{array}$$
$$\begin{bmatrix} \times & \times & \times & & \\ \times & \times & \times & \times & \\ & \times & \times & \times & \times \\ & & \times & \times & \times \\ & & & \times & \times \end{bmatrix}. \tag{91}$$

The triangular resolution of these forms has already been discussed in Chapter 3. If there are two boundary conditions at the first point, L has three non-zero elements in each row and U has two such elements: if the position of the boundary positions is reversed, L will have two non-zero elements and U three in each row. This corresponds to the representation of the differential operator of third order as the product of one of second order and one of first order: in the first case we effectively integrate forward with the second-order operator, using the two given boundary conditions, and backward with the first-order operator using the single boundary condition, and in the second case the circumstances are reversed.

Use of Taylor's series for the boundary conditions

34. We saw in Chapter 4 that boundary conditions involving derivatives were often conveniently treated by the use of Taylor's series. This usually improved the accuracy of first approximations, in that no corrections were needed to satisfy the boundary conditions; it also gave matrices of convenient shape.

For equations of third and fourth orders, similarly, this method can often be used with considerable advantage. For a fourth-order equation we have discussed the general finite-difference process in the first three sections of this chapter. In particular we noted that the approximate finite-difference form of (1), applied at boundary point 0, introduced two external pivotal values, for which approximate finite-difference forms of general boundary conditions like (2) provided two extra equations.

By using Taylor's series we can make the last two equations as exact as we please. For, with the notation of Fig. 18, we have

$$\left.\begin{aligned} y_{\pm1} &= y_0 \pm hy_0' + \frac{h^2}{2!}y_0'' \pm \frac{h^3}{3!}y_0''' + \frac{h^4}{4!}y_0^{\mathrm{iv}} \pm \cdots \\ y_{\pm2} &= y_0 \pm (2h)y_0' + \frac{(2h)^2}{2!}y_0'' \pm \frac{(2h)^3}{3!}y_0''' + \frac{(2h)^4}{4!}y_0^{\mathrm{iv}} \pm \cdots \end{aligned}\right\}. \tag{92}$$

In these equations the fourth derivative can be calculated in terms of y_0, y_0', y_0'', and y_0''' from the differential equation (1), and successive higher derivatives from the successive differentiated forms of (1). We can then write

$$y_r = a_r + b_r y_0 + c_r y_0' + d_r y_0'' + e_r y_0''', \qquad r = \pm1, \pm2, \tag{93}$$

in which the coefficients can be calculated as accurately as we please. Any three of the four equations (93) suffice for the expression of y_0', y_0'', y_0''' in terms of y_{-2}, y_{-1}, y_0, y_1, and y_2, and substitution in the two boundary conditions of type (2) provides the two extra required equations, involving exact coefficients and no difference corrections. If desired y_{-2} and y_{-1} can be eliminated, and we are left with a matrix of a standard convenient form.

35. As an illustration we consider again the example of § 7, in which the chief error of the first approximation arose through neglect of the difference correction involved in the satisfaction of the boundary conditions.

The basic finite-difference equation is (13), which we need apply only at the pivotal points 0·2(0·2)1·0. Equations corresponding to (93) are here given by

$$y(\pm h) = \left(1+\frac{h^4}{4!}+\frac{h^8}{8!}+\ldots\right)y(0)\pm\left(h+\frac{h^5}{5!}+\frac{h^9}{9!}+\ldots\right)y'(0)+$$
$$+\left(\frac{h^2}{2!}+\frac{h^6}{6!}+\frac{h^{10}}{10!}+\ldots\right)y''(0)\pm\left(\frac{h^3}{3!}+\frac{h^7}{7!}+\frac{h^{11}}{11!}+\ldots\right)y'''(0), \quad (94)$$

so that we have

$$\left.\begin{aligned}\left(\frac{h^2}{2!}+\frac{h^6}{6!}+\frac{h^{10}}{10!}+\ldots\right)y''(0) &= \frac{1}{2}\left\{y(h)+y(-h)-\left(1+\frac{h^4}{4!}+\frac{h^8}{8!}+\ldots\right)2y(0)\right\}\\ \left(\frac{h^3}{3!}+\frac{h^7}{7!}+\frac{h^{11}}{11!}+\ldots\right)y'''(0) &= \frac{1}{2}\left\{y(h)-y(-h)-\left(h+\frac{h^5}{5!}+\frac{h^9}{9!}+\ldots\right)2y'(0)\right\}\end{aligned}\right\}. \quad (95)$$

At the origin y and y' are specified so that, using (95) for h and $2h$, we deduce finally the equation

$$y(-0{\cdot}2) = -2{\cdot}099744+3{\cdot}000152y(0{\cdot}2)-0{\cdot}500005y(0{\cdot}4), \quad (96)$$

so that the first equation, representing (13) at $x = 0{\cdot}2$, can be written as

$$8{\cdot}998552y(0{\cdot}2)-4{\cdot}500005y(0{\cdot}4)+y(0{\cdot}6) = 6{\cdot}099744-Cy(0{\cdot}2). \quad (97)$$

The next two equations are given by

$$\left.\begin{aligned}-4y(0{\cdot}2)+5{\cdot}9984y(0{\cdot}4)-4y(0{\cdot}6)+y(0{\cdot}8) &= -1-Cy(0{\cdot}4)\\ y(0{\cdot}2)-4y(0{\cdot}4)+5{\cdot}9984y(0{\cdot}6)-4y(0{\cdot}8)+y(1{\cdot}0) &= -Cy(0{\cdot}6)\end{aligned}\right\}, \quad (98)$$

and the last two equations, representing (13) at $x = 0{\cdot}8$ and $1{\cdot}0$, include external values which we eliminate from the boundary conditions at $x = 1$.

From the equations corresponding to (95), with $y'''(1) = y'(1)$, $y''(1) = y(1)$ as boundary conditions, we derive the equations

$$\left.\begin{aligned}y(1{\cdot}4)-y(0{\cdot}6) &= 2{\cdot}040134\{y(1{\cdot}2)-y(0{\cdot}8)\}\\ y(1{\cdot}2)+y(0{\cdot}8) &= 2{\cdot}040134y(1{\cdot}0)\end{aligned}\right\}, \quad (99)$$

and the last two equations, to be used with (97) and (98) to produce the first approximate solution, are then given by

$$\left.\begin{aligned}y(0{\cdot}4)-4y(0{\cdot}6)+4{\cdot}9984y(0{\cdot}8)-1{\cdot}959866y(1{\cdot}0) &= -Cy(0{\cdot}8)\\ 2y(0{\cdot}6)-4{\cdot}080267y(0{\cdot}8)+2{\cdot}000011y(1{\cdot}0) &= -Cy(1{\cdot}0)\end{aligned}\right\}. \quad (100)$$

The solution of these equations gives an approximate solution which differs from e^x by amounts 0·0001, 0·0004, 0·0008, 0·0013, 0·0019, at $x = 0{\cdot}2(0{\cdot}2)1{\cdot}0$, discrepancies equal to only one-quarter of those resulting from the previous method. There is no difficulty, of course, in extending the solution outside the range, calculating Cy from the second of (13), and obtaining a correction.

The application of this method to other types of boundary condition, and to third-order differential equations, is straightforward and needs no further illustration.

Additional notes and bibliography

§ 17, etc. By 'oscillation' we mean here the rapid alternation in sign caused by a factor $(-1)^r$ multiplying a well-behaved function. We had an example of this in Chapter 5, where in § 4 we observed that the general solution of the difference equation

$$y_{r+1}-2hy_r-y_{r-1} = 0$$

was given by $\quad y_r = Pe^{r\theta}+(-1)^r Qe^{-r\theta}, \qquad h = \sinh\theta.$

Only the function $e^{r\theta}$ corresponds to a genuine solution of $y' = y$.

The second-order difference equation

$$y_{r+1}-(2-h^2)y_r+y_{r-1} = 0$$

has the general solution

$$y_r = Ae^{ir\theta}+Be^{-ir\theta}, \qquad h = 2\sin\tfrac{1}{2}\theta.$$

The two solutions involved are of trigonometric type and therefore oscillate, but this oscillation is not undesirable; both functions are well behaved at a small enough interval, and correspond to genuine solutions of the differential equation $y'' = y$.

§ 20. With general q_r in the third of (47), we write that equation as

$$y'''+(p_1 y)''+(p_2 y)'+p_3 y = 0,$$

where $p_1 = q_1$, $p_2 = q_2-2q_1'$, $p_3 = q_3-q_2'+q_1''$. In (53) q_r is replaced by p_r, and equations (54) and (55) become

$$\{(1+\tfrac{1}{2}hp_1+\tfrac{1}{12}h^2p_2)y\}_{r+2}-\{(3+\tfrac{1}{2}hp_1-\tfrac{3}{4}h^2p_2-\tfrac{1}{2}h^3p_3)y\}_{r+1}+$$
$$+\{(3-\tfrac{1}{2}hp_1-\tfrac{3}{4}h^2p_2+\tfrac{1}{2}h^3p_3)y\}_r-\{(1-\tfrac{1}{2}hp_1+\tfrac{1}{12}h^2p_2)y\}_{r-1}+Cy_{r+\frac{1}{2}} = 0,$$
$$Cy_{r+\frac{1}{2}} = \{-\tfrac{1}{12}h\mu\delta^4(p_1 y)-\tfrac{1}{120}h^2\delta^5(p_2 y)+\tfrac{1}{90}\mu\delta^6(p_1 y)+\ldots\}_{r+\frac{1}{2}}.$$

§ 23. The more general equation

$$y'''+yy''+\beta\{y'^2-1\} = 0,$$

with $y(0) = y'(0) = 0$, $y'(\infty) = 1$, was first solved by Hartree, using the differential analyser, for a range of values of β. He showed that the solution is not unique for negative values of β (including the example of § 23), and that the condition at $x = \infty$ should be replaced by

$$y'(\infty) \to 1 \text{ as rapidly as possible,}$$

to give the physically acceptable solution. The wanted solution for $p = y'$ behaves at infinity like $1+O(e^{-x^2})$, other possible solutions like $1+O(x^{-n})$. It is interesting to note that the numerical method of solution outlined converges to the required solution, even when a first approximation is taken deliberately to be very close to an unwanted solution.

Further information on this problem is given by

HARTREE, D. R., 1937, *Proc. Camb. Phil. Soc.* **33,** 223–39.

FOX, L., 1949, The solution by relaxation methods of ordinary differential equations. *Proc. Camb. Phil. Soc.* **45,** 50–68.

7

Eigenvalue problems

Introduction

†**1.** It is well known that the differential system

$$y''+p^2y = 0, \qquad y(0) = y(1) = 0, \tag{1}$$

has non-trivial solutions only for certain values of p, given by

$$p = m\pi, \qquad m = 1, 2, 3,\dots . \tag{2}$$

The corresponding solutions are then

$$y = k\sin px = k\sin m\pi x. \tag{3}$$

The problem leading to (1) is called an eigenvalue or characteristic-value problem, $p^2 = m^2\pi^2$ is an eigenvalue or characteristic value of the solution, and $y = \sin m\pi x$ is an eigenfunction or characteristic function. We shall in this book use the German 'eigen' rather than the English 'characteristic': there seems to be no real distinction, and the former is shorter and in common use. We shall also refer to the combination of eigenvalue and eigenfunction as 'eigensolution'.

A differential equation of eigenvalue type is generally linear and homogeneous, with homogeneous boundary conditions given at distinct points in the range of integration. We have therefore a problem of true boundary-value type. If y is an eigenfunction, any constant multiple of y is also a solution, and the constant corresponding to k in (3) is indeterminate unless some additional condition is imposed. We might, for example, require y or its first derivative to have a specific non-zero value at some convenient point in the range, or choose the constant so that the integral of y^2, taken between the two end points, has a given value. In this chapter we shall concentrate on methods of finding one or more eigenvalues and corresponding eigenfunctions. The subsequent determination of the constant k to satisfy an extra condition is generally relatively trivial.

Not all differential equations involving an eigenvalue have real solutions. We shall not often be concerned with the determination of complex solutions, but we may need to decide whether a given problem has a real solution or not.

There is a considerable literature on the mathematical theory of eigenvalue problems, but we shall concentrate mainly on direct numerical methods. In this chapter we consider boundary-value techniques: the use of initial-value methods is discussed in Chapter 8.

2. Replacement of the differential equation in (1) by simple finite-difference equations, in which the difference correction is neglected, leads to a set of linear algebraic equations typified by

$$y_{r-1}-(2-p^2h^2)y_r+y_{r+1} = 0, \tag{4}$$

in which the first and last equations may have slightly different forms when the boundary conditions are incorporated. We can express the aggregate of equations (4) by the matrix equation

$$(A-I\lambda)y = 0, \tag{5}$$

in which A is a square matrix of rather simple form, y is the vector comprising the pivotal values, and λ is $-p^2h^2$.

The differential problem has thus been replaced by an algebraic problem, that of determining the latent roots λ and latent vectors y of the matrix A in (5). It is not unreasonable to suppose that some solutions of the algebraic problem, for sufficiently small values of h, are reasonably close approximations to solutions of the differential problem.

We shall therefore consider first a selection of methods of solving the algebraic problem defined by (5), without restricting the matrix A to have any special form associated with differential equations.

Properties of latent roots and vectors of matrices

3. Interesting and useful properties of the latent roots and vectors of matrices are easily established. We consider a more general problem defined by the equation

$$(A-B\lambda)y = 0, \tag{6}$$

where A and B are arbitrary real matrices of order n.

The latent roots are the roots of the determinantal equation

$$|A-B\lambda| = 0, \tag{7}$$

in which the left-hand side can be expressed as a polynomial of degree n in λ. To each root, real or complex, there corresponds a non-trivial solution y, unique except for an arbitrary constant of multiplication. We shall assume that the roots are all distinct, since this is almost always the case in practical problems.

With (6) we consider the allied equation

$$(A'-B'\lambda)z = 0, \tag{8}$$

where the dash denotes matrix transposition. The latent roots of (8) are the same as those of (6), but the latent vectors are generally different.

If λ_r and y_r are solutions of (6), λ_s and z_s solutions of (8), we can write

$$(A-B\lambda_r)y_r = 0, \qquad z_s'(A-B\lambda_s) = 0. \tag{9}$$

Postmultiplying the second of (9) by y_r, and using the first of (9) to eliminate A, we find

$$(\lambda_r-\lambda_s)z_s' By_r = 0, \tag{10}$$

so that, if the roots are distinct,

$$z_s' By_r = y_r' B'z_s = z_s' Ay_r = y_r' A'z_s = 0. \tag{11}$$

These are the orthogonality properties of latent vectors.

When y_r and z_r are solutions corresponding to the same λ_r, a common form of normalization is embodied in the equations

$$z_r' By_r = y_r' B'z_r = 1, \qquad z_r' Ay_r = y_r' A'z_r = \lambda_r. \tag{12}$$

4. If B is the unit matrix, the problem reduces to that of determining the latent roots and vectors of A. The matrices B and B' are replaced by I, and the orthogonality properties and normalization conditions take the simpler forms

$$z_s' y_r = z_s' Ay_r = 0, \qquad z_r' y_r = 1, \qquad z_r' Ay_r = \lambda_r. \tag{13}$$

5. In many problems, for example in the application of Lagrangian equations in the theory of small oscillations of discrete systems, the matrices A and B are symmetric, and B usually positive definite. In this case we need consider only the single equation (6), z in (8) being identical with y in (6). Equations (11) and (12) then become

$$y_r' By_s = y_r' Ay_s = 0, \qquad y_r' By_r = 1, \qquad y_r' Ay_r = \lambda_r. \tag{14}$$

In this case, also, the roots are all real. For if we assume a complex λ, with conjugate $\bar{\lambda}$, and corresponding complex vectors y and $\bar{y}$, we can write

$$y'A = \lambda y'B, \qquad A\bar{y} = \bar{\lambda}B\bar{y}. \tag{15}$$

Premultiplying the second of (15) by y', and using the first, we find

$$(\lambda-\bar{\lambda})y'B\bar{y} = 0. \tag{16}$$

Since $y'B\bar{y}$ is real and non-zero, we have $\lambda = \bar{\lambda}$ and the roots are real. The vector y is therefore also real, and the root can be obtained in the form

$$\lambda = y'Ay/y'By. \tag{17}$$

If A, as well as B, is positive definite, the roots are therefore both real and positive.

6. The quantity in (17) is often called the 'Rayleigh quotient', and has important properties embodied in 'Rayleigh's Principle'. An arbitrary vector y can be expanded in terms of latent vectors y_r in the form

$$y = \sum_1^n \alpha_r y_r, \tag{18}$$

where, by virtue of the orthogonality and normalization relations (14), the constants α_r are given by

$$\alpha_r = y'By_r = y'Ay_r/\lambda_r. \tag{19}$$

Substituting in (17) from (18), and using (14), we find

$$\lambda = \sum_1^n \alpha_r^2 \lambda_r / \sum_1^n \alpha_r^2. \tag{20}$$

We can write (20) in the form

$$\lambda = \lambda_k + \frac{\sum_{r \neq k} (\lambda_r - \lambda_k)\alpha_r^2}{\sum \alpha_r^2}, \tag{21}$$

from which we can establish the various parts of Rayleigh's Principle. First, if λ_k corresponds to the largest eigenvalue, it is clear that $\lambda < \lambda_k$, so that the Rayleigh quotient underestimates the largest eigenvalue, for all arbitrary vectors not identical with the 'largest' eigenvector. Second, if λ_k corresponds to the smallest eigenvalue, it is clear that $\lambda > \lambda_k$, so that the Rayleigh quotient overestimates the smallest eigenvalue, for all arbitrary vectors not identical with the 'smallest' eigenvector. Third, if the vector y is an approximation to any eigenvector y_k, so that, in (18), α_k is much greater than any other α_r, then first-order changes in y give rise to second-order changes in λ, and the eigenvalue is determined more accurately than its eigenvector.

We shall often call (17) the 'Rayleigh quotient', even when the matrices A and B do not have the properties for which the various parts of Rayleigh's Principle are applicable.

Determination of latent roots and vectors of matrices

†7. As in the solution of simultaneous algebraic equations, we have two main methods for the solution of the algebraic problem, the tentative approach by relaxation and more direct methods in which judgement plays a smaller part. We first consider and illustrate the relaxation method.

Relaxation

8. If the matrix A in equation (5) is symmetric and positive definite, we can write the corresponding algebraic equations in the form

$$\left.\begin{aligned} (a_{11}-\lambda)y_1+a_{12}y_2+...+a_{1n}y_n &= 0 \\ a_{12}y_1+(a_{22}-\lambda)y_2+...+a_{2n}y_n &= 0 \\ \cdot \quad \cdot \quad \cdot \quad \cdot \quad \cdot \quad \cdot \quad \cdot \quad \cdot \quad & \cdot \\ a_{1n}y_1+a_{2n}y_2+...+(a_{nn}-\lambda)y_n &= 0 \end{aligned}\right\}, \tag{22}$$

where $y_1, y_2,..., y_n$ are components of the vector y, and a_{rs} is a coefficient of A. We know that the roots are real and positive and that Rayleigh's Principle is applicable.

In the relaxation method we first make a guess at the values of the components of a required vector y, then calculate Ay and the Rayleigh quotient $y'Ay/y'y$. This tentative value of λ is inserted in the coefficients of equations (22) and the residuals, defined as the left-hand sides of these equations, and written in matrix notation as

$$R = (A-I\lambda)y, \tag{23}$$

can then be computed.

We now try to reduce these residuals by the standard relaxation process, using the acceleration devices described in Chapter 3. There is, however, a new feature not present in our previous problems. Unless the calculated λ is an eigenvalue the residuals cannot all be liquidated: we can perhaps reduce them by a certain amount, after which there is no further improvement. At this stage we take the new vector so obtained, recalculate λ from Rayleigh's quotient, and repeat the process until y and λ show no further change, and the residuals are negligible.

9. There are some interesting computational points. First, there exists for any λ the trivial solution $y = 0$, and the relaxation process used by an inexperienced computer will converge to this solution. We guard against this by keeping fairly constant the largest vector component, using appropriate multiplying factors if this is becoming small during the relaxation process. The reduction in the residual can then be properly assessed.

Second, the relaxation is now aimed not so much at liquidating particular residuals, but at arranging their signs and magnitudes so that a subsequent change in λ has good effect everywhere. Consider in this connexion the effect on the residuals of altering λ. From (23) we have

$$\delta R_r = -y_r \delta\lambda. \tag{24}$$

This equation shows that if we can arrange the residuals, by suitable changes in the y's, to have a distribution such that R_r/y_r is constant, then the vector needs no further alteration and the residuals will all disappear if we increase λ by the value of this constant. This ideal effect may be difficult to achieve, but we can usually approach it by arranging our increments in y so that each residual R_r and its corresponding vector component y_r are either all of the same sign or all of opposite sign. With skill we can often arrange also that their ratio is fairly constant.

Example

10. As a simple example we take the equations

$$\left.\begin{aligned} (5-\lambda)y_1-y_2-2y_3 &= 0 \\ -y_1+(3-\lambda)y_2-2y_3 &= 0 \\ -2y_1-2y_2+(5-\lambda)y_3 &= 0 \end{aligned}\right\}, \tag{25}$$

and try to determine a solution for λ and y.

Starting rather arbitrarily with the guessed vector 1, 1, 1 we find the values 2, 0, 1 for the components of Ay, and the Rayleigh quotient gives $\lambda = 1$. Inserting this value in the equations we calculate the residuals, found to be 1, -1, 0, and try to liquidate them by suitable changes in y. At this stage the 'operations table' is as shown in Table 1.

TABLE 1

	R_1	R_2	R_3
$\delta y_1 = 1$	4	-1	-2
$\delta y_2 = 1$	-1	2	-2
$\delta y_3 = 1$	-2	-2	4

As in previous relaxation processes, we first decide the best operation for liquidating each residual. Clearly a change in y_r is the best single operation for the liquidation of R_r, but we might often do better by forming group operations. In this case there is no obvious satisfactory group-operator.

The first stage in the relaxation process is shown in Table 2.

TABLE 2

y_1	y_2	y_3	R_1	R_2	R_3
1·00	1·00	1·00	1·00	-1·00	0·00
	0·50		0·50	0·00	-1·00
		0·20	0·10	-0·40	-0·20
-0·05			-0·10	-0·35	-0·10
	0·05		-0·15	-0·25	-0·20
0·95	1·55	1·20			

The first row of Table 2 shows the initial guess and the initial residuals. We liquidate first R_2, since it needs the greatest increment in y, and the most appropriate single change is $\delta y_2 = 0{\cdot}50$, giving new residuals as shown. R_3 is now largest, and we reduce it by the change $\delta y_3 = 0{\cdot}20$. The residuals are now generally smaller than their initial values, and since we have actually increased the size of the vector our process has been effective.

We might now recalculate λ from the Rayleigh quotient and continue with the relaxation, but it is best first to make suitable changes, no longer designed to reduce residuals but to arrange them in a form such that a change in λ is everywhere beneficial. The vector is one-signed, so that we want the residuals to be one-signed also. The easiest way to effect this is to make a change in y_1, shown in the fourth row of the table. At this stage the signs are all negative, but we can do still better by making the final change, which makes fairly constant the ratio of remaining residuals to corresponding vector components.

We now take the new vector and recalculate λ, obtaining the value 0·8377. The diagonal terms in the operations table change to the values 4·1623, 2·1623, 4·1623, and the new initial residuals are given by 0·0042, 0·0016, —0·0052. The liquidation process is repeated as shown in Table 3.

TABLE 3

y_1	y_2	y_3	R_1	R_2	R_3
0·95	1·55	1·20	0·0042	0·0016	−0·0052
		0·0010	0·0022	−0·0004	−0·0010
−0·0005			0·0001	0·0001	0·0000
0·9495	1·5500	1·2010			

The process is now very rapid, and the final result of Table 3 is correct in its last figure. The previous λ, calculated at the end of Table 2, was even then correct to four decimals. The Rayleigh quotient given by the final result of Table 3 is in fact correct to eight decimal places, a demonstration of the power of Rayleigh's Principle.

Production of a second solution

11. The solution obtained is that belonging to the smallest eigenvalue. This is by no means inevitable, and if the relaxation process is performed carefully we shall usually converge to the solution nearest in 'shape' to the original guessed vector. To produce a second solution, therefore, we have only to start with a vector sufficiently unlike the first calculated eigenvector, and the method is then unchanged.

If for any reason there is difficulty in avoiding convergence to the solution already obtained, we can at any stage remove from our current vector any component of known eigenvectors by means of the orthogonality relation. Thus if y is a guessed vector, y_1 a known eigenvector, we can take as a new guess the vector

$$z = y - \alpha y_1, \tag{26}$$

and make z orthogonal to y_1 by choosing α so that

$$\alpha = y_1' y / y_1' y_1. \tag{27}$$

The choice of z as initial guess is then likely to cause convergence to the eigenvector of largest component in z, that of y_1 being now negligible.

For the example of § 10 we have already one eigenvector y_1. Starting with a second arbitrary vector 1, 0, 0, we find $\alpha = 0{\cdot}2000$ from equation (27) and a new first guess with components 0·8101, −0·3100, −0·2402. The 'Rayleigh' quotient is 5·976, a value roughly midway between the remaining two eigenvalues 5·0 and 7·2, so that in the relaxation process we shall almost certainly converge to one or other of these solutions.

12. There is little change in the relaxation method for the solution of the more general equation (6). The Rayleigh quotient is given by (17), and both Ay and By have to be calculated for use in (17) and for the determination of the residual vector

$$R = (A - B\lambda)y. \tag{28}$$

Every coefficient in the operations table now changes with λ, and the relaxation process is a little more difficult. In particular we now wish to arrange the sign and size of residuals so that each is proportional to the corresponding element of the vector By, a criterion again more difficult than that of the simpler case.

13. If the original equations are unsymmetric the roots may not all be real. Relaxation is a difficult process for the production of complex solutions, but any existing real solutions may be obtained by the techniques outlined for the symmetric case. Rayleigh's Principle no longer applies, but the Rayleigh quotient may still be used to give an estimate of the root corresponding to a trial vector.

For the determination of a second solution the simple orthogonality property no longer holds, but is replaced by the first pair of equations (13), and involves the determination of the corresponding vector of the transposed matrix A'. Though the latter is not very difficult since the eigenvalue is already known, it is probably quicker to produce a second solution by starting with a vector sufficiently dissimilar from the first, without assistance from the orthogonality property.

Other methods

14. Of other methods for the solution of the general matrix eigenvalue problem, defined by equation (6), the most obvious is the expansion of the determinant $|A-B\lambda|$, to form the polynomial equation (called the characteristic equation when B is I) whose roots are the eigenvalues. These can then be evaluated by standard methods, and the set of linear equations solved for each root. This method is rather laborious, not always conducive to great accuracy, and somewhat uneconomic if only a few roots are required.

Most of the satisfactory methods are of an iterative nature, and prefer to treat the simpler equation (5). This form can always be produced from (6) by premultiplication with the matrix B^{-1}. When both A and B are symmetric, however, it is usually better to split B into its triangles L, L' as explained in Chapter 3, expressing (6) in the form

$$\{L^{-1}A(L^{-1})'-I\lambda\}L'y = 0. \tag{29}$$

This process preserves symmetry, which the other method destroys, but needs further work to produce the required vector y from the calculated $L'y$.

Matrix iteration. Real roots

15. If we take an arbitrary vector y and premultiply it repeatedly with the matrix A, equation (18) shows that the vector obtained after s steps is given by

$$A^s y = \sum_1^n \alpha_r A^s y_r = \sum_1^n \alpha_r \lambda_r^s y_r = \lambda_1^s \sum_1^n \alpha_r \left(\frac{\lambda_r}{\lambda_1}\right)^s y_r. \tag{30}$$

If all the roots are real, and λ_1 that of largest modulus, the right-hand side of (30) reduces ultimately to the dominant term $\lambda_1^s \alpha_1 y_1$. The vector $A^s y$ therefore tends to a multiple of the latent vector y_1, and λ_1 is ultimately the ratio of corresponding elements of $A^{s+1}y$ and $A^s y$.

For the calculation of the root of smallest modulus and the corresponding vector, we can apply the same technique to the equation

$$(A^{-1}-\lambda^{-1}I)y = 0, \tag{31}$$

which involves the preliminary computation of A^{-1}. It is worth noting that in the case of equation (6) we can produce the form (31) by decomposing A into its triangles L, L', and expressing the equation in the form

$$\{L^{-1}B(L')^{-1}-\lambda^{-1}I\}L'y = 0, \tag{32}$$

Matrix iteration. Complex roots

16. If the root of largest modulus is complex, its conjugate is also a root with the same modulus, the corresponding vectors form a conjugate complex pair, and the corresponding terms on the right of (18) are $\alpha_1 y_1 + \bar{\alpha}_1 \bar{y}_1$. In equation (30) there are now two dominant terms and we can write

$$A^s y \sim \alpha_1 \lambda_1^s y_1 + \bar{\alpha}_1 \bar{\lambda}_1^s \bar{y}_1. \tag{33}$$

It is easy to see that this quantity oscillates with varying s, indicating the presence of dominant complex roots.

If we now write $z_s = A^s y$, and consider corresponding elements of z_s, z_{s+1}, and z_{s+2}, it is seen from (33) that for sufficiently large s they satisfy the equation

$$z_{s+2} + uz_{s+1} + vz_s = 0, \tag{34}$$

where $\lambda \pm i\mu = \lambda_1, \bar{\lambda}_1$ are roots of the quadratic equation

$$x^2 + ux + v = 0. \tag{35}$$

The quantities u and v are determinable from any two equations of the set (34), and hence λ_1 and $\bar{\lambda}_1$ can be calculated.

For the latent vectors, we write $\alpha_1 \lambda_1^s y_1 = \xi + i\eta$, a multiple of the latent vector y_1. Then

$$\left.\begin{aligned} z_s &\sim 2\xi \\ z_{s+1} &\sim 2(\lambda\xi - \mu\eta) \end{aligned}\right\}, \tag{36}$$

from which ξ and η, and hence the required y_1, can be obtained when λ and μ have been determined.

Acceleration devices

17. These iterative processes may converge only slowly when there are two or more roots nearly equal in magnitude. There are several known methods for increasing the rate of convergence.

If in equation (30) we use the matrix $(A - pI)$ instead of A, that equation becomes

$$(A - pI)^s y = \sum_1^n \alpha_r (\lambda_r - p)^s y_r, \tag{37}$$

and the right-hand side reduces to a multiple of the vector for which $(\lambda_r - p)$ has largest modulus. If the roots $\lambda_1, \lambda_2, \ldots, \lambda_n$ are real and arranged in order of decreasing algebraic value, then for convergence to λ_1 the best choice of p would be that for which

$$(\lambda_2 - p) = -(\lambda_n - p), \tag{38}$$

and for convergence to λ_n the best choice of p would be that for which

$$(\lambda_1 - p) = -(\lambda_{n-1} - p), \tag{39}$$

no root other than the largest or smallest being calculable directly by this iterative process. In particular, this method can separate roots of equal modulus but opposite sign, when these are the largest or smallest.

In the case of complex roots we may iterate with the matrix $(A-ipI)$, which separates the conjugate complex pair and gives the root and vector for which $(\lambda \pm i\mu - ip)$ has greatest modulus. Successive iterates converge to the corresponding complex vector.

For any root, in the real case, we can iterate with the matrix $(A-pI)^{-1}$, giving

$$(A-pI)^{-s}y = \sum_1^n \alpha_r(\lambda_r - p)^{-s} y_r, \tag{40}$$

and the right-hand side becomes a multiple of the vector corresponding to the smallest arithmetic value of $(\lambda_r - p)$.

With all these devices some trial and error may be necessary to determine a good value for p. In the case of (40), the selected value of p must be an approximation to the required root λ_r.

18. Another acceleration device is that of Aitken, first used in this context. If convergence is slow, it may often be improved by considering three successive iterated vectors $A^s y$, $A^{s+p} y$, $A^{s+2p} y$, with p not necessarily equal to unity. If these vectors are normalized so that one corresponding element is unity, and any other component is denoted by z, then, with an obvious notation, the quantity

$$z = \frac{z_s z_{s+2p} - z_{s+p}^2}{z_s + z_{s+2p} - 2z_{s+p}} = z_s - \frac{(z_{s+p} - z_s)^2}{z_s + z_{s+2p} - 2z_{s+p}}, \tag{41}$$

is a better approximation to that vector component. A similar formula can be used to improve successive approximations to the corresponding eigenvalue. The method is applicable in the case of real roots, and in some cases of complex roots.

Examples

19. To illustrate these techniques we consider first the example of equation (25), whose roots are all real. The matrix involved is given by

$$A = \begin{bmatrix} 5 & -1 & -2 \\ -1 & 3 & -2 \\ -2 & -2 & 5 \end{bmatrix}. \tag{42}$$

First we start with the arbitrary vector 1, 1, 1, and calculate successive vectors in the sequence $A^s y$, as shown in Table 4. In each case the vector is multiplied by a constant so that its first component is unity. This constant is the reciprocal of the corresponding current estimate of λ, obtained as the ratio of successive unnormalized first components, and shown in the last row of Table 4.

TABLE 4

s	0	1	2	3	4	5
$A^s y$	1 1 1	1 0 0·5	1 −0·5 0·125	1 −0·5238 −0·0714	1 −0·4286 −0·2311	1 −0·3096 −0·3902
λ		2	4	5·250	5·6666	5·8908

s	6	7	8	9	10
$A^s y$	1 −0·1886 −0·5471	1 −0·0751 −0·6937	1 0·0251 −0·8229	1 0·1089 −0·9311	1 0·1760 −1·0178
λ	6·0900	6·2828	6·4625	6·6207	6·7533

Convergence is rather slow, partly because the starting vector is a poor approximation to the eigenvector corresponding to the largest root, and partly because the latter is not much larger than its neighbour. If we apply Aitken's process to the last three steps, we produce the vector 1·0000, 0·4456, −1·3674, compared with the correct result 1·0000, 0·3675, −1·2649: the improvement is quite striking. The same process applied to the last three estimates of λ gives the result 7·4401, compared with the true result 7·1623. The Rayleigh quotient obtained from our improved vector is 7·1570, again demonstrating the power of Rayleigh's Principle.

20. We now try to find the smallest root by the method associated with equation (37). The roots are known to be all positive, and the two largest are fairly equal, so that the choice $p = 7$ should ensure fairly rapid convergence to the smallest root. We then take

$$A-7I = \begin{bmatrix} -2 & -1 & -2 \\ -1 & -4 & -2 \\ -2 & -2 & -2 \end{bmatrix}, \tag{43}$$

start with the same vector as before, and calculate the successive vectors $(A-7I)^s y$, normalized so that the second and largest component is unity, and shown in Table 5.

TABLE 5

s	0	1	2	3	4	5	6
$(A-7I)^s y$	1 1 1	0·7143 1·0000 0·8571	0·6444 1·0000 0·8000	0·6228 1·0000 0·7829	0·6159 1·0000 0·7775	0·6137 1·0000 0·7757	0·6129 1·0000 0·7751
$\lambda-7$		−7	−6·4285	−6·2444	−6·1886	−6·1709	−6·1651

At the last stage of Table 5, the eigenvector is correct to within 3 units in the last figure, but $\lambda = 0{\cdot}8349$ compares less accurately with the correct result 0·8377223. Again Rayleigh's quotient applied to our best vector gives the much better result 0·8377225.

21. We might otherwise find the smallest root by the method indicated by equation (31). This can be done either by evaluating A^{-1} and calculating directly $y_1 = A^{-1}y_0$, $y_2 = A^{-1}y_1$, etc., or by solving the simultaneous equations $Ay_1 = y_0$, $Ay_2 = y_1$, etc. The former method is usually quicker, but the latter may have advantage when A has special form, for instance that associated with the finite-difference equations representing a second-order differential equation. In our example the inverse A^{-1} is given by

$$A^{-1} = \tfrac{1}{30}\begin{bmatrix} 11 & 9 & 8 \\ 9 & 21 & 12 \\ 8 & 12 & 14 \end{bmatrix}, \tag{44}$$

and if we apply this successively to the vector 1, 1, 1, we find the successive vectors, normalized so that their second component is unity, shown in Table 6.

TABLE 6

s	0	1	2	3	4	5
$A^{-s}y$	1 1 1	0·6667 1·0000 0·8095	0·6213 1·0000 0·7808	0·6140 1·0000 0·7759	0·6128 1·0000 0·7750	0·6126 1·0000 0·7749
λ^{-1}		1·4000	1·2238	1·1987	1·1946	1·1938

At the last stage of Table 6 the vector is correct to within one unit in the fourth decimal; the corresponding value of λ is 0·8377, and is correct in all four decimals.

22. As a final illustration of this type of process for real roots we demonstrate the application of equation (40) to find the middle root. The sum of the roots is the sum of the diagonal terms of the matrix, and we have obtained approximations to the largest and smallest roots. The remaining root has therefore the approximate value $13-7-1 = 5$. In fact the middle latent root has the exact value 5, and the matrix $(A-5I)$ is singular. We will therefore, for the purpose of illustration, take for p a number close to 5 but sufficiently different to permit convenient and reasonably accurate computation of the inverse. Taking $p = 4{\cdot}9$ we find the matrix

$$(A-4{\cdot}9I)^{-1} = \begin{bmatrix} 4{\cdot}5593 & -4{\cdot}4614 & 1{\cdot}9587 \\ -4{\cdot}4614 & 4{\cdot}3417 & -2{\cdot}3939 \\ 1{\cdot}9587 & -2{\cdot}3939 & 1{\cdot}2949 \end{bmatrix}. \tag{45}$$

Application of this matrix to the initial vector 1, 1, 1 produces the successive normalized vectors as shown in Table 7.

TABLE 7

s	0	1	2	3
$(A-4{\cdot}9I)^{-s}y$	1 1 1	−0·8182 1·0000 −0·3420	−1·0058 1·0000 −0·5039	−0·9999 1·0000 −0·4999
$(\lambda-4{\cdot}9)^{-1}$		−2·5136	8·8107	10·0353

Convergence to the correct vector −1, 1, −0·5 is here extremely rapid, notwithstanding the very poor starting approximation. The latent root, given at the last stage by $\lambda = 4{\cdot}9+(10{\cdot}0353)^{-1}$, is very nearly 5, the correct result.

23. To illustrate the procedure for complex roots we consider the equations

$$\left.\begin{aligned}(3-\lambda)y_1+y_2+y_3 &= 0\\ 2y_1+(1-\lambda)y_2+y_3 &= 0\\ -y_1-y_2+(3-\lambda)y_3 &= 0\end{aligned}\right\}. \tag{46}$$

The results of successively premultiplying the arbitrary vector 1, 1, 1 by the matrix associated with (46) are shown in Table 8.

TABLE 8

s	0	1	2	3	4	5	6	7	8	9
$A^s y$	1 1 1	5 4 1	20 15 −6	69 49 −53	203 134 −277	466 263 −1168	493 27 −4233	− 2727 − 3220 −13219	−24620 −21893 −33710	−129463 −104843 − 54617

The ratios of successive corresponding elements of $A^s y$ are not becoming constant but are oscillating, indicating a pair of complex roots. From equation (34), with s taking its largest value 7, we see that the required latent roots are roots of the quadratic equation $x^2+ux+v = 0$, where u, v may be found from any two of the equations

$$\left.\begin{aligned}-129463-24620u-2727v &= 0\\ -104843-21893u-3220v &= 0\\ -54617-33710u-13219v &= 0\end{aligned}\right\}. \tag{47}$$

The three possible pairs give the respective results

$$\left.\begin{aligned}u &= -6{\cdot}6906475,\ -6{\cdot}6906475,\ -6{\cdot}6906474\\ v &= 12{\cdot}9302317,\ 12{\cdot}9302312,\ 12{\cdot}9302311\end{aligned}\right\}, \tag{48}$$

the agreement indicating high accuracy and, after only nine steps, a quick rate of convergence. We find $\lambda_1, \bar{\lambda}_1 = +3{\cdot}34532\pm1{\cdot}31873i$.

To produce the corresponding vectors we use equations (36), in which we can take z_s to be any convenient multiple of $A^s y$. For $z_s = 10^{-4}A^s y$ the first of (36), with $s = 8$ in Table 8, gives

$$\xi = -1{\cdot}2310,\ -1{\cdot}0946,\ -1{\cdot}6855,$$

and the second gives

$$\eta = 1{\cdot}7859,\ 1{\cdot}1984,\ -2{\cdot}2049.$$

The vectors $\xi \pm i\eta$ are unnormalized forms of the required y_1, $\bar{y}_1$.

24. An alternative method, which separates the complex pair and gives one root and corresponding vector by direct iteration, is to use $A-ipI$ in place of A. A suitable value for p is not generally known, and we may have to use trial and error. With $p = 1$, for example, we find the successive complex vectors shown to three decimals in Table 9. The initial vector is again 1, 1, 1 and for ease of estimating the rate of convergence and the eigenvalue we again normalize each successive vector so that its first component is $1+0i$.

To this precision the correct result is $\lambda-i = 3{\cdot}345-2{\cdot}319i$, $y = 1$, $0{\cdot}741-0{\cdot}102i$, $-0{\cdot}396-1{\cdot}217i$, so that convergence is rather slow. Convergence here depends on the modulus of the ratio $(\lambda_1-i)/(\bar{\lambda}_1-i)$, about 0·83, so that the degree of separation of the roots achieved with this value of p is not high. We can, however, apply Aitken's process in this case also, the last three approximations yielding the better vector $y = 1$, $0{\cdot}730-0{\cdot}078i$, $-0{\cdot}544-0{\cdot}935i$, and the iterative method can be restarted with this as initial vector. Aitken's process applied to the root gives $\lambda-i = 3{\cdot}667-1{\cdot}937i$, some improvement over the last value of Table 9.

TABLE 9

s	0	1	2	3
$(A-iI)^s y$	1 1 1	1·000 0·808−0·038i 0·231−0·154i	1·000 0·751−0·026i −0·245−0·234i	1·000 0·718−0·030i −0·648−0·356i
$\lambda-i$		5−i	4·039−1·192i	3·506−1·260i

s	4	5	6
$(A-iI)^s y$	1·000 0·687−0·050i −1·040−0·596i	1·000 0·661−0·093i −1·354−1·105i	1·000 0·678−0·160i −1·147−1·903i
$\lambda-i$	3·070−1·386i	2·647−1·646i	2·307−2·198i

Determination of a second solution

25. Only one of the iterative methods, that based on equation (40), enables any required root and vector to be found directly. With every other process we converged to the largest or smallest root, no intermediate solution being calculable directly with or without acceleration devices. If any of these methods is used to obtain an intermediate solution we must therefore take suitable measures to prevent regression to a root already obtained.

Suppose that the largest root λ_1 of the matrix A has already been calculated, and we wish to obtain the next largest. If the matrix is symmetric and the roots therefore real, we can, as suggested for the relaxation method, use the orthogonality relations to eliminate from a new guessed vector any component belonging to λ_1, so that the vector has the form

$$y = \sum_{2}^{n} \alpha_r y_r. \tag{49}$$

In theory we can now iterate as before, converging to the root λ_2 and vector y_2. In practice, however, a rounding error at any stage will re-introduce a small component of y_1. If $\lambda_1 \gg \lambda_2$ this will grow substantially, and, unless this component is again removed at some stage, we shall again converge to λ_1. For the relaxation method this danger was more remote, since our relaxation process is rather more tentative, gentle, and flexible.

For unsymmetric matrices the simple orthogonality property does not hold, but if the eigensolution of the transposed matrix is known or calculated, we can of course use the relevant orthogonality relations given in § 3.

In either case we can eliminate from a guessed vector any component of y_1 by first premultiplying by the matrix $(A-\lambda_1 I)$. We can then proceed to iterate with A, applying $(A-\lambda_1 I)$ from time to time to prevent the reintroduction of λ_1.

26. Better methods involve the modification of the matrix so that the basic iterative method will converge to λ_2 with no additional assistance. For symmetric matrices, we form

$$A_2 = A-\lambda_1 y_1 y_1'/y_1' y_1, \tag{50}$$

a symmetric matrix whose roots and vectors are the same as those of A, except that the vector y_1 now corresponds to a root $\lambda = 0$ of A_2. Iteration with A_2 therefore converges to λ_2. For unsymmetric matrices we can use a similar method if the vector z_1, corresponding to the root λ_1, has been calculated for the transposed matrix A'. Equation (50) is then replaced by

$$A_2 = A-\lambda_1 y_1 z_1'/z_1' y_1, \tag{51}$$

and convergence to λ_2 is assured.

27. In the methods of § 26 the order of the matrix is unaltered. Another class of methods reduces the order of the matrix by unity at each 'root-removal' operation. Suppose first that, in the unsymmetric case, the latent root λ_1 and the corresponding vectors of both A and A' have been calculated, and that the vectors have been normalized so that their first component is unity. If the matrix is partitioned by lines separating the first row and column from the rest, and the vectors likewise partitioned, we can write

$$\begin{bmatrix} a_{11} & R_1' \\ C_1 & B \end{bmatrix}\begin{bmatrix} 1 \\ \eta \end{bmatrix} = \lambda\begin{bmatrix} 1 \\ \eta \end{bmatrix}, \qquad \begin{bmatrix} a_{11} & C_1' \\ R_1 & B' \end{bmatrix}\begin{bmatrix} 1 \\ \zeta \end{bmatrix} = \lambda\begin{bmatrix} 1 \\ \zeta \end{bmatrix}, \tag{52}$$

where R_1', C_1 are respectively the first row and column of the original matrix with the first element a_{11} excluded, B is the remaining square matrix of order $(n-1)$, and the vectors y and z of A and A' are written in the respective partitioned forms $\begin{bmatrix} 1 \\ \eta \end{bmatrix}$, $\begin{bmatrix} 1 \\ \zeta \end{bmatrix}$. In particular the quantities λ_1, η_1, and ζ_1 have already been computed. Then among the four equations represented by (52) we have, in particular,

$$C_1 + B\eta = \lambda\eta. \tag{53}$$

The orthogonality relation given in (11) can here be written as

$$\zeta_1'\eta = -1, \tag{54}$$

for any η other than η_1. We can therefore write (53) in the form

$$(B - C_1\zeta_1')\eta = \lambda\eta, \tag{55}$$

showing that the latent roots of the matrix $(B - C_1\zeta_1')$ are $\lambda_2, \lambda_3, \ldots, \lambda_n$, and the latent vectors, of order $(n-1)$, are proportional to the last $(n-1)$ components of those of the original matrix. The constant of proportionality comes from another equation of the set (52), given by

$$a_{11} + R_1'\eta = \lambda. \tag{56}$$

The roots of the transposed of $(B - C_1\zeta_1')$ are also $\lambda_2, \lambda_3, \ldots, \lambda_n$, but the vectors are not $\zeta_2, \zeta_3, \ldots, \zeta_n$, though closely related to those vectors. This relation becomes obvious when we consider a second method, in which the transposed matrix is not used.

28. Partitioning as before we have, from the first of (52), the two equations

$$\left.\begin{aligned} a_{11} + R_1'\eta &= \lambda \\ C_1 + B\eta &= \lambda\eta \end{aligned}\right\}, \tag{57}$$

which are satisfied also by η_1 and λ_1. Eliminating a_{11} and λ_1 from the four equations so obtained we find

$$(B - \eta_1 R_1')(\eta - \eta_1) = \lambda(\eta - \eta_1). \tag{58}$$

This shows that the latent roots of $(B-\eta_1 R_1')$ are $\lambda_2, \lambda_3,..., \lambda_n$, and their vectors ξ, say, are given by $k(\eta-\eta_1)$ where k is a constant. The corresponding vectors of the original matrix are thus given by $\begin{bmatrix} 1 \\ \cdots \\ \eta \end{bmatrix}$, where

$$\eta = \eta_1 + \frac{1}{k}\xi, \tag{59}$$

and, from the first of (57), we find

$$\frac{1}{k} = \frac{\lambda - a_{11} - R_1'\eta_1}{R_1'\xi} = \frac{\lambda-\lambda_1}{R_1'\xi}. \tag{60}$$

In the method of § 27 the relation between the vectors of the reduced matrix and the transposed of the original matrix is the same as that between the reduced and original matrices in this alternative process.

29. If in the last method the root λ_1 is complex, it has a complex conjugate of the same modulus, and the corresponding vectors are $\begin{bmatrix} 1 \\ \cdots \\ p \pm iq \end{bmatrix}$, say. Then one solution of (58) is known immediately, for one root is the complex conjugate of λ_1, and the corresponding vector $\eta-\eta_1$ is $2iq$. We can therefore remove this solution directly, thus reducing the order of the original matrix by two.

This description of iterative methods and accelerating devices does not exhaust the list of known methods for this algebraic problem, but a knowledge of them suffices for all practical problems. We proceed now to consider the differential problem, its connexion with the algebraic problem, and methods for its solution.

Mathematical connexions

†30. There are several analogies with the results derived for matrices and those obtained in the mathematical theory of differential equations. For example, a second-order differential equation of the form

$$fy''+gy'+ky = 0, \tag{61}$$

in which f, g, and k are functions of the independent variable x, has an associated 'adjoint' equation given by

$$(fz)''-(gz)'+kz = 0. \tag{62}$$

The general simple and approximate finite-difference form of (61) is given by

$$y_{r-1}(f_r-\tfrac{1}{2}hg_r)-y_r(2f_r-h^2k_r)+y_{r+1}(f_r+\tfrac{1}{2}hg_r) = 0, \tag{63}$$

that of (62) by

$$z_{r-1}(f_{r-1}+\tfrac{1}{2}hg_{r-1})-z_r(2f_r-h^2k_r)+z_{r+1}(f_{r+1}-\tfrac{1}{2}hg_{r+1}) = 0, \tag{64}$$

and the first and last equations of types (63) and (64) are affected by the boundary conditions. The relation between the boundary conditions of the system and its adjoint is in general complicated, but if we take the simplest case, in which both boundary values vanish, the adjoint boundary conditions are of identical form, so that the first of (63) and (64), with $r = 1$, have the respective forms

$$\left.\begin{aligned} -y_1(2f_1-h^2k_1)+y_2(f_1+\tfrac{1}{2}hg_1) &= 0 \\ -z_1(2f_1-h^2k_1)+z_2(f_2-\tfrac{1}{2}hg_2) &= 0 \end{aligned}\right\}, \tag{65}$$

and the corresponding final equations of the sets (63) and (64) are given by

$$\left.\begin{aligned} y_{n-2}(f_{n-1}-\tfrac{1}{2}hg_{n-1})-y_{n-1}(2f_{n-1}-h^2k_{n-1}) &= 0 \\ z_{n-2}(f_{n-2}+\tfrac{1}{2}hg_{n-2})-z_{n-1}(2f_{n-1}-h^2k_{n-1}) &= 0 \end{aligned}\right\}. \tag{66}$$

We then see that the complete matrix corresponding to the equations in z is the transposed of that corresponding to the equations in y, so that a differential system and its adjoint are analogous to a matrix and its transposed.

The system is self-adjoint if the adjoint system is identical with the original. The boundary conditions are already identical, and the differential equations are the same if

$$g = f'. \tag{67}$$

The corresponding relation obtained from the matrices is that

$$f_r+\tfrac{1}{2}hg_r = f_{r+1}-\tfrac{1}{2}hg_{r+1}. \tag{68}$$

Expansion of f_{r+1} and g_{r+1}, by means of the Taylor's series with origin at x_r, shows that, with neglect of terms of order h^2, (68) is equivalent to the mathematical relation (67), so that a self-adjoint differential system is analogous to a symmetric matrix.

For more complicated boundary conditions, the first analogy also depends on the neglect of some power of h.

31. If $g = f'$ in (61), we can write that equation in the form

$$\frac{d}{dx}\left(f\frac{dy}{dx}\right)+ky = 0, \tag{69}$$

and, when $g \neq f'$, we can produce from (61) the equivalent equation

$$\frac{d}{dx}\left\{\left(\exp\int\frac{g}{f}\,dx\right)\frac{dy}{dx}\right\}+\frac{k}{f}\left(\exp\int\frac{g}{f}\,dx\right)y = 0, \tag{70}$$

so that the form (69) is no less general than (61). The second-order system then fails to be self-adjoint only for the most general boundary conditions. If the boundary conditions, at the end points a and b, are of the fairly general form

$$\alpha_1 y(a)+\beta_1 y'(a) = 0, \qquad \alpha_2 y(b)+\beta_2 y'(b) = 0, \tag{71}$$

so that each end point is involved in only one condition, it is true that the second-order system is self-adjoint.

32. There are also mathematical analogies with the orthogonal properties of eigenvectors. For example, a system comprising the equation

$$fy''+gy'+(k+\lambda l)y = 0, \tag{72}$$

and associated boundary conditions in which λ does not appear, has solutions y_r for an infinite number of discrete values λ_r. The adjoint system has solutions z_r for the same values of λ_r. It can then be proved that, in the case of unequal roots,

$$\int\limits_a^b ly_r z_s\,dx = \int\limits_a^b ly_s z_r\,dx = 0 \quad (r \neq s), \tag{73}$$

which is analogous to the orthogonality property of equation (11) for unsymmetric matrices. If the system is self-adjoint, $y = z$ and (73) becomes

$$\int\limits_a^b ly_r y_s\,dx = 0 \quad (r \neq s), \tag{74}$$

corresponding to the orthogonality property of equation (14) for symmetric matrice. In this case the vectors may also be normalized according to the equation

$$\int\limits_a^b ly_r^2\,dx = 1, \tag{75}$$

corresponding to the normalization condition in (14).

The result (74) is easily proved for the self-adjoint system given by

$$\frac{d}{dx}\left(f\frac{dy}{dx}\right)+(k+\lambda l)y = 0, \tag{76}$$

with boundary conditions of type (71). For we have

$$\frac{d}{dx}\{f(y_r y_s'-y_s y_r')\} = (\lambda_r-\lambda_s)ly_r y_s, \tag{77}$$

and the quantity in braces vanishes, by virtue of the boundary conditions, at the two end points.

In the self-adjoint case, also, it is easy to prove that the roots are all real if l is everywhere positive, corresponding to the result of § 5 for a symmetric matrix A and positive-definite symmetric matrix B.

†33. Finally, we have a Rayleigh's Principle analogous to that for matrices. For equation (76) the Rayleigh quotient is given by

$$\lambda = -\int_a^b y\left\{\frac{d}{dx}\left(f\frac{dy}{dx}\right)+ky\right\}dx \Big/ \int_a^b ly^2\,dx, \tag{78}$$

where y is a function which satisfies the boundary conditions, but is otherwise arbitrary. Such a function can be expanded as a linear combination of the eigenfunctions y_r in the form

$$y = \sum_1^\infty \alpha_r y_r. \tag{79}$$

Substituting in (78), and using the orthogonality and normalization relations (74) and (75), we find

$$\lambda = \sum_1^\infty \alpha_r^2 \lambda_r \Big/ \sum_1^\infty \alpha_r^2. \tag{80}$$

As in § 6, we see that if y is an approximation to an eigenfunction y_k, so that α_k is much greater than other α_r, then first-order changes in y give rise to second-order changes in λ, and the eigenvalue is determined more accurately than the eigenfunction.

If all the λ_r are positive, which is the case in many physical problems, and corresponds to some structure of the differential system corresponding to a positive-definite matrix A in § 5, we can also assert that the Rayleigh quotient (78) overestimates the smallest eigenvalue. It certainly underestimates the largest, since this eigenvalue is generally infinite.

Interpretation of finite-difference solutions

34. When a differential system of eigenvalue type is replaced by a set of finite-difference equations, as in the case of equations (4) which replace equation (1), we are led to consider the solution of the matrix equation of type (5) or (6). The order of the matrices involved is n, the number of chosen pivotal points, so that, whereas the differential problem has an infinity of solutions, the algebraic problem gives only n solutions. A natural question is: to which of the infinity of true solutions do these n approximate solutions correspond, and what has happened to all the others?

The answers to these questions are not obvious, and not always known or easily ascertainable. It is, however, reasonable to suppose that we get approximations to those solutions of the differential problem for which the finite-difference equations provide a fair approximation at the selected interval.

Consider, in this connexion, the solution of the finite-difference equations (4), with $y_0 = y_n = 0$, which approximate to the differential system (1). We shall calculate the values of p^2 obtained at different intervals. With $h = \frac{1}{2}$ there is one internal point, and just one eigenvalue $h^2p^2 = 2$, or $p^2 = 8$. This is an approximation to the smallest eigenvalue, $p^2 = \pi^2$, of the differential system. Taking two internal points, with $h = \frac{1}{3}$, we find the two eigenvalues $p_1^2 = 9$, $p_2^2 = 27$, of which the first is a better approximation to the value π^2, the second a poor approximation to $4\pi^2$, the second eigenvalue of the differential system. With $h = \frac{1}{4}$ we find the three values $p_1^2 = 9{\cdot}4$, $p_2^2 = 32$, $p_3^2 = 54{\cdot}6$. The first is a still closer approximation to π^2, the second rather nearer than its predecessor to $4\pi^2$, and the third a poor approximation to $9\pi^2$, the third eigenvalue of the differential system.

At any stage it appears that the introduction of an extra pivotal value provides a new solution, corresponding to the next largest eigenvalue, and improves previous approximations to all the smaller eigenvalues. This is in agreement with our conjecture concerning the adequacy of finite-difference approximations. In general, successively larger eigenvalues of the differential system are associated with eigenfunctions containing an increasing number of zeros in the total range, so that an increasing number of pivotal points is necessary for adequate representation of the 'larger' eigenfunctions.

Algebraic treatment of differential systems

†35. In solving differential systems our main approach is to use finite-difference equations, transforming the differential problem into an algebraic problem. As in previous chapters, however, it is desirable, for accuracy and economy of labour, to keep as small as possible the order of the matrices involved in the algebraic problem. This involves a large interval and a non-negligible difference correction in the finite-difference equations. The introduction of the difference correction, to improve approximate solutions obtained with its neglect, constitutes a new problem.

There are other considerations. First, the matrices stemming from a differential system are of a specially simple form so that, as in the corresponding treatment of the two-point boundary problems of previous chapters, both relaxation and more direct methods are relatively easy to use. Second, the interesting eigenvalues of differential systems are usually the smallest, or those of smallest modulus, so that the simplest iterative method based on (30) will not give directly a useful solution, and equations like (31) or (40), involving the initial computation of an inverse matrix, or of (37), involving a trial-and-error process to determine a suitable p, are more appropriate.

For the rest of this chapter we illustrate the application of relaxation and other methods, including the treatment of the difference correction, to obtain the required solutions of differential systems.

Relaxation. Simple examples

36. We first illustrate the relaxation method applied to the simple problem defined by the system

$$y''+\lambda y = 0, \qquad y(0) = y(1) = 0, \tag{81}$$

for which the simplest finite-difference equations have the form

$$y_{r+1}-(2-h^2\lambda)y_r+y_{r-1}+Cy_r = 0, \tag{82}$$

with $y_0 = y(0) = 0$, $y_n = y(1) = 0$, and the difference-correction operator is given by

$$C = -\tfrac{1}{12}\delta^4+\tfrac{1}{90}\delta^6-\dots. \tag{83}$$

The treatment of the difference correction is similar to that of previous chapters. First neglecting Cy we find a solution for both λ and y. The difference corrections are then calculated and used to effect an improvement. Details will become clear as the example proceeds.

With the interval $h = \frac{1}{4}$, with Cy neglected, and with the boundary values inserted, we have the following three simultaneous equations, in which $\nu = h^2\lambda = \frac{1}{16}\lambda$:

$$\left.\begin{aligned} -(2-\nu)y(\tfrac{1}{4})+y(\tfrac{1}{2}) &= 0\\ y(\tfrac{1}{4})-(2-\nu)y(\tfrac{1}{2})+y(\tfrac{3}{4}) &= 0\\ y(\tfrac{1}{2})-(2-\nu)y(\tfrac{3}{4}) &= 0 \end{aligned}\right\}. \tag{84}$$

Using the relaxation method, we first select a trial vector. Suspecting that there is an eigensolution with no internal zero, we take $y(\frac{1}{4}) = y(\frac{3}{4}) = \frac{1}{2}$, $y(\frac{1}{2}) = 1$ for our first guess. Since the matrix is symmetric we can certainly approximate to a corresponding ν by using the Rayleigh quotient, which gives $\nu = \frac{2}{3}$. The initial residuals are then calculated, their values being $+\frac{1}{3}$, $-\frac{1}{3}$, $+\frac{1}{3}$, and with ν inserted in equations (84) we can proceed to attempt their liquidation.

The solution with no internal zero is symmetrical about $x = \frac{1}{2}$, as is obvious from equations (84), so that we need consider only half the range, and the operations table is symbolized by Figs. 24 and 25.

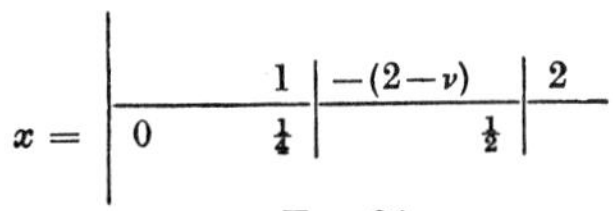

FIG. 24

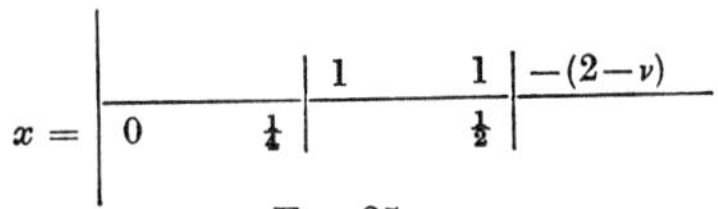

FIG. 25

The relaxation process is merely a suitable rearrangement of the two residuals. We clearly want to make the residuals one-signed, and we do this either by a positive change in y at $x = \frac{1}{4}$, or a negative change at $x = \frac{1}{2}$. So that the vector shall not be reduced in size we choose the former, and make such a change that the ratio of residual to vector is fairly constant. Since there are only two points this can be done precisely, the change δy at $x = \frac{1}{4}$ being a root of the equation

$$\frac{\frac{1}{2}+\delta y}{\frac{1}{3}-\frac{4}{3}\delta y} = \frac{1}{2\delta y-\frac{1}{3}}, \tag{85}$$

so that $\delta y = -\frac{1}{2}(1\pm\sqrt{2})$. We obviously take the negative sign so that the next approximation is given by $y(\frac{1}{4}) = y(\frac{3}{4}) = 1/\sqrt{2}$, $y(\frac{1}{2}) = 1$, and the corresponding value of ν obtained from Rayleigh's quotient is $2-\sqrt{2}$. The new residuals are of course zero, so that our first approximate solution has been found.

37. We now try to correct this solution by applying the difference correction. To compute the latter we can as in previous problems extend our approximate solution to external points by the standard extrapolation process. The difference table, using four decimals in y and ν, is shown in Table 10, in which only half the range need be considered. It is found that there is skew symmetry about $x = 0$ so that we can fill in as many differences as required in a compact manner.

TABLE 10

x	y		δ^2		δ^4		δ^6		δ^8		δ^{10}		δ^{12}
+0	+0·0000		0		0		0		0		0		0
		+7071		−4142		+2426		−1420		+828		−472	
$+\frac{1}{4}$	+0·7071		−4142		+2426		−1420		+ 828		−472		+232
		+2929		−1716		+1006		− 592		+356		−240	
$+\frac{1}{2}$	+1·0000		−5858		+3432		−2012		+1184		−712		+480

This is an interesting table. The differences remain large for some time, but they all have a smooth flow and there is clearly no mistake in the function y, the largest binomial coefficient of order twelve having the value 924. We can therefore calculate the difference corrections with confidence. Retaining differences up to the tenth we find

$$Cy(\tfrac{1}{4}) = -0{\cdot}02196, \qquad Cy(\tfrac{1}{2}) = -0{\cdot}03107.$$

We now calculate a new value of λ from the modified Rayleigh quotient

$$h^2\lambda = -\frac{y'(Ay+Cy)}{y'y}, \tag{86}$$

where Cy is the vector of difference corrections and y' is the transposed of the column vector y. The change $\delta\nu$ is clearly given by $\delta\nu = -y'Cy/y'y$, which is found to have the value 0·031063. This gives $\lambda = 9{\cdot}8696$, agreeing with π^2 in all the figures quoted. The new residuals are negligible, so that there is no correction to be made to the vector y.

†38. We note some interesting points, computational and otherwise. The fact that in this case the first approximation to y needs no change can be deduced by examining the differential equation. Our first approximation is a solution of the difference equation

$$\delta^2 y = -\nu y, \tag{87}$$

and we wish to solve

$$D^2 y = -\lambda y. \tag{88}$$

By using the expansion for D^2 in terms of δ^2, given by

$$h^2D^2 = \delta^2 - \tfrac{1}{12}\delta^4 + \tfrac{1}{90}\delta^6 - \tfrac{1}{560}\delta^8 + \tfrac{1}{3150}\delta^{10} - \ldots, \tag{89}$$

we see that (87) and (88) are identical provided that

$$h^2\lambda = \nu + \tfrac{1}{12}\nu^2 + \tfrac{1}{90}\nu^3 + \tfrac{1}{560}\nu^4 + \tfrac{1}{3150}\nu^5 + \ldots, \tag{90}$$

since

$$\delta^{2n} y = -\nu\delta^{2n-2} y = \nu^2\delta^{2n-4} y \ldots = (-)^n \nu^n y. \tag{91}$$

Substitution in (90), of the value of ν obtained in § 36 and the known corresponding eigenvalue λ of the differential system, provides a peculiar expansion for π^2. We found, at the interval $h = \frac{1}{4}$, that ν is exactly $2-\sqrt{2}$, so that we have

$$\tfrac{1}{16}\pi^2 = (2-\sqrt{2}) + \tfrac{1}{12}(2-\sqrt{2})^2 + \tfrac{1}{90}(2-\sqrt{2})^3 + \ldots. \tag{92}$$

If we try to find a solution for $h = \frac{1}{2}$, with only one internal point, we find $\nu = 2$, and the expansion corresponding to (92) is

$$\tfrac{1}{4}\pi^2 = 2 + \tfrac{1}{12}(2^2) + \tfrac{1}{90}(2^3) + \ldots. \tag{93}$$

This series converges, and we can therefore solve the differential equation at an interval corresponding to 90° in $\sin x$.

Not all differential equations will have the property that the solution y, obtained with neglect of the difference corrections, also satisfies the differential equation, but there is a tendency for the eigenvector to have a more accurate first approximation than the eigenvalue.

A second interesting point is that equation (86) is not the true Rayleigh quotient for the differential equation, which is given by

$$\lambda = -\frac{\int\limits_0^1 y''y\,dx}{\int\limits_0^1 y^2\,dx} = -\frac{\int\limits_0^1 (Ay+Cy)y\,dx}{h^2\int\limits_0^1 y^2\,dx}. \tag{94}$$

We have replaced the integrals by sums, and though this is clearly good enough to provide a reasonably accurate λ, we might get more accurate figures by using accurate quadrature formulae. For the approximation ν, on the other hand, when the difference correction is neglected, the use of sums is a correct application of Rayleigh's Principle for the solution of the corresponding finite matrix.

39. As a second example we consider the evaluation of an eigenvalue and eigenfunction of Mathieu's equation

$$y''+(\lambda-2q\cos 2x)y = 0, \tag{95}$$

where q is a constant and λ the eigenvalue. This equation is known to have solutions of periods π and 2π for certain values of λ, and we choose for consideration that solution of period 2π which vanishes at $x = 0$ and π and is symmetric about $x = \frac{1}{2}\pi$. The boundary conditions associated with (95) are therefore given by

$$y(0) = 0, \qquad y(\tfrac{1}{2}\pi-x) = y(\tfrac{1}{2}\pi+x), \tag{96}$$

and we take the case $q = 1$.

With neglect of difference corrections the finite-difference equation corresponding to (95) has the form

$$y_{r+1}+y_{r-1}-(2+2h^2\cos 2x_r)y_r = -\lambda h^2 y_r = -\nu y_r. \tag{97}$$

At the interval $h = \frac{1}{4}\pi$ we can obtain a solution, but we find that the differences of the approximate y do not enable corrections to be made with any confidence. We therefore take the smaller interval $h = \frac{1}{8}\pi$, and the four independent equations corresponding to (97) are then as follows:

$$\left.\begin{aligned} -2{\cdot}2181y_1+y_2 &= -\nu y_1 \\ y_1-2y_2+y_3 &= -\nu y_2 \\ y_2-1{\cdot}7819y_3+y_4 &= -\nu y_3 \\ 2y_3-1{\cdot}6916y_4 &= -\nu y_4 \end{aligned}\right\}, \tag{98}$$

where $y_r = y(\frac{1}{8}r\pi)$.

Assuming that there is a solution with no internal zero, we take the trial vector $y = 2x/\pi$ and use Rayleigh's quotient, with the last of (98) divided by 2 to preserve symmetry, for the corresponding approximation to ν, found to be $-0{\cdot}0096$. This value is inserted in (98), and the relaxation process is carried out as shown in Table 11, in which the residuals are given in units of the fourth decimal. The diagonal terms of the operations table are at this stage $-2{\cdot}2277$, $-2{\cdot}0096$, $-1{\cdot}7915$, and $-1{\cdot}7012$.

TABLE 11

Row	y_1	y_2	y_3	y_4	R_1	R_2	R_3	R_4
1	0·25	0·50	0·75	1·00	−569	− 48	+1564	−2012
2			0·10		−569	+952	− 228	− 12
3		0·05			− 69	− 53	+ 272	− 12
4			0·01		− 69	+ 47	+ 93	+ 188
5	−0·004				+ 20	+ 7	+ 93	+ 188
6				0·005	+ 20	+ 7	+ 143	+ 103
7			0·002		+ 20	+ 27	+ 107	+ 143
	0·2460	0·5500	0·8620	1·0050	+ 20	+ 27	+ 107	+ 143

The first five rows of Table 11 give operations designed to reduce the size of the residuals, while the last two operations are aimed at giving residuals which are algebraically proportional to the corresponding vector components. In the last row the new vector is assembled and the residuals recalculated as a check on the computation.

In practice we would be more likely to carry out the computation in the more compact form shown in Fig. 26, the numbers beneath the line denoting the diagonal coefficients of the matrix.

				+ 27		+ 107		
				+ 7		+ 143		+ 143
				+ 47	20	+ 93		+ 103
		+ 20		− 53	100	+ 272		+ 188
	−40	− 69	500	+952	1000	− 228	50	− 12
0	2500	−569	5000	− 48	7500	+1564	10000	−2012
	−2·2277		−2·0096		−1·7915		−1·7012	

FIG. 26

We now take the new vector, find from Rayleigh's quotient the new eigenvalue $-0{\cdot}0210$, calculate the new residuals, and perform further relaxation as shown in Fig. 27.

				0		0		
		0		− 2		− 2		
		+ 4		− 6		0		+ 1
	2	+ 6	− 2	−11		+ 9	2	+ 4
	−15	−28	−20	+ 4	5	+29	20	− 6
0	2460	− 8	5500	−36	8620	+ 9	10050	+28
	−2·2391		−2·0210		−1·8029		−1·7126	

FIG. 27

At this stage the residuals are nearly negligible, and we should expect little further change in eigenvalue or vector. We find in fact that the result of Fig. 27, when normalized so that its largest component is 1·0000, is indeed a correct solution, to four decimals, of the algebraic problem and the eigenvalue $-0{\cdot}0210$ is correct to the same number of decimals. We denote this eigenvalue by $\nu^{(1)}$ and the eigenvector by $y^{(1)}$.

40. The next step is the calculation of the difference correction of $y^{(1)}$, given by

$$Cy_r^{(1)} = (-\tfrac{1}{12}\delta^4+\tfrac{1}{90}\delta^6-\tfrac{1}{560}\delta^8+\tfrac{1}{3150}\delta^{10}-\ldots)y_r^{(1)}. \tag{99}$$

When we compute external values by the step-by-step process we find that they exhibit the symmetry which is a quality of the solution of the differential equation. This we can use to obtain as many differences as needed in Table 12.

TABLE 12

x	$y^{(1)}$		δ^2		δ^4		δ^6		δ^8		δ^{10}	$Cy^{(1)}$
$+0$	$+0{\cdot}0000$		0		0		0		0		0	
		+2430		+ 579		−1043		+ 748		+ 752		
$\frac{1}{8}\pi$	·2430		+ 579		−1043		+ 748		+ 752		− 5446	+0·0092
		3009		− 464		− 295		+1500		−4694		
$\frac{1}{4}\pi$	·5439		+ 115		−1338		+2248		−3942		+ 7754	+0·0146
		3124		−1802		+1953		−2442		+3060		
$\frac{3}{8}\pi$	0·8563		−1687		+ 615		− 194		− 882		+ 4470	−0·0050
		+1437		−1187		+1759		−3324		+7530		
$+\frac{1}{2}\pi$	+1·0000		−2874		+2374		−3518		+6648		−15060	−0·0254

These differences are not convergent, in the sense defined in Chapter 2, but they have a fairly definite 'flow', even in differences as high as the tenth, and we should expect their use to give a substantial improvement on the previous approximation. Lack of complete confidence can always be removed by using a smaller interval later, and here our work will be materially lightened if a good approximation is already known.

The twelfth difference contributes to $Cy^{(1)}$ a maximum of 2 units in the fourth decimal, and since the sign of this contribution varies at different pivotal points, we should expect its combined effect to be small. We therefore retain differences up to order ten, and compute from (99) the difference corrections, whose values are given on the right of Table 12. From these values we use Rayleigh's quotient in the form (86), to compute a correction to the eigenvalue, given by

$$\delta\nu = -\frac{\sum y^{(1)}Cy^{(1)}}{\sum \{y^{(1)}\}^2} = 0{\cdot}0043. \tag{100}$$

The resulting residuals, given by $Cy^{(1)}+y^{(1)}\delta\nu$, are then computed, and relaxation proceeds as shown in Fig. 28, the diagonal terms of the operations table having the new values given below the line.

41. In the relaxation process we deliberately refrain from making changes to the largest pivotal value, at $x = \frac{1}{2}\pi$, so that the increments in the other pivotal values constitute a definite alteration to the initial approximation $y^{(1)}$, and are not contaminated by any multiple of the latter, which would involve a similar change in $Cy^{(1)}$. If it is inconvenient to avoid making a change at $x = \frac{1}{2}\pi$ we could achieve an equivalent effect by adding our increments to $y^{(1)}$ and normalizing the resulting vector to the value unity at $x = \frac{1}{2}\pi$.

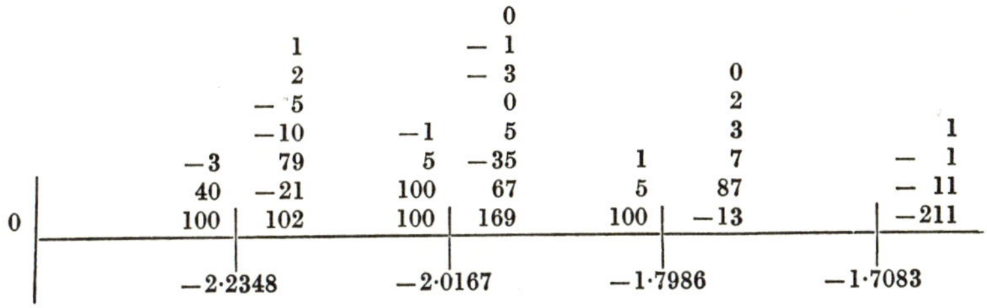

FIG. 28

42. The fact that we have been able to effect almost complete liquidation of residuals means that the alteration in the vector is unaccompanied by appreciable further alteration to the eigenvalue, and we should now have the correct solution if the difference corrections of these changes in y are negligible. This, however, is not the case, the difference correction of the changes in the vector having the values $+8$, -28, -12, and $+55$ in units of the fourth decimal. These give rise to a change $-0{\cdot}0002$ in ν, and the residuals can again be liquidated completely. The new difference correction causes no further change in ν, but four more steps are necessary before further corrections to y are negligible. When all the increments are collected together we carry out a final check, to take account of possible errors and the probable accumulation of the various almost negligible residuals. The final Rayleigh quotient

$$\lambda = -\frac{\sum (y''-2y\cos 2x)y}{\sum y^2}, \tag{101}$$

with y'' calculated as accurately as possible using all significant differences, gives our accepted eigenvalue. The results are shown in Table 13.

The residuals, given with an extra figure in Table 13, indicate that the pivotal values at $x = \frac{1}{8}\pi$, $x = \frac{1}{2}\pi$ could with advantage have respective negative and positive increments of one unit in the fourth decimal. The results, compared with a solution computed by other means and normalized to have the same value unity at $x = \frac{1}{2}\pi$, then have maximum errors

Table 13

x	y		δ^2		δ^4		δ^6		δ^8		δ^{10}	$h^2(y''-2y\cos 2x)$	R
$+0$	$+0{\cdot}0000$		0		0		0		0		0		
		$+2556$		$+504$		-1034		$+940$		-16			
$\frac{1}{8}\pi$	$\cdot2556$		$+504$		-1034		$+940$		-16		-3136	$+0{\cdot}00422$	$-0{\cdot}00013$
		3060		-530		-94		$+924$		-3152			
$\frac{1}{4}\pi$	$\cdot5616$		-26		-1128		$+1864$		-3168		$+6164$	$+0{\cdot}00963$	$+0{\cdot}00008$
		3034		-1658		$+1770$		-2244		$+3012$			
$\frac{3}{8}\pi$	$0{\cdot}8650$		-1684		$+642$		-380		-156		$+1944$	$+0{\cdot}01456$	$-0{\cdot}00015$
		$+1350$		-1016		$+1390$		-2400		$+4956$			
$+\frac{1}{2}\pi$	$+1{\cdot}0000$		-2700		$+2032$		-2780		$+4800$		-9912	$+0{\cdot}01723$	$+0{\cdot}00022$

$$\nu = -0{\cdot}017008, \qquad \lambda = -0{\cdot}11029$$

of one unit in the fourth decimal. The correct eigenvalue is $-0{\cdot}11025$, differing from our value by less than four parts in 10,000.

43. This example shows that we need not worry unduly about large differences of high order. On the other hand we cannot have complete confidence in the result, since the differences are not converging, and Cy may not be correct in its fourth decimal. We can, however, be sure that our calculated pivotal values are correct to three decimals.

As in the previous example, we notice that when the difference correction is neglected the approximate eigenvector is more accurate than the approximate eigenvalue, the errors being about 5 and 25 per cent. respectively. Subsequent correction to y, however, entailed more labour than that for λ, in accordance with the stationary property of Rayleigh's Principle. Indeed it would have been preferable to avoid the necessity for computing the last four changes in y by using a smaller interval after the calculations embodied in equation (100) and Fig. 28, at which stage the eigenvector was correct to two decimals and the eigenvalue to three decimals.

General description of method

44. Since this procedure for eigenvalue problems is rather more complicated than that of previous chapters, we repeat in a symbolic form the steps in the calculation for a more general linear differential equation. The latter is first expressed in the form

$$Ly+Cy = \lambda f(x)y, \tag{102}$$

where Ly is a linear combination of pivotal values, Cy the difference correction, λ the eigenvalue, and f a known function of x only. The computing steps are then as follow.

(i) Neglecting Cy, we obtain a first approximation $\lambda^{(1)}$ and $y^{(1)}$. In this part of the work, the standard 'latent-root' problem for a finite matrix, the Rayleigh quotient is given by

$$\lambda = \frac{\sum (Ly)y}{\sum f(x)y^2}, \tag{103}$$

and $\lambda = \lambda^{(1)}$ when $y = y^{(1)}$. This quotient may or may not have all the qualities associated with Rayleigh's Principle, but (103) will in any case be a useful formula for estimating the eigenvalue corresponding to an approximate vector. At any stage in the relaxation process the residuals are given by

$$R = Ly-\lambda f(x)y, \tag{104}$$

and the aim of relaxation is to make each typical residual R_r algebraically proportional to the corresponding quantity $f_r y_r$.

(ii) When $\lambda^{(1)}$ and $y^{(1)}$ have been calculated, so that R is negligible in (104), we proceed to difference $y^{(1)}$ and calculate the difference corrections $Cy^{(1)}$. We now wish to obtain λ and y to satisfy the equations

$$Ly+Cy^{(1)} = \lambda f(x)y. \tag{105}$$

As a first approximation we take $y = y^{(1)}$, and obtain an estimate for λ from the modified Rayleigh quotient

$$\lambda = \frac{\sum \{Ly^{(1)}+Cy^{(1)}\}y^{(1)}}{\sum f\{y^{(1)}\}^2}. \tag{106}$$

In the previous example this procedure was modified slightly, and a correction $\delta\lambda = \lambda-\lambda^{(1)}$ was calculated from

$$\delta\lambda = \frac{\sum \{Cy^{(1)}\}y^{(1)}}{\sum f\{y^{(1)}\}^2}. \tag{107}$$

If the solution at stage (i) is exact, the procedures are clearly identical, and the calculation of (107) is easier than that of (106).

The residuals are defined by

$$R = Ly+Cy^{(1)}-\lambda fy, \tag{108}$$

where λ is given by (106). The initial residuals are obtained from (108) with $y = y^{(1)}$, or by the equation

$$R = Cy^{(1)}-\delta\lambda fy^{(1)}, \tag{109}$$

which is equivalent to (108) if the results of stage (i) are exact.

(iii) Equation (105) is apparently no longer homogeneous, the matrix corresponding to $Ly-\lambda f(x)y$ is not singular when $\lambda \neq \lambda^{(1)}$, and it would appear possible to solve (105) exactly for any $\lambda \neq \lambda^{(1)}$. Such a solution, however, is meaningless. For λ is not likely to differ much from $\lambda^{(1)}$, so that the inverse matrix has large coefficients, large increments are necessary in y to effect a solution, and $Cy^{(1)}$ is then no longer a close approximation to Cy. From another point of view, we might consider increments $ky^{(1)}$ to the first approximation $y^{(1)}$. The effect on the residuals is given by

$$\delta R = kLy^{(1)}-\lambda fky^{(1)} = (\lambda_1-\lambda)fky^{(1)}, \tag{110}$$

which might be quite small even for large k. With such an increment, however, we must associate a change $kCy^{(1)}$ in the non-homogeneous part of the equation, and these quantities may be large.

We must therefore proceed tentatively, just as in the relaxation process of stage (i). The 'shape' of the vector is important, its magnitude immaterial and indeterminate. Increments to the first approximation $y^{(1)}$ should therefore be small, the size of y should be kept constant, preferably so that its largest element is the same as that of $y^{(1)}$, so that the 'constants' $Cy^{(1)}$ still furnish a good approximation to the required Cy. In these circumstances we may not be able to effect complete liquidation of residuals without making another change in λ, the new value being given by (106), with the new vector y replacing $y^{(1)}$ except in the term $Cy^{(1)}$.

(iv) When λ and y have been found by this process to satisfy (105), we recompute the difference correction and repeat the process until the original equation (102) is satisfied.

(v) Finally, we might obtain a more accurate value of λ by using integrals instead of sums in (106), with the final solution y replacing $y^{(1)}$, whenever the differential equation is of a form for which all the qualities of Rayleigh's Principle are applicable.

Though this description of method may appear involved it is fairly easy in practice, and the necessary computational skill is not difficult to acquire.

Harder examples

†**45.** Our differential equations have so far involved no first derivative and the matrices were inevitably symmetric. We consider now an example in which the differential equation contains the first derivative.

A problem of an oscillating narrow lake of elliptical plan form involves the solution of the differential equation

$$y''+\frac{x}{1-x^2}y'+\lambda y = 0, \tag{111}$$

with $y = 0$ at $x = \pm 1$. The finite-difference equation is here given by

$$\left.\begin{aligned} y_{r+1}\left(1+\tfrac{1}{2}h\frac{x_r}{1-x_r^2}\right)+y_{r-1}\left(1-\tfrac{1}{2}h\frac{x_r}{1-x_r^2}\right)-2y_r+Cy_r = -\lambda h^2 y_r = -\nu y_r \\ Cy_r = \left\{-\tfrac{1}{12}\delta^4+\tfrac{1}{90}\delta^6-\ldots+\frac{hx_r}{1-x_r^2}(-\tfrac{1}{6}\mu\delta^3+\tfrac{1}{30}\mu\delta^5-\ldots)\right\}y_r \end{aligned}\right\}. \tag{112}$$

At least one solution is clearly symmetric about the origin, so that with the interval $h = 0{\cdot}2$ we neglect the difference corrections and consider the following five independent equations, in which $y_r = y(rh)$:

$$\left.\begin{aligned} -2y_0+2y_1 &= -\nu y_0 \\ 0{\cdot}9792y_0-2y_1+1{\cdot}0208y_2 &= -\nu y_1 \\ 0{\cdot}9524y_1-2y_2+1{\cdot}0476y_3 &= -\nu y_2 \\ 0{\cdot}9062y_2-2y_3+1{\cdot}0938y_4 &= -\nu y_3 \\ 0{\cdot}7778y_3-2y_4 &= -\nu y_4 \end{aligned}\right\}. \tag{113}$$

Assuming as a first guess the trial values $y = 1-x$, for $x \geqslant 0$, and symmetry about $x = 0$, we find the value of ν given by Rayleigh's quotient to be 0·1475, and the initial residuals are calculated. These are the first values on the right of the pivotal points in Fig. 29, which shows the relaxation process up to the point where the residuals cannot be further reduced, and have been made roughly proportional to the pivotal values. The numbers beneath the line in Fig. 29 are the coefficients in the finite-difference equations (113), with the term νy included. The arrangement of the numbers accords with a suggestion of § 19 in Chapter 3.

			97				64			
	108		134		78		249		40	
−400	68		526		59		140	100	−38	
−1000	−673	20	118	400	−46	100	−222	−400	147	
10000	−2525	8000	1097	6000	695	4000	215	2000	−594	0
	2	0·9792	1·0208	0·9524	1·0476	0·9062	1·0938	0·7778		
−1·8525		−1·8525		−1·8525		−1·8525		−1·8525		

Fig. 29

A repetition of the process, using the new vector, produces the value $\nu = 0{\cdot}1347$, and with this value the residuals can be reduced to a maximum of 2 units. A final step gives $\nu = 0{\cdot}1345$, negligible residuals, and the vector shown in Table 14.

46. We now wish to calculate the difference corrections of this first approximation, and a difficulty is encountered immediately when we try to extend the pivotal values beyond $x = 1$. This is clearly possible only if all the derivatives of the function are finite, and we can see easily from the differential equation that if y vanishes at $x = 1$ then all the derivatives, if finite, are also zero, giving the trivial solution. The only other solution which vanishes at $x = 1$ behaves there like $(1-x)^{\frac{3}{2}}$, with infinite derivatives of order two and higher, so that we cannot extend the tabulation beyond $x = 1$.

We cannot therefore compute the difference corrections near $x = 1$ from the second of (112), and we use the alternative method of regarding our first solution as an approximation to the solution of the differential equation, from which we compute the derivatives by the best formulae available. The latter are taken from the groups 1 and 2 in Table 7 of Chapter 2, which give first and second derivatives of y_r in terms of the backward differences of y_{r+s}. The accuracy of this process is uncertain, and even the convergence is doubtful of some of the formulae for derivatives near $x = 1$. We must therefore regard this method not as one for producing accurate results at this interval, but for providing a good approximation for use at a smaller interval.

TABLE 14

x	y		δ^2		δ^4		δ^6		δ^8	h^2y''	hy'	$F(x)$
+0·0	+0·8603		−1158		+250		−38		+64	−0·1179	−0·0000	−0·1179
		− 579		+125		−19		+32				
·2	·8024		1033		231		− 6		+31	·1052	·1137	·1099
		1612		356		−25		+63				
·4	·6412		677		206		+57			·0694	·2027	·0887
		2289		562		+32						
·6	·4123		− 115		+238					− ·0133	·2457	·0594
		2404		+800								
0·8	·1719		+ 685							+0·0653	−0·2220	−0·0334
		−1719										
+1·0	+0·0000											

TABLE 15

x	y		δ^2		δ^4		δ^6		δ^8	h^2y''	hy'	$F(x)$
+0·0	+0·8604		−1184		+260		−26		+16	−0·1206	−0·0000	−0·1206
		− 592		+130		−13		+ 8				
·2	·8012		1054		247		−18		+87	·1075	·1162	·1123
		1646		377		−31		+95				
·4	·6366		677		216		+77			·0694	·2066	·0891
		2323		593		+46						
·6	·4043		− 84		+262					− ·0103	·2481	·0568
		2407		+855								
0·8	·1636		+ 771							+0·0729	−0·2195	−0·0247
		−1636										
+1·0	+0·0000											

Table 14 shows the differences of this first approximation and the computed values of h^2y'' and hy', in which differences of order eight were retained. The column labelled $F(x)$ gives the values of

$$F(x) = h^2\left(y''+\frac{x}{1-x^2}y'\right). \tag{114}$$

The next step is a new calculation of ν from the modified Rayleigh quotient

$$\nu = -\frac{\sum Fy}{\sum y^2}, \tag{115}$$

which produces the result $\nu = 0{\cdot}1391$. The new residuals, given by $F+\nu y$, are calculated and further relaxation performed. For this purpose, of course, we use the operations given by equations (113), being careful to make little or no change in the size of the vector. The residuals cannot be liquidated without a further change of 0·0009 in ν, and at this stage we repeat, as shown in Table 15, the operations of Table 14.

Using (115) we find the new value 0·1403 for ν, and the new residuals can be liquidated by changes in y never greater than 13 units in the fourth decimal, and a change $+0{\cdot}0002$ in ν. The changes in y have a negligible difference correction, so that there is no further change in ν or y. The final result is shown in Table 16, which gives also the final residuals. The latter suggest that the values at $x = 0$ and $x = 0{\cdot}8$ should receive respective positive and negative increments of amount 0·0001.

47. It is tempting to remove the first derivative from the differential equation (111) by using the substitution

$$y = (1-x^2)^{\frac{1}{4}}z. \tag{116}$$

The function z then satisfies the differential equation

$$z''-\left\{\frac{1}{2(1-x^2)}+\frac{5x^2}{4(1-x^2)^2}\right\}z+\lambda z = 0. \tag{117}$$

The arithmetic involved in the solution of this equation is not substantially different from that already expended, and the difference correction might be expected to be smaller. In fact, however, the function z behaves like $(1-x)^{\frac{3}{4}}$ near $x = 1$, so that the infinity in the second derivative is of larger order than that of y, and a smaller interval would be required than before for a solution of comparable accuracy.

On the other hand the Rayleigh quotient applied to (117) has all the qualities of Rayleigh's Principle, since the kinetic and potential energies

TABLE 16

x	y		δ^2		δ^4		δ^6		δ^8	h^2y''	hy'	$F(x)$	R
+0·0	+0·8604		−1186		+258		−18		− 2	−0·1208	−0·0000	−0·1208	+0·0002
		− 593		+129		− 9		− 1					
·2	·8011		1057		249		−19		+94	·1078	·1165	·1127	− ·0001
		1650		378		−28		+93					
·4	·6361		679		221		+74			·0697	·2072	·0894	·0000
		2329		599		+46							
·6	·4032		− 80		+267					− ·0100	·2487	·0566	+ ·0001
		2409		+866									
0·8	·1623		+ 786							+0·0744	−0·2191	−0·0230	−0·0002
		−1623											
+1·0	+0·0000												

$\nu = 0{\cdot}1406,$
$\lambda = 3{\cdot}51.$

are respectively given by

$$T \propto \int_{-1}^{1} \frac{y^2}{(1-x^2)^{\frac{1}{2}}}\,dx, \qquad V \propto \int_{-1}^{1} \frac{y'^2}{(1-x^2)^{\frac{1}{2}}}\,dx, \tag{118}$$

and Rayleigh's Principle here gives

$$\lambda = \int_{-1}^{1} \frac{y'^2}{(1-x^2)^{\frac{1}{2}}}\,dx \bigg/ \int_{-1}^{1} \frac{y^2}{(1-x^2)^{\frac{1}{2}}}\,dx. \tag{119}$$

Substituting for y from (116) we find that (119) can be expressed in the form

$$\lambda = -\frac{\int_{-1}^{1} z[z''-z\{(1/2(1-x^2))+(5x^2/4(1-x^2)^2)\}]\,dx}{\int_{-1}^{1} z^2\,dx}, \tag{120}$$

which corresponds to our definition of the Rayleigh quotient applied to the differential equation (117).

48. It is better, however, to remove the singularity completely from the numerical work. It can be shown that at $x = \pm 1$ the function y has the form

$$y = (1-x^2)^{\frac{1}{4}}\{a(1-x^2)+b(1-x^2)^2+...\}. \tag{121}$$

If we write $y = (1-x^2)^{\frac{1}{4}}z$, then z satisfies the differential equation

$$z''-\frac{x}{1-x^2}z'-\frac{1+x^2}{(1-x^2)^2}z+\lambda z = 0, \tag{122}$$

is symmetric about the origin, vanishes at $x = \pm 1$, and has there the form

$$z = a(1-x^2)+b(1-x^2)^2+... . \tag{123}$$

At $x = \pm 1$, (122) becomes

$$z''-\frac{x(1-x^2)z'+(1+x^2)z}{(1-x^2)^2} = 0. \tag{124}$$

The second term in (124) is an 'undetermined form', but standard treatment produces, for use at $x = \pm 1$, the equations

$$z''\pm\tfrac{1}{5}z' = 0. \tag{125}$$

With this equation and subsequent use of (122) we can extend z beyond $x = \pm 1$ and use central-difference formulae for difference corrections.

The simple finite-difference equation corresponding to (122) is given by

$$z_{r+1}\left(1-\frac{hx_r}{2(1-x_r^2)}\right)+z_{r-1}\left(1+\frac{hx_r}{2(1-x_r^2)}\right)-z_r\left\{2+\frac{h^2(1+x_r^2)}{(1-x_r^2)^2}\right\} = -\nu z_r, \tag{126}$$

and at $x = 1$ this can be replaced by

$$z_{r+1}(1+\tfrac{1}{10}h)+z_{r-1}(1-\tfrac{1}{10}h) = 0. \tag{127}$$

With the same interval $h = 0{\cdot}2$ we compute as before a first approximation, obtaining $\nu = 0{\cdot}1420$ and the vector shown in Table 17. The value at $x = 1{\cdot}2$ is obtained from (127), and successive values from the application of (126) at $x = 1{\cdot}2$ and successive points.

TABLE 17

x	z		δ^2		δ^4		δ^6	$F(x)$	R
+0·0	+0·8710		−888		+104		−14	−0·1245	− 6
		− 444		+ 52		− 7			
·2	·8266		836		97		9	·1180	5
		1280		149		16			
·4	·6986		687		81		9	·0996	− 3
		1967		230		25			
·6	·5019		457		+ 56		−44	·0709	+ 5
		2424		286		69			
0·8	+ ·2595		−171		− 13		+66	−0·0346	+23
		2595		273		3			
1·0	·0000		+102		16		+ 1		
		2493		257		2			
1·2	− ·2493		359		18		−64		
		2134		239		66			
1·4	·4627		598		84		+54		
		1536		155		−12			
1·6	·6163		753		− 96				
		− 783		+ 59					
1·8	·6946		+812						
		+ 29							
+2·0	−0·6917								

It would appear from the differences that the interval is slightly too big near $x = 1$, but the bump in the differences is clearly not representative of a singularity, being more reminiscent of the similar phenomenon in the differences of a function of the form $(1+x^2)^{-1}$. We can clearly compute the difference corrections, neglecting differences of order greater than six, and have full confidence in three-figure accuracy, while the fourth should not be greatly in error. It is just as easy to compute accurate values of the terms not involving λ in (122), and these are given multiplied by h^2 in the column labelled $F(x)$. The Rayleigh quotient $\sum Fz/\sum z^2$ then gives the new value 0·1422 for ν, and from this we can calculate the residuals given by $R = F+\nu z$, given in the table in units of the fourth decimal.

The liquidation of these residuals necessitates a change of about 0·0001 in ν, and the following changes in the vector in units of the fourth decimal:

x	0·0	0·2	0·4	0·6	0·8	1·0
δz	0	+2	+9	+16	+19	0

The difference correction of this change in z is small, certainly negligible to three decimals, so that our result is correct to this number of figures. The final eigenvalue $\nu = 0{\cdot}1423$ agrees to all the figures given with a result computed by other workers.

It is interesting to compare this result with that obtained by the previous method. The ν of the latter is in error by nearly 0·002, showing the danger of relying on backward-difference formulae near a singularity, which were used in the production of Table 16. The vector also shows considerable change, particularly near $x = 1$; the computed values of y are shown below, normalized to have the same value at the origin as we obtained previously.

x	0·0	0·2	0·4	0·6	0·8	1·0
y	0·8604	0·8002	0·6333	0·3979	0·1549	0·0000

Non-homogeneous eigenvalue problem

49. The previous substitution suggests a further possibility. If we write $y = (1-x^2)^{\frac{3}{2}}z$, then the differential equation for z is given by

$$z'' - \frac{1}{1-x^2}(5xz'+3z)+\lambda z = 0, \tag{128}$$

and z is still symmetric about $x = 0$, but now has finite non-zero boundary values at $x = \pm 1$. The finite-difference equations are no longer homogeneous, and can be solved for any value of λ. It would appear at first sight that λ is not therefore determinable uniquely, but equation (128) shows that another condition must be satisfied if z is to be well behaved, an essential quality. This condition is given by

$$5xz'+3z = 0 \text{ at } x = \pm 1. \tag{129}$$

There are now two possible procedures. First, we can select various values of λ and solve the simultaneous finite-difference equations, choosing that λ for which (129) is satisfied. This is clearly equivalent to choosing λ so that the function has smooth differences, since z'' is infinite at $x = 1$ unless (129) is satisfied: the problem is therefore in a sense similar to that of the boundary-value treatment of first-order equations. The easiest numerical method is to use a step-by-step process starting from $x = 0$. At this point we can take z to have a fixed value, say unity. Equation (128) is replaced by a three-term finite-difference equation, with difference corrections neglected. Because of symmetry the equation at the origin contains only $y(0)$ and $y(h)$, so that the latter value is calculable directly, and successive values $y(2h)$, $y(3h)$,... are obtained from the three-term formula.

The behaviour of z is very sensitive to a small change in λ, as shown in the results of Table 18. The result $h^2\lambda = 0{\cdot}143$ is clearly a close approximation to the truth, though the differences of the corresponding solution are not very satisfactory near $x = 1$.

TABLE 18

	$h^2\lambda = 0{\cdot}14$	$h^2\lambda = 0{\cdot}15$	$h^2\lambda = 0{\cdot}145$	$h^2\lambda = 0{\cdot}143$
x	z	z	z	z
+0·0	+1·0000	+1·0000	+1·0000	+1·0000
·2	0·9900	0·9850	0·9875	0·9885
·4	·9611	·9390	·9500	·9545
·6	·9177	·8555	·8864	·8991
0·8	·8798	0·6851	0·7815	·8213
+1·0	+0·0691	+2·7927	+1·4502	+0·8929

50. The second possibility is to use (129) as a boundary condition, replacing the condition that the boundary value should be finite and non-zero and the solution well behaved. By this process we recover the homogeneous nature of the finite-difference equations, and have a true eigenvalue problem. To be able to use central-difference formulae for all derivatives we would also like to make use of the differential equation at $x = 1$. By the rule for indeterminate forms the middle term in equation (128) is equivalent to $4z'+\frac{5}{2}z''$ at $x = 1$. We therefore have two equations reducible to

$$\left.\begin{aligned} \tfrac{7}{2}z''-\tfrac{12}{5}z+\lambda z &= 0 \\ 5z'+3z &= 0 \end{aligned}\right\} \qquad (130)$$

at $x = 1$. Using the simplest central-difference formulae for the derivatives and ignoring the difference corrections, we find that the pivotal value $y(1+h)$ can be eliminated, and at the interval $h = 0{\cdot}2$ we can produce the equation

$$2z(0{\cdot}8)-2{\cdot}2674z(1{\cdot}0) = -0{\cdot}2857\nu z(1{\cdot}0). \qquad (131)$$

The finite difference equations corresponding to (128) at internal pivotal points are as follows:

$$\left.\begin{aligned} -2{\cdot}1200z(0)+2{\cdot}0000z(0{\cdot}2) &= -\nu z(0) \\ 1{\cdot}1042z(0)-2{\cdot}1250z(0{\cdot}2)+0{\cdot}8958z(0{\cdot}4) &= -\nu z(0{\cdot}2) \\ 1{\cdot}2381z(0{\cdot}2)-2{\cdot}1429z(0{\cdot}4)+0{\cdot}7619z(0{\cdot}6) &= -\nu z(0{\cdot}4) \\ 1{\cdot}4687z(0{\cdot}4)-2{\cdot}1875z(0{\cdot}6)+0{\cdot}5313z(0{\cdot}8) &= -\nu z(0{\cdot}6) \\ 2{\cdot}1111z(0{\cdot}6)-2{\cdot}3333z(0{\cdot}8)-0{\cdot}1111z(1{\cdot}0) &= -\nu z(0{\cdot}8) \end{aligned}\right\}. \qquad (132)$$

The solution of these equations gives $\nu = 0{\cdot}1424$ and the vector shown in Table 19, including some calculated external values. We see that the differences are now very smooth and much smaller than before. The difference correction is not quite negligible, but its calculation and the subsequent correction of the first approximation are clearly almost trivial.

The relaxation process is interesting. We find that the coefficients of the equation are of such a nature that large increments to z near $x = 1$ have only small effect at smaller values of x; large residuals near the top end of the range are therefore relatively unimportant in their effect on both z and ν.

TABLE 19

x	z		δ^2		δ^4		δ^6
+0·0	+1·0000		−224		+10		+4
		− 112		+ 5		+2	
·2	0·9888		219		12		−3
		331		17		−1	
·4	·9557		202		11		−4
		533		28		−5	
·6	·9024		174		6		+6
		707		34		+1	
0·8	·8317		140		7		−4
		847		41		−3	
1·0	·7470		99		+ 4		
		946		+45			
1·2	·6524		− 54				
		−1000					
+1·4	+0·5524						

Boundary condition at infinity

51. Not uncommonly one of the boundaries is at infinity. The methods used for the case when no eigenvalue is involved, illustrated in §§ 27, 28 of Chapter 4, are applicable with only obvious modifications. As an example we take the differential equation

$$y''+\left(\frac{2}{x}-\lambda\right)y = 0, \tag{133}$$

and look for a well-behaved solution which vanishes at the origin and also as $x \to \infty$.

One possible method is to satisfy the second boundary condition at some finite value of x. Choosing $x = 4$ to represent infinity, and the interval $h = \frac{1}{2}$, we replace (133) by the approximate finite-difference equation

$$y_{r+1}+y_{r-1}-\left(2-\frac{2h^2}{x_r}\right)y_r = h^2\lambda y_r = \nu y_r, \tag{134}$$

and obtain a solution by methods already illustrated. For purposes of calculating the subsequent difference correction we can extrapolate in the usual way for $x > 4$, while at the origin we replace (133) by the equivalent differential equation

$$y''+2y' = 0, \tag{135}$$

and (134) by the corresponding finite-difference equation

$$y_1(1+h)+y_{-1}(1-h) = 0. \tag{136}$$

Table 20 shows the approximate solution, with $\nu = 0{\cdot}226$ or $\lambda = 0{\cdot}904$.

TABLE 20

x	y		δ^2		δ^4		δ^6
$-0{\cdot}5$	$-2{\cdot}400$						
		$+2400$					
$0{\cdot}0$	$+0{\cdot}000$		-1600				
		800		$+981$			
$+0{\cdot}5$	$\cdot800$		619		-630		
		$+\ 181$		351		$+450$	
$1{\cdot}0$	$\cdot981$		268		180		-360
		$-\ \ 87$		171		90	
$1{\cdot}5$	$\cdot894$		97		90		52
		184		81		38	
$2{\cdot}0$	$\cdot710$		$-\ \ 16$		52		9
		200		29		29	
$2{\cdot}5$	$\cdot510$		$+\ \ 13$		23		$-\ \ 18$
		187		$+\ \ 6$		$+\ \ 11$	
$3{\cdot}0$	$\cdot323$		19		$-\ \ 12$		
		168		$-\ \ 6$			
$3{\cdot}5$	$\cdot155$		$+\ \ 13$				
		$-\ 155$					
$+4{\cdot}0$	$+0{\cdot}000$						

It is clear that the interval is too large near the origin, and $x = 4$ is not a satisfactory interpretation of infinity. Nevertheless the calculated eigenvalue differs by only 10 per cent. from the true value $\lambda = 1$, and there is little doubt that we could get good results by reducing the interval and increasing the total range. We note, moreover, that as x increases the diagonal term of the matrix also increases, so that it is relatively easy to liquidate residuals: also the eigenvalue is determined mainly by the large values of y near the origin, being relatively unaffected by the shape of the function beyond $x = 4$.

52. A second method is to use knowledge of the behaviour of y as $x \to \infty$. Clearly $y \propto f(x)e^{-x\sqrt{\lambda}}$, where f changes relatively slowly, so that for large enough values of x two successive values of y have the ratio $e^{-h\sqrt{\lambda}}$, and this can be used as a boundary condition. Applying this condition at $x = 3{\cdot}5$ we have

$$y(4{\cdot}0) = e^{-\frac{1}{2}\sqrt{\lambda}}y(3{\cdot}5). \tag{137}$$

This equation is non-linear in λ so that it is difficult to use directly. A less direct method is to fix λ and the value of y at $x = 3{\cdot}5$, solve the resulting equations, extrapolate to $x = 4{\cdot}0$, and repeat with a new λ until (137) is satisfied.

It is usually easy to decide in which direction λ should be changed. In our case, for example, we might conveniently start with the solution of Table 20 with $\nu = 0{\cdot}226$: here $y(4{\cdot}0)$ is zero and so too small for the satisfaction of (137). The finite-difference equation gives

$$y(4{\cdot}0) = (1{\cdot}8571+\nu)y(3{\cdot}5)-y(3{\cdot}0). \tag{138}$$

If we keep $y(3{\cdot}5)$ constant, the change in $y(4{\cdot}0)$ is given by

$$\delta y(4{\cdot}0) = y(3{\cdot}5)\delta\nu-\delta y(3{\cdot}0), \tag{139}$$

where $y(3{\cdot}0)$ changes only with ν. If ν is increased, all the internal residuals are decreased and so all internal values of y are decreased. Thus an increase in ν increases algebraically both terms on the right of (139) and has therefore an overall beneficial effect. The required increase in ν is not likely to be large, since the algebraic equations are ill-conditioned and small residuals necessitate relatively large changes in y. The ill-conditioning is here so bad that it is preferable in this case to keep constant the value $y(0{\cdot}5)$, calculating successive values by a step-by-step process. Making the change $\delta\nu = 0{\cdot}01$, so that $\nu = 0{\cdot}236$, we find the results shown in Fig. 30.

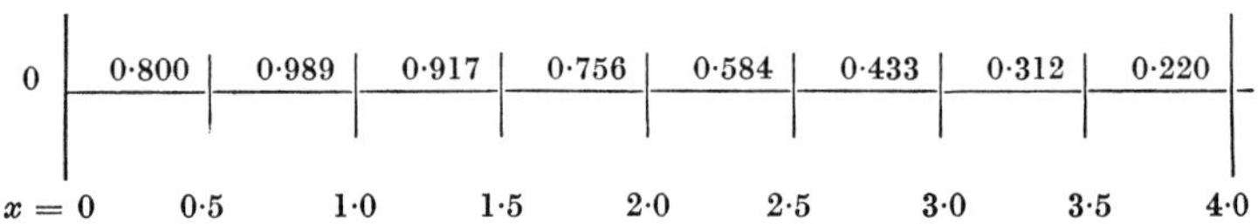

FIG. 30

We find $y(4{\cdot}0)/y(3{\cdot}5) = 0{\cdot}705$, while $e^{-\frac{1}{2}\sqrt{\lambda}} = e^{-\sqrt{\nu}} = 0{\cdot}615$. The ratio, previously too small, is now too large but quite close to the correct result, so that $\lambda = 0{\cdot}944$ is a good approximation to the solution of this problem. The latter is still only an approximation to the original problem, not only because the difference correction has been neglected but also because we have used an approximation to the true boundary condition at ∞. The first source of error can be removed by incorporating the difference correction, though a smaller interval is needed near the origin, and our imposed boundary condition is fairly satisfactory if the ratio of successive pivotal values is the same, and equal to $e^{-\frac{1}{2}\sqrt{\lambda}}$, for several successive points.

The correct solution is a multiple of xe^{-x}, and we give its values in Fig. 31 normalized to agree with our approximate solutions at $x = 0{\cdot}5$. It is interesting to observe that the results of Table 20 are better than we should expect. This is caused by the fact that the difference correction, starting with $-\frac{1}{12}\delta^4 y$, has everywhere positive values, cancelling to some extent the negative residuals introduced by the positive change in ν. This circumstance is somewhat fortuitous.

0	0·800	0·971	0·883	0·714	0·541	0·394	0·279	0·193

Fig. 31

53. Finally, we might try the effect of changing the interval. The results already obtained show that the interval needs to be fairly small near the origin but can increase steadily with increasing x. We take $h = \frac{1}{4}$ from $x = 0$ to 2, $h = \frac{1}{2}$ from $x = 2$ to 4, $h = 1$ from $x = 4$ to 10, at which point we might reasonably expect the function to vanish to three decimals. In the first part of the range, with $h = \frac{1}{4}$, there are seven finite-difference equations, for the points $x = \frac{1}{4}(\frac{1}{4})1\frac{3}{4}$: in the second part $h = \frac{1}{2}$ and there are four equations at the points $x = 2(\frac{1}{2})3\frac{1}{2}$: in the last part $h = 1$ and there are six equations at the points $x = 4(1)9$. The corresponding equations for the three regions are respectively given by

$$\left.\begin{aligned} y_1+y_{-1}-\left(2-\frac{1}{8x}\right)y_0 &= \tfrac{1}{16}\lambda y_0 \\ y_1+y_{-1}-\left(2-\frac{1}{2x}\right)y_0 &= \tfrac{1}{4}\lambda y_0 \\ y_1+y_{-1}-\left(2-\frac{2}{x}\right)y_0 &= \lambda y_0 \end{aligned}\right\}. \tag{140}$$

The technique involves the use of Rayleigh's quotient for calculating a reasonable λ from guessed values of y, and the relaxation method for subsequent liquidation and adjustment of residuals. We have discussed already in §§ 30–34 of Chapter 4 the application of relaxation when the interval is not constant, so that no further description is needed here and we give only the results, shown in Table 21.

The difference corrections are clearly not quite negligible, and would have the effect of adding positive residuals in the first half of the range, negative residuals in the second half. Both the coefficients in the finite-difference equations and the size of the vector components indicate that the former can be liquidated only by an appropriate change in λ, the change being positive in agreement with the analytical solution. The negative residuals in the second half of the range must then be liquidated by making negative changes in y, again in accordance with the known solution. If we use the results of Table 21 to calculate y'' as accurately as

TABLE 21

x	y		δ^2		δ^4
+0·00	+0·000				
		+512			
·25	·512		−224		
		288		+73	
·50	·800		151		−21
		137		52	
0·75	·937		99		15
		+ 38		37	
1·00	·975		62		12
		− 24		25	
1·25	·951		37		8
		61		17	
1·50	·890		20		− 5
		81		+12	
1·75	·809		− 8		
		− 89			
+2·00	+0·720				

x	y		δ^2		δ^4
+2·0	+0·720				
		−172			
2·5	·548		+25		
		147		+6	
3·0	·401		31		−7
		116		−1	
3·5	·285		+30		
		− 86			
+4·0	+0·199				

x	y		δ^2		δ^4
+ 4·0	+0·199				
		−106			
5·0	·093		+55		
		51		−27	
6·0	·042		28		+11
		23		16	
7·0	·019		12		10
		11		6	
8·0	·008		6		+ 2
		5		− 4	
9·0	·003		+ 2		
		− 3			
+10·0	+0·000				

$\lambda = 0{\cdot}98080$

possible, and recalculate λ from the Rayleigh quotient, we obtain the result $\lambda = 0{\cdot}998$.

Finally, it should be noted that in the application of Rayleigh's quotient to equations (140) the first set of these equations should be multiplied by $h^{-2} = 16$ and the second set by 4, so that we approximate more correctly to the true Rayleigh quotient

$$\lambda = \frac{\int\limits_0^\infty \{y''+(2/x)y\}y\,dx}{\int\limits_0^\infty y^2\,dx}. \tag{141}$$

Without this preliminary multiplication the large values in the first part of the range would not make their proper contribution.

Mixed boundary conditions

54. Occasionally the boundary conditions, though homogeneous, involve more than one pivotal value. Consider, for example, the solution of the differential system given by

$$y''+\lambda y = 0, \qquad y(1) = 2y(2), \qquad y'(1) = \tfrac{1}{2}y'(2). \tag{142}$$

Success depends only on our ability to find as many finite-difference equations as the number of unknowns introduced. Taking the interval $h = \frac{1}{4}$, we can express the first of (142), with neglect of difference corrections, in the simple form

$$y_{r+1}+y_{r-1}-(2-\nu)y_r = 0, \qquad \nu = \lambda h^2, \tag{143}$$

and if we apply this at all pivotal points $x = 1(\frac{1}{4})2$, we have introduced the pivotal values at these points and also at $x = \frac{3}{4}, \frac{9}{4}$. The second boundary condition can be expressed approximately in the form

$$y(\tfrac{5}{4})-y(\tfrac{3}{4}) = \tfrac{1}{2}\{y(\tfrac{9}{4})-y(\tfrac{7}{4})\}, \tag{144}$$

and the first boundary condition relates $y(1)$ and $y(2)$. We therefore have, for seven unknowns, the following seven equations:

$$\left.\begin{array}{ccccccccc}
y(\frac{3}{4}) & y(1) & y(\frac{5}{4}) & y(\frac{6}{4}) & y(\frac{7}{4}) & y(2) & y(\frac{9}{4}) & & \\
-1 & & 1 & & \frac{1}{2} & & -\frac{1}{2} & = & 0 \\
 & 1 & & & & -2 & & = & 0 \\
1 & -(2-\nu) & 1 & & & & & = & 0 \\
 & 1 & -(2-\nu) & 1 & & & & = & 0 \\
 & & 1 & -(2-\nu) & 1 & & & = & 0 \\
 & & & 1 & -(2-\nu) & 1 & & = & 0 \\
 & & & & 1 & -(2-\nu) & 1 & = & 0
\end{array}\right\}. \tag{145}$$

By judicious combinations we can reduce the number of equations to four, with the parameter ν present in each equation, as follows:

$$\left.\begin{array}{ccccl} y(1) & y(\frac{5}{4}) & y(\frac{6}{4}) & y(\frac{7}{4}) & \\ -(2-\nu) & 1\cdot 6 & 0 & 0\cdot 8 & =0 \\ 1 & -(2-\nu) & 1 & 0 & =0 \\ & 1 & -(2-\nu) & 1 & =0 \\ 0\cdot 5 & 0 & 1 & -(2-\nu) & =0 \end{array}\right\}. \qquad (146)$$

By our previous methods we have no difficulty in obtaining, to three decimals, the following solutions:

x	1	$\frac{5}{4}$	$\frac{6}{4}$	$\frac{7}{4}$	(2)
y	1·935	1·755	1·530	1·265	(0·968)

$\nu = 0\cdot 0258, \quad \lambda = 0\cdot 413,$

and there is clearly no inherent difficulty in applying the difference corrections, or using more accurate formulae at the boundaries. The analytical solution gives λ as the roots of the equation $\cos\sqrt{\lambda} = 0\cdot 8$, of which the smallest root is $\lambda = 0\cdot 414$.

A fourth-order equation

55. Problems in elasticity relating to the buckling or vibration of elastic beams demand the solution of fourth-order differential equations involving an eigenvalue. The procedure is a straightforward extension of the second-order case, and for illustration we treat the simplest problem defined by the differential system

$$y^{\text{iv}} = \lambda y, \qquad y(0) = y'(0) = y(1) = y'(1) = 0. \qquad (147)$$

The corresponding finite-difference equation at a general point is given by

$$y_{r-2} - 4y_{r-1} + 6y_r - 4y_{r+1} + y_{r+2} + Cy_r = h^4\lambda y_r = \nu y_r, \qquad (148)$$

and at the boundary points we use the further equations

$$y_0 = 0, \qquad -\tfrac{1}{12}y_2 + \tfrac{2}{3}y_1 - \tfrac{2}{3}y_{-1} + \tfrac{1}{12}y_{-2} + C_1 y_0 = 0, \qquad (149)$$

the second of which represents the derivative condition.

If we take the interval $h = 0\cdot 1$, and seek a solution symmetric about the mid-point $x = \frac{1}{2}$, we find the following set of independent equations, from which the difference corrections have been excluded:

$$\left.\begin{array}{rrrrrrrl}
y(-0{\cdot}2) & y(-0{\cdot}1) & y(0{\cdot}1) & y(0{\cdot}2) & y(0{\cdot}3) & y(0{\cdot}4) & y(0{\cdot}5) & \\
-1 & 8 & -8 & 1 & & & & = 0 \\
1 & -4 & -4 & 1 & & & & = 0 \\
 & 1 & 6 & -4 & 1 & & & = \nu y(0{\cdot}1) \\
 & & -4 & 6 & -4 & 1 & & = \nu y(0{\cdot}2) \\
 & & 1 & -4 & 6 & -4 & 1 & = \nu y(0{\cdot}3) \\
 & & & 1 & -4 & 7 & -4 & = \nu y(0{\cdot}4) \\
 & & & & 2 & -8 & 6 & = \nu y(0{\cdot}5)
\end{array}\right\}. \tag{150}$$

The first two equations do not contain ν, and can with convenience be used merely to eliminate the external values $y(-0{\cdot}2)$ and $y(-0{\cdot}1)$, so that we can produce finally the following set of equations:

$$\left.\begin{array}{rl}
9y(0{\cdot}1)-4{\cdot}5y(0{\cdot}2)+\;y(0{\cdot}3) & = \nu y(0{\cdot}1) \\
-4y(0{\cdot}1)+\;6y(0{\cdot}2)-4y(0{\cdot}3)+\;y(0{\cdot}4) & = \nu y(0{\cdot}2) \\
y(0{\cdot}1)-\;4y(0{\cdot}2)+6y(0{\cdot}3)-4y(0{\cdot}4)+\;y(0{\cdot}5) & = \nu y(0{\cdot}3) \\
y(0{\cdot}2)-4y(0{\cdot}3)+7y(0{\cdot}4)-4y(0{\cdot}5) & = \nu y(0{\cdot}4) \\
2y(0{\cdot}3)-8y(0{\cdot}4)+6y(0{\cdot}5) & = \nu y(0{\cdot}5)
\end{array}\right\}. \tag{151}$$

The numerical technique follows familiar lines. With a first approximation to the vector y, we calculate Ay, estimate ν from the quotient $y'Ay/y'y$, calculate the residuals $Ay-\nu y$, and use the relaxation procedure to reduce and distribute the residuals such that R and y are in a fairly constant ratio: ν is then recalculated from the Rayleigh quotient, and the process repeated until all the equations are satisfied. The last equation of (151) should of course be divided by 2 when using the Rayleigh quotient.

As before, we would expect the mode corresponding to the lowest eigenvalue to have no zero in the range, and we make the initial assumption that $y = 2x$, for $0 \leqslant x \leqslant \frac{1}{2}$, and is symmetric about the point $x = \frac{1}{2}$. With this assumption we find the Rayleigh value $\nu = 0{\cdot}12$, calculate the residuals, and proceed with the relaxation process. This is shown in Fig. 32, the first values to the right of the pivotal points giving the initial residuals.

At the final stage of Fig. 32 we decide to recalculate ν, obtaining the value $0{\cdot}0492$, and we can proceed with the relaxation, reducing the residuals, with this value of ν, to a maximum of 6 units in the last figure. A final calculation of ν gives the result shown in Fig. 33, which includes the final residuals, and in which the pivotal values have been normalized so that $y(0{\cdot}5) = 1{\cdot}000$.

					− 60				
			− 43		− 80				
			− 63		0				
			17		−118		− 70		
	− 62		− 63		− 98		10		− 22
	− 82		172		−258		−128		−140
	96		52		−228		− 48		20
− 20	− 84		2		− 28	20	− 8		− 20
− 30	182		202	20	−322	50	−352		380
− 50	132	− 40	152	50	−122	50	−152	20	280
200	576	400	− 48	600	− 72	800	−496	1000	680
$x = 0{\cdot}1$		0·2		0·3		0·4		0·5	

FIG. 32

0·118	−1	0·388	−3	0·689	2	0·916	−1	1·000	1
$x = 0{\cdot}1$		0·2		0·3		0·4		0·5	

$\nu = 0{\cdot}0490$

FIG. 33

56. Our next step is the incorporation of the difference corrections in the calculation, and for this purpose we calculate $y(-0{\cdot}1)$ and $y(-0{\cdot}2)$ from the first two of (150), and $y(-0{\cdot}3)$ by applying (148), with Cy neglected, at $x = -0{\cdot}1$. We now have enough values to enable all difference corrections to be calculated, differences of orders greater than six being negligible. The results are shown on the left of Table 22.

TABLE 22

x	y		δ^2		δ^4		δ^6	*Difference corrections*
−0·3	+1·826							
		−1102						
− ·2	0·724		+538					
		564		−134				
− ·1	·160		404		+ 8			
		− 160		126		− 8		
·0	·000		278		0		+13	
		+ 118		126		+ 5		
+ ·1	·118		152		5		6	−0·0006
		270		121		11		
·2	·388		+ 31		16		+ 9	−0·0015
		301		105		20		
·3	·689		− 74		36		−12	+0·0020
		227		69		8		
·4	0·916		143		44		2	+0·0003
		+ 84		− 25		+ 6		
+0·5	+1·000		−168		+50		−12	+0·0020

Now the main correction Cy in (148) is $-\frac{1}{6}\delta^6 y$, to be added to the left of all but the first equation in (150): the correction to be added to the latter is $-12C_1 y_0 = -\frac{12}{30}\mu\delta^5 y_0$. The corrections to be added to equations (151) are thus $-\frac{1}{6}\delta^6 y_r$ for all save the first equation, and

$$-\tfrac{1}{6}\delta^6 y_1 + (\tfrac{1}{10}\mu\delta^5 + \tfrac{1}{24}\delta^6)y_0$$

in that equation. These values are also given in Table 22.

We can now obtain from Rayleigh's quotient the correction

$$\delta\nu = \sum yCy \,/\, \sum y^2 = 0{\cdot}0010,$$

giving a corrected ν of 0·0500, which compares satisfactorily with the correct value 0·0501, since only three figures have been retained in the pivotal values. To obtain better accuracy in both root and vector we should need to keep extra figures. Indeed in the solution of an eigenvalue problem involving a fourth-order equation we shall always need to retain more guarding figures than in the case of a second-order equation, just as in the more straightforward problems of previous chapters.

Determination of a second solution

57. In all our previous examples the solution obtained has been that corresponding to the smallest eigenvalue, and this is usually the most important solution, and often the only one required. Occasionally, however, we wish to calculate the next highest eigenvalue and the corresponding eigenfunction. This demand creates no inherently new problem, the relaxation process is unchanged, and we have only to guard against slipping back into a solution already obtained. This regression is usually easily avoided, since with most differential equations the shape of any required solution is roughly known, and the number of zeros in the range is often known exactly. If we start with a vector containing this number of zeros, and carry out the relaxation process carefully, making no changes which affect the number of zeros, there is generally no difficulty in obtaining the required solution.

In cases of difficulty we can use the orthogonality properties of eigensolutions, discussed in § 32, to remove from a guessed solution any component of an eigensolution already obtained. In particular, for self-adjoint systems, the guessed function y can be written in the form

$$y = a_1 y_1 + a_2 y_2 + \ldots, \tag{152}$$

where y_r is an eigenfunction. If y_1 has already been computed, we can calculate a_1 by multiplying (152) by y_1 and integrating from one boundary point a to the other b, finding

$$a_1 = \frac{\int_a^b yy_1\,dx}{\int_a^b y_1^2\,dx}, \tag{153}$$

the integrals of $y_1 y_r$ ($r \neq 1$) vanishing in virtue of the orthogonality relation. The function $y-a_1 y_1$ is then used at the start instead of y. At any subsequent stage in the process any component of y_1 inadvertently introduced can be removed by the same process. In (153) it is usually good enough to replace integrals by sums.

More accurate finite-difference equations

58. In our previous examples of differential systems the finite-difference equations have involved the eigenvalue only in their diagonal coefficients, and the matrix form of the algebraic problem was that of equation (5). The more general form (6) is most likely to arise, for differential systems, when the differential equation lacks its first derivative, and we use more accurate finite-difference equations. For example, an equation of the type

$$y''+\{f(x)+\lambda\}y = 0 \tag{154}$$

can be replaced by the finite-difference equation

$$\left.\begin{aligned} y_{r+1}\{1+\tfrac{1}{12}h^2(f_{r+1}+\lambda)\}+y_{r-1}\{1+\tfrac{1}{12}h^2(f_{r-1}+\lambda)\}-&\\ -y_r\{2-\tfrac{5}{6}h^2(f_r+\lambda)\}+Cy_r = 0&\\ C = \tfrac{1}{240}\delta^6-\ldots& \end{aligned}\right\}. \tag{155}$$

The difference correction is now smaller, and we should expect to be able to obtain a better first approximation, though with more labour. We note also that the symmetry of the matrix of these equations has disappeared, even when present in the simpler finite-difference equations.

If we apply this method to the example of Mathieu's equation in § 39, we find, at the interval $h = \frac{1}{8}\pi$, and remembering the symmetry at $x = \frac{1}{2}\pi$, the following four equations, in which $\nu = \frac{1}{12}h^2\lambda$:

$$\left.\begin{array}{ccccc} y\left(\dfrac{\pi}{8}\right) & y\left(\dfrac{2\pi}{8}\right) & y\left(\dfrac{3\pi}{8}\right) & y\left(\dfrac{4\pi}{8}\right) & \\ -2{\cdot}181741+10\nu & 1+\nu & & & = 0\\ 0{\cdot}981826+\nu & -2+10\nu & 1{\cdot}018174+\nu & & = 0\\ & 1+\nu & -1{\cdot}818259+10\nu & 1{\cdot}025702+\nu & = 0\\ & & 2{\cdot}036348+2\nu & -1{\cdot}742979+10\nu & = 0 \end{array}\right\}. \tag{156}$$

We find the results $\lambda = -0{\cdot}1113$, $y = 0, 0{\cdot}2547, 0{\cdot}5602, 0{\cdot}8641, 1{\cdot}0000$, which are much closer to the final results of Table 13 than are those of Table 12, the first approximation by the previous method.

Other methods for differential systems

59. We turn now to an examination of other methods of solving differential systems, analogous to the matrix iteration processes of §§ 15–29 for finite matrices. Any of these techniques can, of course, be applied directly to the matrices obtained from the finite-difference equations. The treatment of the difference correction, however, is not quite obvious.

There are, moreover, certain points in the analogies between matrices and differential systems which merit further consideration. For the moment we concentrate on problems with real solutions.

60. For a differential equation of the form

$$f(D)y = \lambda y, \tag{157}$$

in which $f(D)$ denotes a linear differential operator, and with suitable two-point boundary conditions, the number of eigenvalues is generally infinite, and any arbitrary function y_0 can be expressed as a convergent series of eigenfunctions in the form

$$y_0 = \sum_1^\infty \alpha_r y_r. \tag{158}$$

Then, in analogy with the basic iteration procedure for matrices, given by equation (30), we might consider the formal expression

$$\{f(D)\}^n y_0 = \sum_1^\infty \alpha_r \lambda_r^n y_r, \tag{159}$$

and the right-hand side would appear to tend ultimately to a multiple of the 'largest' eigenfunction. The analogy, however, is only apparent, since the infinite series on the right of (159) usually diverges. Similarly the equations corresponding to (37), one of the 'acceleration devices', and given for the differential system by

$$\{f(D)-p\}y = (\lambda-p)y, \tag{160}$$

and

$$\{f(D)-p\}^n y_0 = \sum \alpha_r(\lambda_r-p)^n y_r, \tag{161}$$

have little meaning, the series in (161) again diverging.

The divergence corresponds to the fact that a Fourier series or a series of other orthogonal functions may not converge when differentiated term by term. We can, however, integrate such series, so that we should expect to be able to use the equivalent of equations (31) and the accelerating device (40), in which inverse matrices are used. Thus the iterative procedure given by

$$f(D)y_n = y_{n-1} \tag{162}$$

will provide a series of functions y which converge to the 'smallest' eigenfunction, the corresponding eigenvalue being given by the ratio $y_{n-1}(x)/y_n(x)$ for any x. Convergence can be accelerated by using the iterative scheme

$$\{f(D)-p\}y_n = y_{n-1}, \tag{163}$$

where p is an approximation to the required eigenvalue. In both these methods the initially-chosen function y_0 need not necessarily satisfy the boundary conditions of the problem, but each successive step involves the solution of a differential equation of standard boundary-value type, the boundary conditions being always satisfied.

We find then that there is a direct analogy between differential equations, using what we might call 'inverse differential operators', and the use of inverse matrices, and this we proceed to develop. We return later to a consideration of the direct matrix iteration process.

Inverse differential operators. Examples

61. As a simple example, consider the differential system given by

$$y''-\lambda y = 0, \qquad y(-1) = y(1) = 0. \tag{164}$$

In this case the solution of each differential equation of the set (162) is simply equivalent to a double integration, introducing two arbitrary constants whose determination comes from the satisfaction of the boundary conditions. Starting with $y_0 = 1$, the simplest even polynomial, which incidentally does not satisfy the boundary conditions, we find

$$y_1 = \tfrac{1}{2}(x^2-1), \qquad y_2 = \tfrac{1}{24}(x^4-6x^2+5),$$
$$y_3 = \tfrac{1}{720}(x^6-15x^4+75x^2-61), \tag{165}$$

and the ratios of successive values of y at the origin are -2, $-2{\cdot}4$, $-2{\cdot}459$, converging fairly quickly to the true value

$$\lambda = -\pi^2/4 = -2{\cdot}467\ldots.$$

For this example equation (163) becomes

$$(D^2-p)y_n = y_{n-1}, \tag{166}$$

and with $y_0 = 1$ and the choice $p = -2$, we find

$$y_1 = \tfrac{1}{2}(1-\sec\sqrt{2}\cos x\sqrt{2}),$$

which has the value $-2{\cdot}706$ at $x = 0$. The estimate for λ at this stage is then given by $\lambda = -2-1/2{\cdot}706 = -2{\cdot}37$, so that we would expect rapid convergence to the correct result.

62. We cannot use $y_0 = 1$ as a start for the determination of the next largest eigenvalue, since the corresponding eigenfunction is an odd function of x, and therefore already orthogonal to y_0 and all functions derived from y_0 by our iterative process. We could try $y_0 = 1$ as a start for the third eigensolution, but for convergence we must either remove from it any component of the first solution present, or use the iterative scheme corresponding to (166), where p is now taken to be a reasonable approximation to the required solution $\lambda = -9\pi^2/4$.

63. The application to more complicated differential equations is straightforward. For example, for the equation

$$f(D)y = \lambda g(D)y, \tag{167}$$

where $g(D)$ is a differential operator of order not greater than that of $f(D)$, we use the iterative scheme

$$f(D)y_n = g(D)y_{n-1}, \tag{168}$$

and, ultimately, $y_n \to \lambda_1^{-1}y_{n-1}$, where λ_1 is here the numerically smallest eigenvalue, and y_n tends to the corresponding eigenvector. For convergence to a root close to p, we use

$$\{f(D)-pg(D)\}y_n = g(D)y_{n-1}, \tag{169}$$

whence, ultimately, $y_n \to (\lambda_s-p)^{-1}y_{n-1}$, where (λ_s-p) is smaller than any other (λ_r-p), and y_n becomes the eigenvector corresponding to λ_s.

Finite-difference methods for inverse differential operators

64. For most problems the solutions of the successive differential equations of type (162) or (168) cannot be obtained analytically, but we can use our numerical methods for this purpose. The differential equation (162), for the determination of y_n, can be represented by a set of finite-difference equations, and we can write these equations in the matrix form

$$Ay_n + Cy_n = b, \tag{170}$$

where Cy is the difference correction and b depends on the known y_{n-1}. Our usual process of solution is to introduce the difference correction in an iterative way, and we can write

$$y_n^{(r)} = A^{-1}\{b - Cy_n^{(r-1)}\}. \tag{171}$$

Having obtained an accurate y_n, we proceed to compute y_{n+1} from a similar formula, the matrix A being unchanged.

It is not necessary to invert A. The linear equations can be solved each time; for a second-order equation A will have the well-known advantageous form for this purpose, and the decomposition into triangles has to be made once only. However, the inversion may be desirable when the process needs several steps for convergence. It is also not necessary to introduce the difference correction until convergence is well established.

In a similar manner, equation (168) can be replaced by the finite-difference matrix equation

$$Ay_n + C_1 y_n = By_{n-1} + C_2 y_{n-1}, \tag{172}$$

and the solution of this equation is computed iteratively from

$$y_n^{(r)} = A^{-1}\{By_{n-1} + C_2 y_{n-1} - C_1 y_n^{(r-1)}\}. \tag{173}$$

Since y_{n-1} is known, however, we can calculate $g(D)y_{n-1}$ accurately, and replace (173) by

$$y_n^{(r)} = A^{-1}\{g(D)y_{n-1} - C_1 y_n^{(r-1)}\}. \tag{174}$$

65. As an example we consider the determination of the smallest eigenvalue and corresponding eigenfunction of the differential system

$$y'' - \frac{\lambda y}{x} = 0, \qquad y(0) = y(4) = 0. \tag{175}$$

In this case we solve the successive equations typified by

$$y_n'' = \frac{y_{n-1}}{x}, \tag{176}$$

and the smallest eigenvalue is the ultimate ratio of y_n to y_{n-1} for any value of x in the range at which y does not vanish.

The simplest finite-difference equation corresponding to (176) is given by

$$\left.\begin{aligned} \{y_1 + y_{-1} - 2y_0 + Cy_0\}_n &= \left(\frac{h^2}{x_0} y_0\right)_{n-1} \\ C &= -\tfrac{1}{12}\delta^4 + \tfrac{1}{90}\delta^6 - \dots \end{aligned}\right\}. \tag{177}$$

At the interval $h = \frac{1}{2}$ there are seven internal pivotal points, and the inverse of the matrix A of coefficients of the left-hand sides of the finite-difference equations is given by

$$-8A^{-1} = \begin{bmatrix} 7 & 6 & 5 & 4 & 3 & 2 & 1 \\ 6 & 12 & 10 & 8 & 6 & 4 & 2 \\ 5 & 10 & 15 & 12 & 9 & 6 & 3 \\ 4 & 8 & 12 & 16 & 12 & 8 & 4 \\ 3 & 6 & 9 & 12 & 15 & 10 & 5 \\ 2 & 4 & 6 & 8 & 10 & 12 & 6 \\ 1 & 2 & 3 & 4 & 5 & 6 & 7 \end{bmatrix}. \tag{178}$$

Neglecting the difference correction for the time being, we start with the arbitrary function $y_0 = 1$ at all pivotal points, and calculate the next iterate y_1 by applying the matrix (178) to the vector given by

$h^2/x = 1/4x$. The resulting vector is normalized to form the vector z_1 in which the largest component, at $x = 1{\cdot}5$, is unity. The normalizing factor is clearly an estimate of λ_1^{-1}. Repeating this process, we form $z_2,\ldots, z_5$, at which stage convergence is reasonably well established. To avoid complication with the introduction of the difference correction, we now identify z_5 with a new starting vector y_5, and form the true y_6. The ratio of y_6 to y_5 at $x = 1{\cdot}5$ is close to the previous λ^{-1}, so that the final approximate solution has been obtained, and the approximate λ is

$$(-1{\cdot}0846)^{-1} = -0{\cdot}9220.$$

The results of this computation, together with the differences and difference corrections of y_6, needed for further improvements, are shown in Table 23.

In the calculation of the difference correction, differences higher than the fourth are clearly negligible, and the 'missing' values can be estimated with fair confidence.

66. At this stage we have an approximation $y_6^{(1)}$ to the solution of $y_6'' = x^{-1}y_5$. We improve it by adding the correcting term $A^{-1}\{-Cy_6^{(1)}\}$, which, in units of the fourth decimal, is found to have the values -27, -45, -53, -54, -49, -37, -21 at the respective pivotal points. The difference correction of this increment is negligible, so that an accurate solution of $y_6'' = x^{-1}y_5$ is given by

x	0·0	0·5	1·0	1·5	2·0	2·5	3·0	3·5	4·0
$-y_6$	0·0000	0·6322	0·9732	1·0899	1·0392	0·8688	0·6181	0·3197	0·0000

At $x = 1{\cdot}5$ the ratio of y_6 to y_5 is now $-1{\cdot}0899$, and the corresponding value of λ is $-0{\cdot}9175$. If y_6 is normalized in the same form as y_5, we have the result

x	0·0	0·5	1·0	1·5	2·0	2·5	3·0	3·5	4·0
y_6	0·0000	0·5801	0·8929	1·0000	0·9535	0·7971	0·5671	0·2933	0·0000

This is not quite identical with y_5, so that we should carry out one more step, calculating an accurate y_7 from y_6, just as y_6 was obtained from y_5, and evaluating the ratio y_7/y_6.

This problem, which has the analytical solution

$$y = \surd x J_1\{2\surd(-\lambda x)\}, \qquad J_1\{4\surd(-\lambda)\} = 0, \tag{179}$$

giving $\lambda = -0{\cdot}9176$, could of course be solved by the method of § 61, since the integration of (176) can be performed analytically. Nevertheless it serves to illustrate the matrix method, differing from the general case only in the simplicity of the inverse matrix (178).

TABLE 23

x	y_0	z_1	z_2	z_3	z_4	$z_5 = y_5$	$-y_6$		δ^2		δ^4		δ^6	$-Cy_6$
+0·0	+1	+0·0000	+0·0000	+0·0000	+0·0000	+0·0000	+0·0000							
								+6295						
0·5	1	·6474	·5944	·5837	·5812	·5805	·6295		−2903					(+10)
								3392		+670				
1·0	1	0·9179	0·8996	0·8950	0·8937	0·8932	0·9687		2233		−104			9
								+1159		566		+14		
1·5	1	1·0000	1·0000	1·0000	1·0000	1·0000	1·0846		1667		90		−5	7
								− 508		476		9		
2·0	1	0·9565	0·9515	0·9523	0·9529	0·9531	1·0338		1191		81		0	7
								1699		395		9		
2·5	1	·8187	·7963	·7955	·7961	·7964	0·8639		796		72		+5	6
								2495		323		+14		
3·0	1	·6056	·5679	·5657	·5661	·5663	·6144		473		− 58			5
								2968		+265				
3·5	1	·3297	·2945	·2925	·2927	·2928	·3176		− 208					(+ 4)
								−3176						
+4·0	+1	+0·0000	+0·0000	+0·0000	+0·0000	+0·0000	+0·0000							
λ^{-1}		−1·3268	−1·1198	−1·0907	−1·0858	−1·0849								

67. This simplicity disappears if we apply the faster process, taking $p = -1$ as an estimate for λ, and using the iterative scheme

$$y_n'' - \frac{p}{x} y_n = \frac{1}{x} y_{n-1}. \tag{180}$$

The inverse matrix corresponding to the finite-difference form of the left-hand side of (180) is given by

$$A^{-1} = \begin{bmatrix} 5{\cdot}255 & 8{\cdot}882 & 10{\cdot}288 & 9{\cdot}980 & 8{\cdot}423 & 6{\cdot}025 & 3{\cdot}124 \\ 8{\cdot}882 & 13{\cdot}323 & 15{\cdot}432 & 14{\cdot}970 & 12{\cdot}635 & 9{\cdot}037 & 4{\cdot}686 \\ 10{\cdot}288 & 15{\cdot}432 & 16{\cdot}717 & 16{\cdot}216 & 13{\cdot}687 & 9{\cdot}789 & 5{\cdot}076 \\ 9{\cdot}980 & 14{\cdot}970 & 16{\cdot}216 & 14{\cdot}760 & 12{\cdot}458 & 8{\cdot}910 & 4{\cdot}620 \\ 8{\cdot}423 & 12{\cdot}635 & 13{\cdot}687 & 12{\cdot}458 & 9{\cdot}671 & 6{\cdot}917 & 3{\cdot}586 \\ 6{\cdot}025 & 9{\cdot}037 & 9{\cdot}789 & 8{\cdot}910 & 6{\cdot}917 & 4{\cdot}232 & 2{\cdot}194 \\ 3{\cdot}124 & 4{\cdot}686 & 5{\cdot}076 & 4{\cdot}620 & 3{\cdot}586 & 2{\cdot}194 & 0{\cdot}619 \end{bmatrix}, \tag{181}$$

and application to successive vectors, starting with the same vector as before, produces the vectors y_1 and y_2 as shown in Table 24, the ratios of y_1 to y_2 being also recorded at the various pivotal points. These are so nearly constant that we can assert that a good approximation to the eigenvalue, with difference corrections neglected, is given by

$$\lambda = -1+0{\cdot}078 = -0{\cdot}922.$$

TABLE 24

y_0	1	1	1	1	1	1	1
y_1	9·378	14·566	16·362	15·599	13·008	9·218	4·743
y_2	121·14	186·40	208·69	198·93	166·23	118·22	61·12
y_1/y_2	0·0774	0·0781	0·0784	0·0784	0·0783	0·0780	0·0776

This value of λ agrees closely with the approximate value obtained by the previous method, and is produced in many fewer steps. The vector y_2 of Table 24, when adjusted so that its largest value is the same as that of y_5 of Table 23, differs from y_5 at other points by a maximum of 2 units in the fourth decimal. As in the previous method, we should now proceed to incorporate the difference correction and get a better solution to the differential equation.

68. Finally, we may note that there is no difficulty in using more accurate finite-difference equations. For example, if the $f(D)$ of equation (168) is of second order, but lacking the first derivative, the technique of operating on both sides with $1+\frac{1}{12}\delta^2$ produces a smaller difference correction on the left, and on the right the quantities $(1+\frac{1}{12}\delta^2)g(D)y_{n-1}$ are calculable with all the accuracy obtainable from values of y_{n-1}

known to a given precision. In equation (169), however, this method is applicable, for second-order equations, only if $g(D)$ also contains no first derivative.

Determination of a second solution

69. To calculate a second solution of our last example, we need only repeat the method of equation (169) with a different value of p, sufficiently close to the required eigenvalue. This will often be known in advance from mathematical or physical considerations.

Otherwise we can easily find a sufficiently good approximation to p by the simpler process based on equation (176), for which the inverse matrix is already available. By starting with a vector from which the component of the first eigenvector has been removed by use of the orthogonality relations, the method of equation (176) will in theory give convergence to the next highest eigenvalue. In practice the process will introduce, through rounding error, small components of the unwanted first eigensolution which will then proceed to grow without limit. Before this growth becomes serious, however, we shall have a reasonable approximation to the eigensolutions which can then be used in the faster process of equation (169).

Consider, for example, the determination of the next highest eigensolution of the system (175). Since in the first approximation the difference correction is ignored we shall want to start with a vector orthogonal to the first solution of the original matrix equation, in which the difference correction was also neglected. This first solution is labelled y_5 in Table 23. Starting with an arbitrary vector z_0, equal to 1 at $x = 2$ and zero elsewhere, we wish to deduce a vector y_0 such that

$$y_0 = z_0 - \alpha y_5 \tag{182}$$

is orthogonal to y_5. Using the orthogonality relation (11) for solutions of the matrix equation $(A - B\lambda)y = 0$, we have

$$\alpha = \frac{y_5' B z_0}{y_5' B y_5}, \tag{183}$$

where B is here a diagonal matrix with elements x^{-1}, for

$$x = 0{\cdot}5(0{\cdot}5)3{\cdot}5.$$

The resulting vector y_0 is shown in Table 25, together with the next four vectors obtained by premultiplication of $h^2 y_r/x$ with the matrix A^{-1} of equation (178). The vector is in each case normalized so that its largest component is unity, this normalizing factor giving an estimate of the corresponding λ_2.

The last column of Table 25 has been obtained by removing from y_4 any component reintroduced of the first eigensolution, and comparison of y_4 and y_5 shows how quickly this infiltration becomes manifest.

The value of λ obtained from y_3 and y_4 is about $-3{\cdot}1$, and the choice of $p = -3$ in equation (180) and the starting vector given by y_4 of Table 25 will cause rapid convergence to a first approximation to the second eigensolution, subsequently to be corrected by use of the difference correction in the usual way.

TABLE 25

x	y_0	$-y_1$	y_2	$-y_3$	y_4	y_5
+0·0	−0·0000	−0·0000	−0·0000	−0·0000	−0·0000	−0·0000
0·5	·0929	·4518	·7181	·7160	·6836	·7377
1·0	·1429	− ·3545	− ·3881	− ·3275	− ·2760	− ·3296
1·5	− ·1600	+0·1651	+ ·3530	+ ·3596	+ ·3810	+ ·3475
2·0	+ ·8475	1·0000	0·9665	0·8656	0·8554	0·8424
2·5	− ·1274	0·5825	1·0000	1·0000	1·0000	1·0000
3·0	·0906	·3157	0·7633	0·8267	0·8399	0·8449
3·5	·0468	·1381	·4045	·4578	·4698	·4737
+4·0	−0·0000	+0·0000	+0·0000	+0·0000	+0·0000	+0·0000

Direct differential operators

70. Returning now to a further discussion of methods not involving inverse matrices, we have already remarked that the direct *analytical* approach, based on equation (159), gives no useful result. The direct *matrix* process, however, is not quite analogous to the direct analytical process, since the matrix involved is a finite matrix, and the number of its solutions is finite. Moreover, the form of the finite matrix ensures that successive vectors satisfy the correct boundary conditions. We might therefore expect that the direct iterative process represented by (159), but carried out numerically as represented by equation (30), would converge to the largest eigensolution representable by the chosen number of pivotal points. As previously indicated, however, the approximate solution, obtained with neglect of the difference correction, is likely to be very poor for the largest eigenvalue.

To produce the smallest eigensolution we must use (37), the matrix equivalent of (161), with p chosen so that $(\lambda-p)$, for the smallest λ, is greater than that for any other λ. The approximate solution for the smallest eigenvalue should be closer to the truth than that for any other.

The iterative process represented by (159), with the difference correction included, can be written in the form

$$Ay_n+Cy_n = h^2y_{n+1}, \tag{184}$$

and that of (161) in the form

$$(A-h^2p)y_n+Cy_n = h^2y_{n+1}, \tag{185}$$

and the technique of including the difference correction, when convergence is well established, is fairly simple.

71. As an illustration we consider again the example of § 65. At the same interval $h = \frac{1}{2}$, the differential operator h^2xy'' is represented by the matrix

$$A = \begin{array}{c} \begin{matrix} y(0{\cdot}5) & y(1{\cdot}0) & y(1{\cdot}5) & y(2{\cdot}0) & y(2{\cdot}5) & y(3{\cdot}0) & y(3{\cdot}5) \end{matrix} \\ \begin{bmatrix} -1 & 0{\cdot}5 & & & & & \\ 1 & -2 & 1 & & & & \\ & 1{\cdot}5 & -3 & 1{\cdot}5 & & & \\ & & 2 & -4 & 2 & & \\ & & & 2{\cdot}5 & -5 & 2{\cdot}5 & \\ & & & & 3 & -6 & 3 \\ & & & & & 3{\cdot}5 & -7 \end{bmatrix} \end{array}, \tag{186}$$

there being no entries corresponding to $y(0)$ and $y(4)$, whose values are specified as zero.

Using the iterative scheme (184), and starting with the arbitrary vector 1, 1, 1, 1, 1, 1, 1, we find the successive vectors recorded in Table 26, with the normalizing factors, giving estimates of the largest eigenvalue, arranged so that the last component, at $x = 3{\cdot}5$, is unity.

Convergence is rather slow, but if we apply Aitken's acceleration device to produce the major components of a new vector y_9, estimating the less important values at $x = 0{\cdot}5$, $1{\cdot}0$, and $1{\cdot}5$, and repeat the iteration once to provide y_{10}, we find the following results:

x	0·5	1·0	1·5	2·0	2·5	3·0	3·5	
y_9	0·0000	−0·0030	+0·0400	−0·1845	+0·5373	−1·0111	+1·0000	$\lambda = -42{\cdot}1554$
y_{10}	+0·0001	−0·0044	+0·0381	−0·1796	+0·5385	−1·0133	+1·0000	

To two figures the vector is fairly constant, and the eigenvalue is about −42. This corresponds to the seventh eigensolution of the differential equation, with the true eigenvalue −32·37... . We cannot, however, correct our approximation, since the function oscillates so wildly that calculation of the difference correction is impossible.

72. If we wish to calculate the smallest eigensolution without the use of an inverse matrix, we must use instead of A the matrix $(A-h^2pI)$, with a suitably chosen value of h^2p. We might take the value −28, equal to the sum of the diagonal terms of A, but quicker convergence is assured if we take $h^2p = -10$, an approximation, already obtained in § 71, to

TABLE 26

x	y_0	y_1	y_2	y_3	y_4	y_5	y_6	y_7	y_8
+0·5	+1	+ 0·1429	+ 0·0204	+ 0·0036	+ 0·0008	+ 0·0002	+ 0·0001	+ 0·0000	+ 0·0001
1·0	1	·0000	− ·0204	− ·0072	− ·0023	− ·0008	− ·0007	− ·0012	− ·0019
1·5	1	·0000	·0000	+ ·0036	+ ·0023	+ ·0054	+ ·0110	+ ·0169	+ ·0220
2·0	1	·0000	·0000	·0000	− ·0279	− ·0620	− ·0924	− ·1163	− ·1340
2·5	1	·0000	·0000	+ ·1261	+ ·2442	+ ·3340	+ ·3978	+ ·4420	+ ·4722
3·0	1	0·0000	− 0·4286	− 0·6555	− 0·7866	− 0·8666	− 0·9171	− 0·9497	− 0·9710
+3·5	+1	+ 1·0000	+ 1·0000	+ 1·0000	+ 1·0000	+ 1·0000	+ 1·0000	+ 1·0000	+ 1·0000
λ		−14·0000	−28·0000	−34·0004	−37·1770	−39·0124	−40·1324	−40·8394	−41·2958

the largest root of A. Convergence is likely to be fairly slow, however, for any value of p, since the relevant values of $-h^2\lambda$ for the differential equation are 0·23, 0·77, 1·62, 2·77, 4·24, 6·01, 8·09, and no h^2p exists for which $h^2(\lambda-p)$ is much larger for the smallest λ than for any other.

To demonstrate the method of applying the difference correction, however, we take as a start the best vector y_6 of Table 23, to which we should undoubtedly converge ultimately, and apply to it the matrix $(A+10I)$, where A is given in equation (186). The vector y_6 of Table 23, normalized so that its largest component is unity, the result of pre-multiplication by $(A+10I)$, and the new normalizing factor, are shown in Table 27.

TABLE 27

x	0·5	1·0	1·5	2·0	2·5	3·0	3·5	*Normalizing factor*
y	0·5804	0·8931	1·0000	0·9532	0·7965	0·5665	0·2928	
$(A+10I)y$	0·5804	0·8931	1·0000	0·9532	0·7965	0·5664	0·2929	9·7695

For the solution of the matrix equation the vector is clearly correct in its last figure, and the eigenvalue is given by $h^2\lambda = 9{\cdot}7695-10$, or $\lambda = -0{\cdot}9220$, in close agreement with previous results.

We now proceed to incorporate the difference correction, given here by

$$Cy = x(-\tfrac{1}{12}\delta^4+\tfrac{1}{90}\delta^6-\ldots)y, \tag{187}$$

and whose values, at the respective pivotal points, are found to be

(6) 8, 10, 12, 14, 15, (15).

When we include these values on the left of equation (185), computing a more accurate h^2y_{n+1} and normalizing as before, we find the normalizing factor is now 9·7705, giving $h^2\lambda = -0{\cdot}2295$, incorrect by only 1 unit in the fourth decimal. With work of this precision further accuracy could not be expected.

In many of these examples, of course, the Rayleigh quotient could be used finally to produce a more accurate eigenvalue from the finally-computed eigenvector.

Harder problems. Simultaneous equations

†73. So far we have not encountered complex solutions, and have used none of the root-removal processes of §§ 26–29. These methods will be required, however, in problems for which the qualitative nature of the solutions cannot be assessed in advance. It may happen, for example, that an equation or system of equations has few or no real solutions, and it is required to determine those real solutions, if any, which exist.

This problem may arise particularly in the solution of simultaneous differential equations involving an eigenvalue. Consider, for example, a pair of such equations of the form

$$\left.\begin{aligned} f_1(D)y+f_2(D)z+\lambda\{f_3(D)y+f_4(D)z\} &= 0 \\ g_1(D)y+g_2(D)z+\lambda\{g_3(D)y+g_4(D)z\} &= 0 \end{aligned}\right\}, \tag{188}$$

with suitable homogeneous boundary conditions.

We might try to solve (188) by eliminating either y or z, obtaining a single equation of higher order in the other. The parameter λ will almost certainly occur then in a non-linear form, and the basic methods of this chapter no longer apply. An alternative procedure is to express (188) in matrix form by the usual method of replacing the differential equations by finite-difference equations. With the boundary conditions suitably incorporated, we can then represent (188) by the matrix equations

$$\left.\begin{aligned} (F_1+C_1)y+(F_2+C_2)z+\lambda\{(F_3+C_3)y+(F_4+C_4)z\} &= 0 \\ (G_1+C_1')y+(G_2+C_2')z+\lambda\{(G_3+C_3')y+(G_4+C_4')z\} &= 0 \end{aligned}\right\}, \tag{189}$$

in which the F and G of (189) are matrices stemming from truncated finite-difference expressions of the corresponding f_r, g_r of (188), the C and C' are effectively diagonal matrices representing the difference corrections, and y and z are vectors, whose components represent the pivotal values of the required solutions $y(x)$, $z(x)$.

We can now combine the two equations (189) into a single equation of the form

$$\left(\begin{array}{c:c} F_1+C_1+\lambda(F_3+C_3) & F_2+C_2+\lambda(F_4+C_4) \\ \hdashline G_1+C_1'+\lambda(G_3+C_3') & G_2+C_2'+\lambda(G_4+C_4') \end{array}\right)\begin{pmatrix} y \\ z \end{pmatrix} = 0 \tag{190}$$

and, removing the partitions, we have a square matrix and vector of order twice the original, the vector comprising the unknown pivotal values of both y and z.

The answer to questions on the existence and results for real solutions of (188) is thus reduced to the same problem for (190), and it is reasonable, for the existence theorem, to ignore the difference corrections C and C'. We can write (190) in the standard form

$$(A-B\lambda)\eta = 0, \tag{191}$$

and by premultiplication with A^{-1} or B^{-1} can reduce it to one or other of the easier forms

$$(A^{-1}B-I\lambda^{-1})\eta = 0, \qquad (B^{-1}A-I\lambda)\eta = 0. \tag{192}$$

Suppose we are asked to determine the largest (smallest) real root λ, or possibly the largest (smallest) real and positive root. A reasonable procedure is to apply the direct matrix iteration methods to the second (first) of (192), calculating first the root or roots of largest modulus, using if necessary any of the accelerating devices of §§ 17 and 18, and removing each root as it is obtained by one of the methods of §§ 26–29.

If by these methods we find a root with the required qualities, and the associated vector is reasonably 'smooth', so that the finite-difference equations have some meaning, we can assert the existence of a solution with the required qualities. If the vector has no smoothness at all, or if no real root is found, we can assert that, if a real root exists, it is associated with a vector of considerable oscillation. If this information does not suffice for the physical problem, it would be necessary to use a smaller interval and repeat the process.

Occasionally rounding errors introduced in the root-removal process reduce the accuracy of the first approximation as a solution of the matrix equation, but this can be overcome by choosing a p reasonably close to the approximate eigenvalue for use in the 'inverse' process represented by equation (40). Again the chosen p should not be so near the approximate λ that the matrix is nearly singular, and not so far removed from the required λ that we converge to a different root or complex pair.

74. If a satisfactory approximate solution has been obtained, the final question is: can this be improved by inclusion of the difference corrections? With the inverse process there is clearly no difficulty, for (191) is an approximation to the accurate equation

$$\{A+C-\lambda(B+C')\}\eta = 0, \tag{193}$$

where C and C' are larger matrices formed from the C_r, C'_r of (190). The accurate inverse process is represented by the equation

$$\{A+C-p(B+C')\}\eta_n = (B+C')\eta_{n-1}, \tag{194}$$

in which, in the first approximation, C and C' are neglected. The incorporation of the difference correction is performed as for the simpler problems, by the equation

$$\eta_n^{(r)} = (A-pB)^{-1}\{(B+C')\eta_{n-1}-(C-pC')\eta_n^{(r-1)}\}. \tag{195}$$

The direct process cannot easily include the difference correction, since it necessitates 'root-removal' operations, and the connexion between the original differential equations and the 'reduced' matrix equations is not obvious.

75. The arithmetical labour involved in the solution of simultaneous differential equations can be very substantial. A set of twelve simultaneous equations, using ten pivotal points, gives a final matrix of order

120. Operations of linear algebra on a matrix of this size are hardly practicable with desk-computing machines, and the use of a modern high-speed electronic computer is almost essential.

Additional notes and bibliography

§ 1. The mathematical theory of differential systems of eigenvalue type has been given by

INCE, E. L., 1944, *Ordinary differential equations*. Dover.

§§ 7–29. In this review of methods for solving the algebraic problem, we have concentrated on those thought to be most useful in application to the differential problem and on those of which we have first-hand experience. Another consideration is the type of calculating machine available. Relaxation is quite useful, in the hands of a skilled worker, for calculating one or two solutions for a matrix of relatively low order, say not greater than twelve, particularly when the matrix is symmetric and of simple construction. For larger problems, and for unsymmetric matrices, one of the other methods may have to be used, and in this respect an automatic high-speed computer is a great asset, desk-machine work being very laborious for this problem.

For one or other of these reasons we have discarded any method of constructing the characteristic equation. In the general algebraic problem, however, when most of the roots are required, this method, assisted by a high-speed computer and with expert attention to detail, is often quite practicable and useful.

Another method not featured in the text, which finds all the roots of a symmetric matrix, is that known as the method of 'rotations'. The basis of this method is the calculation of an orthogonal matrix T such that

$$T'AT = D,$$

the diagonal matrix D having the latent roots of A as its components, and the latent vectors of A are the corresponding columns of T. The orthogonal matrix T is composed of the successive products of simple orthogonal matrices constructed in a definite manner depending on the results of earlier steps in the process.

In this method the number of steps in the construction of the matrix T is not known in advance, and the process is continued until the matrix D has negligible coefficients except in its diagonal. A somewhat similar method produces the form

$$T'AT = B,$$

where B is a matrix of 'co-diagonal form', or a 'band' matrix of width three. Here the number of steps is at most $\frac{1}{2}(n-1)(n-2)$, and the determination of the latent roots of B is a relatively simple operation. The method, particularly suitable for high-speed computation, has been described by

GIVENS, W., 1953, A method of computing eigenvalues and eigenvectors suggested by classical results on symmetric matrices. *National Bureau of Standards Applied Mathematics Series*, **29**. Washington.

With regard to the devices for accelerating an iterative method, the trial and error associated with the determination of a suitable value for p, in equations like (37) and (40), makes the use of a high-speed machine almost essential.

Of the root-removal processes that based on equation (50), for symmetric matrices, is the one most convenient if relaxation methods are used. The symmetry is preserved, so that the important properties of Rayleigh's Principle, very valuable for this method, are still applicable for the transformed matrix. With other methods, not using or depending on Rayleigh's Principle, and with which high-speed machines are an asset, methods which depress the order of the matrix are generally preferred, the loss of symmetry being an almost negligible disadvantage.

Most of the iterative methods mentioned have been discussed, with particular reference to high-speed computation, by

WILKINSON, J. H., 1954, The calculation of the latent roots and vectors of matrices on the Pilot Model of the ACE. *Proc. Camb. Phil. Soc.* **50**, 536–66.

Many root-removal processes, including one for symmetric matrices which depresses the order while still preserving symmetry, though necessarily more complicated than that of equation (58), have been given by

FELLER, W., and FORSYTHE, G. E., 1951, New matrix transformations for obtaining characteristic vectors. *Quart. Appl. Math.* **8**, 325–31.

§§ 30–32. The properties of solutions of differential systems of various forms have been discussed by

INCE, E. L., 1944, *Ordinary differential equations.* Dover, and by
COURANT, R., and HILBERT, D., 1953, *Methods of mathematical physics.* Interscience Publishers, Inc., New York.

§ 33. Rayleigh's Principle was originally applied to vibrating systems, discrete or continuous, and was based on the fact that at any time the mean kinetic and potential energies are equal.

Consider, for example, the problem of a vibrating string fixed at its ends $x = 0$ and 1. If the transverse displacement has amplitude $y(x)$ and the frequency of vibration is p, the displacement is given by $y(x)\cos pt$, and y satisfies the differential equation

$$d^2y/dx^2+p^2y = 0,$$

in which all constants such as the density and tension of the string have been suppressed for simplicity. The mean kinetic and potential energies are given by

$$T \propto \int_0^1 \left(\frac{dy}{dx}\right)^2 dx, \qquad V \propto p^2 \int_0^1 y^2\, dx,$$

and Rayleigh's Principle is embodied in the equation

$$p^2 = \int_0^1 \left(\frac{dy}{dx}\right)^2 dx \bigg/ \int_0^1 y^2\, dx.$$

Since y vanishes at $x = 0$ and 1, we can express the numerator in a different form by integrating by parts, obtaining

$$p^2 = -\int_0^1 y\frac{d^2y}{dx^2}\, dx \bigg/ \int_0^1 y^2\, dx,$$

a form analogous to that of equation (78). For systems of this kind, of course, the roots are all positive, and the complete Rayleigh Principle holds. Its applications to vibrating systems have been described by

TEMPLE, G., and BICKLEY, W. G., 1933, *Rayleigh's Principle.* Oxford University Press.

§ 35. A method, different in principle from those of this chapter, which uses finite differences to produce an approximation to the characteristic equation, whose roots are approximations to some eigenvalues of the differential system, has been given by

MILNE, W. E., 1953, *Numerical solution of differential equations*. Wiley, New York.

§ 38. If in equation (85) we take the alternative solution for δy, the eigensolution is exact, and corresponds to the third eigenvalue of the differential equation. Rayleigh's quotient gives $\nu = 2+\surd 2$, and we find, corresponding to (92), the expansion

$$\tfrac{9}{16}\pi^2 = (2+\surd 2)+\tfrac{1}{12}(2+\surd 2)^2+\tfrac{1}{90}(2+\surd 2)^3+\dots.$$

§ 45. This problem has been treated by different methods by

JEFFREYS, H., 1924, *Proc. Lond. Math. Soc.* (2) **23,** 455–76, and
GOLDSTEIN, S., 1928, *Proc. Lond. Math. Soc.* (2) **28,** 91–101.

§ 73. A problem of this kind has been treated by

BULLARD, E. C., and GELLMAN, H., 1954, Homogeneous dynamos and terrestrial magnetism. *Phil. Trans.* A, **247** 213–78.

8

Initial-value Techniques for Boundary-value Problems

Introduction

1. ALL boundary-value problems can be solved by initial-value techniques, with varying degrees of accuracy and convenience. The basic process solves several initial-value problems, each with some arbitrary initial conditions, combining the solutions to satisfy all the given boundary conditions. In this chapter we illustrate the method for equations of second and higher orders, both linear and non-linear, and for eigenvalue problems.

Linear second-order equations. Specified boundary values

2. Consider first the simplest problem, that of solving a linear second-order equation

$$y''+f(x)y'+g(x)y = k(x), \tag{1}$$

with given values of y at the two end points of the range. We can replace (1) approximately in the usual way by a set of algebraic equations of the form

$$a_r y_{r-1}+b_r y_r+c_r y_{r+1} = d_r, \tag{2}$$

where a, b, c, and d depend on x only, y_r is the pivotal value at a general pivotal point x_r, and we have ignored temporarily the difference correction. In our problem the boundary values y_0 and y_n are known, and the standard technique is to treat and solve (2) as a set of simultaneous equations. If, on the other hand, y_1 or y_{n-1} were also known, we could use a standard step-by-step process.

Suppose that y_1 is given an arbitrary value. Then from (2) we can recur forward, using the known y_0 and arbitrary y_1 to produce all pivotal values including y_n. Except by chance the latter will not agree with the given boundary value at x_n. If we repeat the process with a second guessed y_1, we can produce a second tentative y_n. Since the equation is linear we can combine these two solutions linearly to satisfy the given condition at x_n. If the first trial solution is labelled $y^{(1)}$ and the second $y^{(2)}$, then $\alpha_1 y^{(1)}+\alpha_2 y^{(2)}$ is the required solution to (2) if

$$\left.\begin{aligned} \alpha_1 y_n^{(1)}+\alpha_2 y_n^{(2)} &= y_n \\ \alpha_1+\alpha_2 &= 1 \end{aligned}\right\}. \tag{3}$$

The second of (3) ensures that the general equations (2) are satisfied and that the solution has the specific value y_0 at x_0, and the first of (3) ensures that the solution has the specified value y_n at x_n.

3. As an example of this simple process we consider the equation

$$y''-y-x = 0, \tag{4}$$

with $y(0) = 0$, $y(1) = 1$, solved approximately in §§ 4 and 5 of Chapter 4. There we used the interval $h = \frac{1}{4}$, and the approximate finite-difference equations were given by

$$\left.\begin{aligned} y(\tfrac{1}{4})+y(-\tfrac{1}{4})-2{\cdot}0625y(0) \qquad\qquad &= 0\\ y(\tfrac{1}{2})+y(0)-2{\cdot}0625y(\tfrac{1}{4})-0{\cdot}015625 &= 0\\ y(\tfrac{3}{4})+y(\tfrac{1}{4})-2{\cdot}0625y(\tfrac{1}{2})-0{\cdot}031250 &= 0\\ y(1)+y(\tfrac{1}{2})-2{\cdot}0625y(\tfrac{3}{4})-0{\cdot}046875 &= 0\\ y(\tfrac{5}{4})+y(\tfrac{3}{4})-2{\cdot}0625y(1)-0{\cdot}062500 &= 0 \end{aligned}\right\}, \tag{5}$$

in which $y(0)$ and $y(1)$ are known, and the first and last of (5) are included for use in later investigations.

Taking the given value $y(0) = 0$ and choosing $y(\frac{1}{4}) = 0$ for a first solution $y^{(1)}$, $y(\frac{1}{4}) = 1$ for a second solution $y^{(2)}$, and using the relevant equations of (5) in a forward recurrence process, we produce the trial solutions $y^{(1)}$ and $y^{(2)}$ shown in Table 1.

TABLE 1

x	$y^{(1)}$	$y^{(2)}$	y
+0·00	+0·000000	+0·000000	+0·000000
·25	·000000	1·000000	·180229
·50	·015625	2·078125	·387347
0·75	·063477	3·317383	·649925
+1·00	+0·162171	+4·810852	0·999998

The multipliers in the final solution $\alpha_1 y^{(1)}+\alpha_2 y^{(2)}$ satisfy the equations

$$\left.\begin{aligned} 0{\cdot}162171\alpha_1+4{\cdot}810852\alpha_2 &= 1\\ \alpha_1+\qquad\qquad \alpha_2 &= 1 \end{aligned}\right\}, \tag{6}$$

giving $\alpha_1 = 0{\cdot}819771$, $\alpha_2 = 0{\cdot}180229$, and the final y shown in the last column of Table 1. This agrees to five decimals with the corresponding result in Chapter 4.

4. We could also recur in the opposite direction, fixing arbitrary values at $x = \frac{3}{4}$. Corresponding results are shown in Table 2.

TABLE 2

x	$y^{(1)}$	$y^{(2)}$		y
+1·00	+1·000000	+1·000000	($\alpha_1 = 0{\cdot}3500735$)	+1·000000
0·75	0·000000	1·000000	($\alpha_2 = 0{\cdot}6499265$)	0·649927
·50	−0·953125	1·109375		0·387349
·25	1·934570	1·319336		0·180230
+0·00	−3·021301	+1·627380		+0·000000

5. By a slight modification of method we can shorten the work and avoid the necessity for solving simultaneous equations for α_1 and α_2. If we choose our first trial solution as before, to satisfy (2) and the correct initial condition y_0 and with a guessed y_1, and the second trial solution to satisfy (2) with d_r suppressed and with zero initial value and an arbitrary y_1, then the combination $y^{(1)}+\alpha y^{(2)}$ satisfies the differential equation and the correct initial condition. The constant α is then determined directly from the single equation

$$y_n^{(1)}+\alpha y_n^{(2)} = y_n. \tag{7}$$

In the last example we recurred backwards from $x = 1$. Repeating this with the new technique we have the trial solutions shown in Table 3.

TABLE 3

x	$y^{(1)}$	$y^{(2)}$	y
+1·00	+1·000000	+0·000000	+1·000000
0·75	0·000000	1·000000	0·649927
·50	−0·953125	2·062500	·387348
·25	1·934570	3·253906	·180230
+0·00	−3·021301	+4·648681	+0·000000

We have

$$4{\cdot}648681\alpha - 3{\cdot}021301 = 0, \qquad \alpha = 0{\cdot}6499265,$$

and the final result is given in the last column of Table 3.

Linear second-order equations. Other boundary conditions

†**6.** Our process may require modification if the boundary conditions involve a derivative. If only one boundary point is so affected we can recur from the other boundary as before, combining two solutions to satisfy the condition at the other end. Suppose, for example, that in our previous problem we replace the condition at $x = 0$ by

$$y'+y = 1 \text{ at } x = 0, \tag{8}$$

represented by an algebraic equation of which the simplest of reasonable accuracy at interval $h = \frac{1}{4}$ is given by

$$y(\tfrac{1}{4})-y(-\tfrac{1}{4})+0{\cdot}5y(0) = 0{\cdot}5, \tag{9}$$

again neglecting a difference correction.

We now merely extend our trial solutions of Table 3 one step farther to the point $x = -\frac{1}{4}$, using for this purpose the first of equations (5); in this equation the constant term is zero so that it is here used for both $y^{(1)}$ and $y^{(2)}$. We find $y^{(1)}(-\frac{1}{4}) = -4{\cdot}296863$, $y^{(2)}(-\frac{1}{4}) = 6{\cdot}333999$, and $y = y^{(1)}+\alpha y^{(2)}$ satisfies (9) if

$$-0{\cdot}755752\alpha+0{\cdot}851642 = 0{\cdot}5, \text{ giving } \alpha = 0{\cdot}465288,$$

and the required solution can be assembled as before.

It will be noted that in this case we do not need to use Taylor's series or any other device for starting off the recurrence, and that the derivative condition is not used in the production of the trial solutions as it is, in its Lagrangian form, in boundary-value techniques.

†7. If the derivative is present in both boundary conditions, however, the method is less simple. Suppose the condition at our chosen starting-point is given by

$$y'+ay = b. \tag{10}$$

Then by analogy with our previous method we could take two solutions $y^{(1)}$ and $y^{(2)}$, satisfying respectively the equations

$$y''+fy'+gy = k, \qquad y''+fy'+gy = 0, \tag{11}$$

with the respective initial conditions

$$y'+ay = b, \qquad y'+ay = 0, \tag{12}$$

so that $y^{(1)}+\alpha y^{(2)}$ is a solution of the differential equation satisfying the first boundary condition, and α can again be chosen so that the second is satisfied.

The satisfaction of (12) requires now the use of Lagrangian formulae for the boundary condition. As an example we take the differential equation (4), with boundary conditions

$$y'+y = 1 \text{ at } x = 0, \qquad y'-y = 1 \text{ at } x = 1, \tag{13}$$

given in simple Lagrangian form by the equations

$$y(\tfrac{1}{4})-y(-\tfrac{1}{4})+0{\cdot}5y(0) = 0{\cdot}5, \qquad y(\tfrac{5}{4})-y(\tfrac{3}{4})-0{\cdot}5y(1) = 0{\cdot}5. \tag{14}$$

The finite-difference forms of the differential equation are given by (5), and for the first trial solution $y^{(1)}$ we use (5) and (14), for $y^{(2)}$ those equations with all the constant terms suppressed.

If we decide to recur from $x = 1$ we must use the second of (14) and the last of (5) to produce a relation between consecutive values which satisfies both differential equation and boundary condition. Eliminating $y(\frac{3}{4})$, for example, we have

$$2y(\tfrac{5}{4})-2{\cdot}5625y(1) = 0{\cdot}5625 \text{ for } y^{(1)}, = 0 \text{ for } y^{(2)}. \tag{15}$$

We can therefore take $y(1) = 0$, $y(\frac{5}{4}) = 0{\cdot}28125$ for $y^{(1)}$, and $y(1) = 1$, $y(\frac{5}{4}) = 1{\cdot}28125$ for $y^{(2)}$. The resulting solutions are shown in Table 4.

TABLE 4

x	$y^{(1)}$	$y^{(2)}$	y
+1·25	+0·281250	+1·281250	+1·817742
1·00	·000000	1·000000	1·199213
0·75	− ·218750	0·781250	0·718135
·50	·404297	·611328	+ ·328815
·25	·583863	·479614	− ·008704
+ ·00	0·784295	·377876	·331141
−0·25	−1·033745	+0·299755	−0·674275

The first of (14) is then satisfied by $y^{(1)}+\alpha y^{(2)}$ if

$$0{\cdot}368797\alpha+0{\cdot}057734 = 0{\cdot}5, \text{ giving } \alpha = 1{\cdot}199213,$$

and the required y is given in the last column of Table 4.

The method of § 6 is, of course, merely a special case of the last method; in the former one boundary condition, involving the function only, is particularly easy to satisfy.

A more general method

8. The general equation (1), with two boundary conditions of type (10), has the analytical solution

$$y = Ay_1+By_2+y_3, \tag{16}$$

where y_1 and y_2 are solutions of the homogeneous form of (1), with the right-hand side suppressed, and y_3 is a particular integral of (1); y_1 and y_2 are the so-called complementary functions. In theory we could use any y_1, y_2, and y_3, provided they satisfy their respective differential equations; none of them need satisfy the boundary conditions, the latter being used to determine the constants A and B in (16). If we are prepared to perform three recurrence processes there is then no trouble in starting. For the solution of the last example, for instance, we could take y_3 to satisfy the equations (5) with zero values at $x = -\frac{1}{4}$, $x = 0$ (recurring from this end), y_2 to satisfy the homogeneous form of (5) with values 0 at $x = -\frac{1}{4}$, 1 at $x = 0$, and y_1 to satisfy the same equations with values 1 at $x = -\frac{1}{4}$, 0 at $x = 0$. The results of these calculations are shown in Table 5.

TABLE 5

x	y_3	y_2	y_1	y
−0·25	+0·000000	+0·000000	+1·000000	−0·674272
+ ·00	·000000	1·000000	0·000000	·331140
·25	·000000	2·062500	−1·000000	− ·008704
·50	·015625	3·253906	2·062500	+ ·328813
0·75	·063477	4·648681	3·253906	0·718130
1·00	·162171	6·333999	4·648681	1·199206
+1·25	+0·333501	+8·415192	−6·333999	+1·817732

We now have two simultaneous equations for A and B, obtained from the satisfaction of the boundary conditions represented approximately by equations (14), and given by

$$\left.\begin{aligned} 2{\cdot}562500B-2A &= 0{\cdot}5 \\ 0{\cdot}599511B-0{\cdot}755752A+0{\cdot}188939 &= 0{\cdot}5 \end{aligned}\right\}, \tag{17}$$

with solutions $B = -0{\cdot}331140$, $A = -0{\cdot}674272$. The final y is shown in the last column of Table 5.

9. In all our 'arbitrary values' we have taken numbers like 0 or 1. This is of course not essential, and occasionally it might pay to use other values if the arithmetic is made easier. For the complementary functions, however, one of the assumed values must be non-zero, since otherwise the whole solution is null. In the method in which we calculate two complementary functions, moreover, they must be quite independent solutions, that is one must not be merely a constant multiple of the other.

Inclusion of difference corrections

10. At some stage in our calculations we must include the difference corrections for the differential equation, and sometimes for the boundary conditions if the latter involve a derivative. We could, of course, include them in the evaluation of each trial solution, but these solutions may often have much larger difference corrections than the required linear combination, and the inclusion at that stage would be uneconomic. This would most often be true if one of the complementary functions increases much faster than the other, while the boundary conditions ensure that only a small proportion of the large complementary function is present.

It is therefore desirable to include the difference corrections, at least for the differential equation and often for the boundary conditions, at the latest possible stage, that is when the complete first approximation has been obtained. We shall consider first the cases when the boundary values are specified so that there is no difference correction for the boundary conditions.

11. We can take as an example the problem and method of § 5, with the approximate solution y of Table 3. Using equations (5) in the usual way to extend our values beyond the boundaries for the production of all required central differences, and noting that y is an odd function, we produce the differences and difference corrections shown in Table 6.

Differences of order greater than six are clearly negligible for this purpose, the difference correction being given by

$$Cy_r = (-\tfrac{1}{12}\delta^4 + \tfrac{1}{90}\delta^6 - \tfrac{1}{560}\delta^8 + \ldots)y_r. \tag{18}$$

The fact that we are differencing in the opposite direction does not affect the sign of even differences.

TABLE 6

x	y		δ^2		δ^4		δ^6	$10^6 Cy$
+1·50	+2·120463							
		−645390						
1·25	1·475073		+170317					
		475073		−45317				
1·00	1·000000		125000		+7811			
		350073		37506		−2338		
0·75	0·649927		87494		5473		+325	−452·5
		262579		32033		2013		
·50	·387348		55461		3460		238	285·7
		207118		28573		1775		
·25	·180230		+ 26888		+1685		+ 90	−139·4
		−180230		−26888		−1685		
+0·00	+0·000000		0		0		0	

We now effectively repeat our original problem, with Cy included on the left of equations (5). The original constant terms in these equations are suppressed and the boundary value at our point of departure must also be zero, since these quantities are already included in our first trial solution $y^{(1)}$. In our new trial solution $\eta^{(1)}$, say, we therefore take the two initial values to be zero, and by recurrence produce the values shown on the left of Table 7.

TABLE 7

x	$\eta^{(1)}$	η		δ^2		δ^4
+1·00	+0·000000	−0·000000				
			−473			
0·75	·000000	·000473		+422		
			− 51		−169	
·50	·000452	·000524		+253		+36
			+202		−133	
·25	·001218	·000322		+120		
			+322			
+0·00	+0·002200	−0·000000				

The value of our complete correction has to be zero at $x = 0$, whereas $\eta^{(1)}$ gives $+0{\cdot}002200$. Now $y^{(2)}$ of Table 3 satisfies the homogeneous equation with Cy neglected and has a zero initial value, so that we can add any multiple of $y^{(2)}$ to $\eta^{(1)}$ without upsetting the initial condition at $x = 1$. The function $\eta = \eta^{(1)} + \alpha_2 y^{(2)}$ satisfies the other boundary condition if α_2 is chosen so that $\eta(0) = 0$. We find $\alpha_2 = -0{\cdot}0004733$, and the complete correction $\eta = \eta^{(1)} + \alpha_2 y^{(2)}$ is given, with its differences, on the right of Table 7. The function $y+\eta$ is then the solution of the problem if $C\eta$ is negligible.

This solution agrees closely with the corresponding result obtained by boundary-value methods and given to one less figure in § 6 of Chapter 4. To six decimals the difference correction of η is not quite negligible, so that our answer $y+\eta$ is not correct to six figures, and perhaps the fifth figure may be slightly in error (the available fourth difference of η is about one-hundredth of that of y, so that the further correction to η is about one-hundredth of η, a maximum of about 5 units in the sixth decimal place).

12. It may be worth while here to recapitulate the complete sequence of steps in the solution of the problem given by

$$y''+fy'+gy = k, \qquad y(x_0) = p, \qquad y(x_n) = q. \tag{19}$$

(i) Set up the finite-difference equations, given by

$$a_r y_{r-1}+b_r y_r+c_r y_{r+1}+Cy_r = d_r. \tag{20}$$

(ii) Starting at one end of the range, say x_0, calculate $y^{(1)}$ from (20), with Cy neglected, and with $y(x_0) = p$, $y(x_0+h)$ arbitrary ($= 0$, say).

(iii) Calculate $y^{(2)}$ from (20), with Cy neglected and d_r replaced by zero, and with $y(x_0) = 0$, $y(x_0+h)$ arbitrary ($= 1$, say), but not zero.

(iv) Calculate $y = y^{(1)}+\alpha_1 y^{(2)}$, where α_1 is determined so that

$$y_n^{(1)}+\alpha_1 y_n^{(2)} = q.$$

(v) The function y is the solution of (20), with Cy ignored, and satisfying the given boundary conditions. Extend this beyond the boundaries by recurrence, difference, and calculate Cy at all internal points.

(vi) Calculate $\eta^{(1)}$ from (20), with d_r suppressed but with the calculated Cy included, and with initial values zero at x_0, arbitrary ($= 0$, say) at x_1.

(vii) Calculate $\eta = \eta^{(1)}+\alpha_2 y^{(2)}$, where α_2 is determined so that

$$\eta_n^{(1)}+\alpha_2 y_n^{(2)} = 0.$$

(viii) $y+\eta$ is our first corrected solution, and the process can be repeated if necessary. For this purpose, of course, $\eta^{(1)}$ may need calculation at external points and therefore Cy at stage (v) may also need calculation at external points.

The form of the difference correction depends on the finite-difference equations used. In the previous example there is no first derivative in the differential equation and we could of course have used a more accurate three-term formula.

13. We take now the case in which one of the boundary conditions involves a derivative, so that, for example, the last of (19) is replaced by

$$y' + ay = b \text{ at } x = x_n, \tag{21}$$

these coefficients a and b having no relation with those of (20).

The changes in the process of § 12 are few and generally obvious. It is now desirable to start the recurrence from x_0. The boundary condition at x_n is represented by an algebraic equation, of which suitable forms have been discussed in previous chapters. Here we might take the simplest of these, given by

$$\left.\begin{aligned} y_{n+1} - y_{n-1} + 2ahy_n - 2hb + C_1 y_n &= 0 \\ C_1 &= -\tfrac{1}{3}\mu\delta^3 + \tfrac{1}{15}\mu\delta^5 - \tfrac{1}{70}\mu\delta^7 + \dots \end{aligned}\right\}. \tag{22}$$

Steps (i) to (iii) are unchanged, except that we extend the recurrence to one external point x_{n+1}. At step (iv) we calculate α_1 from the equation

$$y_{n+1}^{(1)} - y_{n-1}^{(1)} + 2ahy_n^{(1)} + \alpha_1\{y_{n+1}^{(2)} - y_{n-1}^{(2)} + 2ahy_n^{(2)}\} = 2hb, \tag{23}$$

neglecting the difference correction $C_1 y_n$ in (22). In step (v) we calculate the difference correction $C_1 y_n$ as well as the general Cy. Step (vi) follows as before, but in step (vii) we calculate α_2 from the equation

$$\eta_{n+1}^{(1)} - \eta_{n-1}^{(1)} + 2ah\eta_n^{(1)} + \alpha_2\{y_{n+1}^{(2)} - y_{n-1}^{(2)} + 2ahy_n^{(2)}\} + C_1 y_n = 0, \tag{24}$$

and the method of repetition, if necessary, is an obvious extension of the previous case.

14. If the derivative is involved in both boundary conditions there are two possible methods, those of §§ 7 and 8, for the calculation of approximate solutions. Their subsequent correction involves no inherently new ideas.

With the method of § 7 we have an initial condition similar to (22), replacing n by 0, and we satisfy it and the differential equation at that point by neglecting both types of difference correction, thereby obtaining a relation between the boundary value and first external value. When we come to calculate the correction, these neglected terms must be included in the calculation of the starting values for $\eta^{(1)}$, but, as in §§ 12 and 13, we need no correcting trial solution corresponding to $y^{(2)}$.

Consider, for example, the problem of § 7 with the results of Table 4. We have first to extend our approximate y at both ends of the range to permit calculation of the general difference corrections of type Cy and the special boundary corrections $C_1 y_0$ and $C_1 y_n$. The complete table of differences with the calculated values of the difference corrections is shown in Table 8.

TABLE 8

x	y		δ^2		δ^4		δ^6	10^6Cy	10^6C_1y
+1·75	+3·696268								
		−1068263							
1·50	2·628005		+258000						
		810263		−66266					
1·25	1·817742		191734		+11983				
		618529		54283		−3393			
1·00	1·199213		137451		8590		+539	−710	−16454
		481078		45693		2854			
0·75	0·718135		91758		5736		356	474	
		389320		39957		2498			
·50	+ ·328815		51801		3238		200	268	
		337519		36719		2298			
·25	− ·008704		+ 15082		+ 940		+ 67	− 78	
		322437		35779		2231			
+ ·00	·331141		− 20697		− 1291		− 90	+107	−11990
		343134		37070		−2321			
− ·25	0·674275		− 57767		− 3612				
		400901		−40682					
− ·50	1·075176		− 98449						
		− 499350							
−0·75	−1·574526								

In the calculation of Cy eighth differences are clearly negligible, but for $C_1 y(1)$ seventh differences, which we shall ignore, make a small contribution and should really be included. They could be calculated by further extension of y, or in this case they could easily be estimated. Since we have differenced in the 'wrong' direction, we must remember to change the sign of the odd differences in the calculation of $C_1 y(0)$ and $C_1 y(1)$.

For the initial values of our correcting trial solution $\eta^{(1)}$, we now have the equations

$$\left.\begin{aligned} \eta(1{\cdot}25)+\eta(0{\cdot}75)-2{\cdot}0625\eta(1)-0{\cdot}000710 &= 0 \\ \eta(1{\cdot}25)-\eta(0{\cdot}75)-0{\cdot}5000\eta(1)-0{\cdot}016454 &= 0 \end{aligned}\right\}, \tag{25}$$

which we can satisfy by taking $\eta(1) = 0$, $\eta(1{\cdot}25) = 0{\cdot}008582$. Recurring towards the origin as in step (vi) of § 12, we find $\eta^{(1)}$ as shown on the left of Table 9.

TABLE 9

x	$\eta^{(1)}$	η		δ^2		δ^4		$10^6C\eta$	$10^6C_1\eta$
+1·25	+0·008582	+0·032340							
			−13797		(−1135)		(−65)		
1·00	·000000	+ ·018543		+1869		(+153)		−13	−348
			11928		982		(−67)		
0·75	− ·007872	+ ·006615		+ 887		+ 86		− 7	
			11041		896		−69		
·50	·015762	− ·004426		− 9		+ 17		− 1	
			11050		879		−71		
·25	·024369	·015476		− 888		− 54		+ 5	
			11938		933		(−73)		
+ ·00	·034421	·027414		−1821		(−127)		+11	−327
			−13759		(−1060)		(−75)		
−0·25	−0·046731	−0·041173							

To satisfy the boundary condition at the origin we must add to $\eta^{(1)}$ a multiple α_2 of the $y^{(2)}$ of Table 4, such that

$$\eta^{(1)}(\tfrac{1}{4})-\eta^{(1)}(-\tfrac{1}{4})+0{\cdot}5\eta^{(1)}(0)+C_1 y_0+0{\cdot}368797\alpha_2 = 0, \tag{26}$$

the coefficient of α_2 having already been calculated in the first stage of the work. We find $\alpha_2 = +0{\cdot}018543$, and the correction $\eta = \eta^{(1)}+\alpha_2 y^{(2)}$ is shown, with its differences, on the right of Table 9.

The difference corrections of η are not negligible, and should be used to calculate a second correction. For the calculation of $C\eta$ and $C_1\eta$ we should, for best precision, have values of η at more external points, involving still further extension of Tables 4 and 8. Here it seems reasonably safe to estimate the missing differences, given in brackets in Table 9, and the calculated values of $C\eta$ and $C_1\eta$ then have the values shown in that table. Calculations corresponding to (25) give initial values of a new trial solution $\eta^{(2)}$, whose other values obtained by recurrence are shown in Table 10.

TABLE 10

x	$\eta^{(2)}$	*Second correction*					*Final solution*
+1·25	+0·000180	+0·000842					
			−325				
1·00	·000000	·000517		+45			+1·21827
			280		−23		
0·75	− ·000167	+ ·000237		+22		+1	0·72499
			258		22		
·50	·000337	− ·000021		0		−1	+0·32437
			258		23		
·25	·000527	·000279		−23		+1	−0·02446
			281		−22		
+ ·00	·000755	·000560		−45			−0·35912
			−326				
−0·25	−0·001041	−0·000886					

For a new correction $\eta^{(2)}+\alpha_2 y^{(2)}$ we have, from equation (26), $\alpha_2 = +0{\cdot}000517$, and this correction is given in Table 10, together with its differences. The new difference correction of type Cy is clearly negligible, though those of $C_1 y$ are still significant. The complete solution so far obtained, given in Table 10 to one less figure, differs from the analytical solution by a maximum of 2 units in the fifth decimal.

15. The rather slow convergence in this example is caused by the difference corrections associated with the boundary conditions, which are here much larger than those associated with the differential equation. We can, of course, use more accurate formulae at the ends of the range, either by using some form of backward-difference formula for the derivative, such as

$$hy'_n = (\nabla-\tfrac{1}{2}\nabla^2-\tfrac{1}{6}\nabla^3-\tfrac{1}{12}\nabla^4-\ldots)y_{n+1}, \tag{27}$$

or by extending the recurrence still further beyond the boundary and using a more accurate central-difference formula.

If this method is used at the start of the range of integration we shall have to solve more simultaneous equations than before to get the starting values. Suppose, for example, that we include third differences in the central-difference formula for the first derivative. Then instead of the second of (14) we have

$$-y(\tfrac{3}{2})+8y(\tfrac{5}{4})-8y(\tfrac{3}{4})+y(\tfrac{1}{2})-3y(1) = 3, \tag{28}$$

and must combine this with the finite-difference equations (5) (and another evaluated at $x = \frac{5}{4}$), to provide the required starting relation between two consecutive values for the solution $y^{(1)}$, and the homogeneous form of that relation for $y^{(2)}$.

16. Alternatively we could use the Taylor-series method and satisfy the initial conditions of all our trial solutions with full accuracy. For example, using the Taylor's series at $x = 1$ we find, from the differential equation (4) and the initial condition given by the second of (13), the relation

$$y(\tfrac{3}{4}) = 0{\cdot}778801y(1)-0{\cdot}223811. \tag{29}$$

Thus for the first solution $y^{(1)}$ we can take $y(1) = 0$, $y(\frac{3}{4}) = -0{\cdot}223811$, for $y^{(2)}$ we can take $y(1) = 1$, $y(\frac{3}{4}) = 0{\cdot}778801$, and produce these solutions by recurrence as before. Results are given in Table 11.

TABLE 11

x	$y^{(1)}$	$y^{(2)}$	y		δ^2		δ^4		δ^6	10^6Cy
			+2·671966							
				−824827						
			1·847139		+193571					
				631256		−55078				
+1·00	−0·000000	+1·000000	1·215883		138493		+8655			
				492763		46423		−2901		
0·75	·223811	0·778801	0·723120		92070		5754		+361	−475
				400693		40669		2540		
·50	·414735	·606277	+ ·322427		51401		3214		197	266
				349292		37455		2343		
·25	·600330	·471645	− ·026865		+ 13946		+ 871		+ 57	− 72
				335346		36584		−2286		
+0·00	−0·807821	+0·366491	·362211		− 22638		−1415			
				357984		−37999				
			0·720195		− 60637					
				−418621						
			−1·138816							

We can now satisfy the boundary condition at $x = 0$ by using Lagrangian formulae of varying accuracy, as in previous examples, or the accurate central-difference formula for the derivative, or we can use the Taylor-series method again, finding, from (4) and the first of (13), the relation

$$y(\tfrac{1}{4})-0{\cdot}778801y(0)-0{\cdot}255225 = 0. \tag{30}$$

This is satisfied by $y = y^{(1)}+\alpha y^{(2)}$ if $0{\cdot}028802+0{\cdot}186221\alpha-0{\cdot}255225 = 0$, or $\alpha = 1{\cdot}215883$. The first approximation y is shown in Table 11. It is, as we would expect, much nearer the truth than that of Table 4.

In finding a second approximation we need consider only the difference correction coming from the differential equation. For its calculation, as given in Table 11, we have as usual to extend the first approximation beyond the boundaries. At internal points differences of order greater than the sixth are negligible.

For the correcting trial solution $\eta^{(1)}$ we can take the first two values to be zero, which satisfy (29) with the constant term ignored. The rest of this function is shown in Table 12. To satisfy (30), with the constant ignored, we find we must add to $\eta^{(1)}$ a multiple of $y^{(2)}$ of amount $+0{\cdot}002372$, and the result η is also shown in Table 12, together with its differences.

TABLE 12

x	$\eta^{(1)}$	η		δ^2		δ^4
+1·00	+0·000000	+0·002372				
			−525			
0·75	·000000	·001847		+591		
			+ 66		−205	
·50	·000475	·001913		386		+38
			452		−167	
·25	·001246	·002365		+219		
			+671			
+0·00	+0·002167	+0·003036				

The new difference correction is not quite negligible, and should be included in a further correction.

Finally, it would be expected that if this example were computed using this method to satisfy the boundary conditions, but the more accurate finite-difference formula with a difference correction starting with $\frac{1}{240}\delta^6 y$, only one correction would be necessary.

17. Another possible method is based on § 8, in which we used three trial solutions, none of them satisfying the boundary conditions. If we consider those of Table 5, we can find a combination $Ay^{(1)}+By^{(2)}+y^{(3)}$ to satisfy the boundary conditions. In § 8 we used the simplest central-difference expression for the first derivative, but we could obviously extend our trial solutions further and use more accurate central-difference formulae, or use sloping difference formulae without further extension. Alternatively we could use equations (29) and (30), obtained from the Taylor's series, giving for A and B the equations

$$\left.\begin{aligned} -0{\cdot}062822-0{\cdot}284244B+0{\cdot}366491A+0{\cdot}223811 &= 0 \\ 1{\cdot}283699B-\qquad A-0{\cdot}255225 &= 0 \end{aligned}\right\}, \tag{31}$$

with solutions $B = -0{\cdot}362213$, $A = -0{\cdot}720197$. The resulting first approximation is of course practically identical with that of Table 11.

When we apply the difference correction only the solution $\eta^{(3)}$, corresponding to $y^{(3)}$, needs calculation, and in its equations we ignore constant terms but include any difference corrections neglected earlier. The other solutions $y^{(1)}$ and $y^{(2)}$, and therefore the coefficients of A and B in the new equations of type (31), need not be recalculated.

We pursue this example no further. The method used in any particular problem will be chosen purely for convenience: much will depend on the form of the boundary conditions, the ease of differentiation to find the Taylor's series, and so on. The question of differences in accuracy between the various methods hardly arises, for we should not allow errors through neglect of significant differences, and, since the same algebraic equations are being solved, all methods should give about the same precision.

Equations of third and fourth orders

†18. The same general principles are used in the solution of equations of third, fourth, and higher orders, though the number of trial solutions required naturally increases with the order of the equation.

Consider first a third-order differential equation. The method of § 19 of Chapter 6 produces finite-difference equations involving four adjacent pivotal values, which we can write in the form

$$a_r y_{r-1}+b_r y_r+c_r y_{r+1}+d_r y_{r+2}+Cy_{r+\frac{1}{2}}+f_r = 0, \tag{32}$$

where the coefficients are known functions of x and the equation corresponds to the satisfaction of the differential equation at the point $x_{r+\frac{1}{2}}$.

If the problem is not of true initial-value type, the boundary conditions may have the following form and distribution:

$$\left.\begin{aligned} p_1 y''+q_1 y'+r_1 y+s_1 = 0, \quad p_2 y''+q_2 y'+r_2 y+s_2 = 0 \text{ at } x = x_0 \\ p_3 y''+q_3 y'+r_3 y+s_3 = 0 \text{ at } x = x_n \end{aligned}\right\}. \tag{33}$$

As for second-order equations there are two main procedures. In the first we find a solution in the form

$$y = \alpha_1 y^{(1)}+\alpha_2 y^{(2)}+y^{(3)}. \tag{34}$$

Starting from x_n, our $y^{(3)}$ should satisfy (32) everywhere and the boundary condition at x_n. Both $y^{(1)}$ and $y^{(2)}$ should satisfy the homogeneous forms of this differential equation and boundary condition at $x = x_n$, and should be independent solutions. The constants α_1 and α_2 are then chosen so that the two boundary conditions at x_0 are satisfied by (34).

If we start from x_0 we make our three solutions satisfy one of the relevant boundary conditions at that point, choosing α_1 and α_2 to satisfy the condition at x_n and the second condition at x_0. In each application of the difference corrections only one trial solution, of type $\eta^{(3)}$ corresponding to $y^{(3)}$, is calculated, $y^{(1)}$ and $y^{(2)}$ being used without correction to satisfy the boundary conditions.

In the second method we find a solution of the form

$$y = Ay^{(1)}+By^{(2)}+Cy^{(3)}+y^{(4)}, \tag{35}$$

where $y^{(4)}$ takes the place of $y^{(3)}$ in (34), and $y^{(1)}$, $y^{(2)}$, and $y^{(3)}$ are three independent solutions of the homogeneous equation. Now none of these solutions need satisfy any boundary conditions, the constants A, B, and C being disposable for this purpose. Again difference corrections are incorporated only in corrections of type $\eta^{(4)}$ corresponding to $y^{(4)}$.

There is yet a third method. If in (34) we choose $y^{(3)}$ to satisfy both boundary conditions at $x = x_0$, and choose $y^{(1)}$ to satisfy the homogeneous forms of both these conditions, then the solution $y^{(2)}$ is not independent of the others and can be omitted: there is then one boundary condition at x_n whose satisfaction gives α_1.

19. The boundary conditions can be satisfied either by using finite-difference formulae, with or without a difference correction, or the Taylor's series which gives an exact relation between various pivotal values. Again the Taylor-series method is preferable if the differentiation is not too difficult. Using the differential equation, the Taylor's series, and any one of (33), we can find a formula of the form

$$y(h) = p(h)y(0)+q(h)y'(0)+r(h), \tag{36}$$

and using this for h and $2h$, we can find a relation

$$y_2+ay_1+by_0+c = 0. \tag{37}$$

If we start from x_0 and satisfy both boundary conditions in the first of (33) there are two relations of type (37), so that only one value, say y_0, can be given an arbitrary value in the production of the solution $y^{(3)}$. For $y^{(1)}$ the two relations are homogeneous, and if we fix $y_0^{(1)}$ the other two values are uniquely determined. From the four-term recurrence relation (32) we can extend the trial solutions as far as desired in either direction.

†20. Consider, for example, the problem defined by the differential equation

$$y'''-y''+y'-y = 0, \tag{38}$$

with boundary conditions

$$y(0) = 1, \qquad y'(0) = 1, \qquad y(1) = 2{\cdot}71828, \tag{39}$$

which was treated in Chapter 6. By the methods used there we find the general finite-difference equation

$$\left.\begin{aligned} & y_{r+2}(1-\tfrac{1}{2}h+\tfrac{1}{12}h^2)+y_{r+1}(-3+\tfrac{1}{2}h+\tfrac{3}{4}h^2-\tfrac{1}{2}h^3)+ \\ & \quad +y_r(3+\tfrac{1}{2}h-\tfrac{3}{4}h^2-\tfrac{1}{2}h^3)+y_{r-1}(-1-\tfrac{1}{2}h-\tfrac{1}{12}h^2)+Cy_{r+\frac{1}{2}} = 0 \\ & \quad C = \tfrac{1}{12}h\mu\delta^4-\tfrac{1}{120}h^2\delta^5-\tfrac{1}{90}h\mu\delta^6-(\tfrac{1}{240}-\tfrac{1}{840}h^2)\delta^7+\dots \end{aligned}\right\}. \tag{40}$$

For the Taylor-series method at the origin the boundary conditions do not involve y'', so that the method of the last section needs modification. Equation (36) is replaced by

$$y(h) = p(h)y(0)+q(h)y'(0)+r(h)y''(0),$$

there being no constant term since this is lacking in the differential equation.

In this case we find

$$y(h) = y(0)\left(1+\frac{h^3}{3!}+\frac{h^4}{4!}+\dots\right)+y'(0)\left(h-\frac{h^3}{3!}+\frac{h^5}{5!}-\dots\right)+$$
$$+y''(0)\left(\frac{h^2}{2!}+\frac{h^3}{3!}+\frac{h^6}{6!}+\dots\right). \tag{41}$$

If we want to use the last method of § 18 we satisfy both initial conditions, obtaining

$$y(h) = \left(1+h+\frac{h^4}{4!}+\frac{h^5}{5!}+\dots\right)+y''(0)\left(\frac{h^2}{2!}+\frac{h^3}{3!}+\frac{h^6}{6!}+\dots\right), \tag{42}$$

when we substitute the known values of $y(0)$ and $y'(0)$ in (41). At the interval $h = 0{\cdot}2$ we find

$$\left.\begin{aligned} y(0{\cdot}2) &= 1{\cdot}2000693+0{\cdot}02133342y''(0) \\ y(-0{\cdot}2) &= 0{\cdot}8000640+0{\cdot}01866676y''(0) \end{aligned}\right\}, \tag{43}$$

and eliminating $y''(0)$ we have finally

$$y(0{\cdot}2)-1{\cdot}142856y(-0{\cdot}2) = 0{\cdot}285711. \tag{44}$$

This again is a slight departure, suggested by the form of (42) and arithmetical considerations, from the procedure resulting in (37).

For our trial solution $y^{(3)}$ we can therefore take $y^{(3)}(0) = 1$, $y^{(3)}(-0{\cdot}2) = 0$, $y^{(3)}(0{\cdot}2) = 0{\cdot}285711$ so that (44) is satisfied; for $y^{(1)}$ we take $y^{(1)}(0) = 0$, $y^{(1)}(-0{\cdot}2) = 1$, $y^{(1)}(0{\cdot}2) = 1{\cdot}142856$, so that the homogeneous form of (44) is satisfied.

At the interval $h = 0{\cdot}2$ the approximate form of equation (40) can be written in the convenient form

$$y_{r+2} = 3{\cdot}181550y_{r+1} - 3{\cdot}394096y_r + 1{\cdot}221402y_{r-1} - 1{\cdot}107011Cy_{r+\frac{1}{2}}, \quad (45)$$

and is used for both trial solutions, since there is no constant term. The solutions are shown in Table 13.

TABLE 13

x	$y^{(3)}$	$y^{(1)}$	y		δ^2		δ^4		δ^6	10^6Cy
			+0·670207							
				+148499						
−0·2	+ 0·000000	+ 1·000000	0·818706		+ 32795					
				181294		+ 7286				
+ ·0	1·000000	0·000000	1·000000		40081		+1622			
				221375		8908		+359		+30
·2	+ 0·285711	1·142856	1·221375		48989		1981		+ 78	
				270364		10889		437		36
·4	− 2·485092	4·857456	1·491739		59878		2418		99	
				330242		13307		536		44
·6	7·654773	11·575276	1·821981		73185		2954		+119	
				403427		16261		+655		+54
0·8	15·570434	21·736534	2·225408		89446		+3609			
				492873		+19870				
+1·0	−26·592376	+35·801178	2·718281		+109316					
				+602189						
			+3·320470							

The solution $y = y^{(3)} + \alpha_1 y^{(1)}$ should equal 2·71828 at $x = 1$, so that $\alpha_1 = 0{\cdot}8187065$ and the approximate y is given in Table 13, together with its differences. In the calculation of the difference correction, given by

$$Cy_{r+\frac{1}{2}} = (\tfrac{1}{60}\mu\delta^4 - \tfrac{1}{3000}\delta^5 - \tfrac{1}{450}\mu\delta^6 - \ldots)y_{r+\frac{1}{2}}, \quad (46)$$

the seventh and higher differences are negligible.

TABLE 14

x	$\eta^{(3)}$	η		δ^2		δ^4		$y+\eta$
−0·2	−0·000000	+0·000024						+0·818730
			− 24					
+ ·0	·000000	·000000		+ 52				1·000000
			+ 28		−22			
·2	·000000	·000028		+ 30		−15		1·221403
			+ 58		−37		+ 3	
·4	·000033	·000086		− 7		−12		1·491825
			+ 51		−49		−10	
·6	·000145	·000137		− 56		−22		1·822118
			− 5		−71			
0·8	·000398	·000132		−127				2·225540
			−132					
+1·0	−0·000874	+0·000000						+2·718281

The inclusion of Cy provides a second trial solution $\eta^{(3)}$ similar to $y^{(3)}$, here satisfying (45) (excluding any constant term, in this case already zero), and with initial conditions which we can take to be

$$\eta^{(3)}(0) = \eta^{(3)}(-0{\cdot}2) = \eta^{(3)}(0{\cdot}2) = 0,$$

so that $\eta^{(3)}$ satisfies the homogeneous form of (44). The solution $\eta^{(3)}$, given in Table 14, must be combined with $\alpha_2 y^{(1)}$ so that the sum is zero at $x = 1$. We find $\alpha_2 = +0{\cdot}0000244$, and the combined solution η is also shown in Table 14. The difference correction of η is clearly negligible. Finally η is added to y to produce the final solution: it differs from the analytical solution e^x by a maximum of 1 unit in the last figure.

21. For fourth-order equations we have the same choice of methods. There are four boundary conditions, four complementary functions, and one particular integral. We could therefore have five trial solutions if none of them are required to satisfy boundary conditions. If the particular integral satisfies one boundary condition, and independent trial solutions satisfy the corresponding homogeneous equations, we need four solutions in all. If all trial solutions satisfy two boundary conditions we need only three with two disposable constants, and in the special case when three boundary conditions are given at one point, we could manage with only one complementary function and one particular integral.

Again the boundary conditions can be satisfied either by using extra finite-difference equations or by using Taylor's series, and again all difference corrections are incorporated in corrections to the particular integral only. We need not give any numerical illustration.

Eigenvalue problems

22. In an eigenvalue problem the boundary conditions and differential equation are homogeneous, and solutions exist only for certain values, the eigenvalues, of a parameter λ involved in the differential equation. If an eigenvalue were known for a second-order equation, for example, a straightforward step-by-step process would produce the eigenfunction, for the homogeneity of the problem means that only one initial condition is needed. If the initial value is zero, the next pivotal value can be selected arbitrarily, and if the derivative is involved in the first boundary condition we can turn this into a homogeneous relation between the first two values, and one of them can be chosen arbitrarily. The other boundary condition, however, is not satisfied unless the λ used in the step-by-step process is an eigenvalue.

An obvious method of solution is to find, for various values of λ, solutions which satisfy the initial condition, and choose the one which satisfies the other condition. A linear combination of solutions, on the lines of previous methods of this chapter, is not of course possible, since this type of equation is effectively non-linear, both y and λ being variables. Instead we must use trial and error or, more systematically, a process of inverse interpolation.

23. As an example of this simple process we consider the simplest problem of this kind, given by

$$y''+\lambda y = 0, \qquad y(0) = y(1) = 0, \tag{47}$$

and find approximate solutions, with difference corrections neglected, to the finite-difference equations typified by

$$y_1+y_{-1}-(2-h^2\lambda)y_0 = 0. \tag{48}$$

We take $h = \frac{1}{4}$ and try some values of $h^2\lambda$, in each case taking $y(0) = 0$, $y(\frac{1}{4}) = 1$ and obtaining successive values by recurrence. Some results are shown in Table 15.

TABLE 15

x	$h^2\lambda = 0$	$h^2\lambda = 0{\cdot}2$	$h^2\lambda = 0{\cdot}4$	$h^2\lambda = 0{\cdot}6$	$h^2\lambda = 0{\cdot}8$	$h^2\lambda = 1{\cdot}0$
+0·00	+0·000	+0·000	+0·000	+0·000	+0·000	+0·000
·25	1·000	1·000	1·000	1·000	1·000	1·000
·50	2·000	1·800	1·600	1·400	1·200	1·000
0·75	3·000	2·240	1·560	+0·960	+0·440	+0·000
+1·00	+4·000	+2·232	+0·896	−0·056	−0·672	−1·000

The evaluation of $h^2\lambda$ for which $y(1) = 0$ is a standard problem of inverse interpolation. The value of $y(1)$ is clearly a function of λ, and we can tabulate it with equal intervals in $h^2\lambda$ as shown in Table 16.

TABLE 16

$h^2\lambda$	$y(1)$		δ^2		δ^4	x	$(h^2\lambda = 0{\cdot}5858)$
+0·0	+4·000					+0·00	+0·0000
		−1768					
·2	2·232		+432			0·25	1·0000
		1336		−48			
·4	+0·896		384		0	0·50	1·4142
		952		−48			
·6	−0·056		336		0	0·75	1·0000
		616		−48			
0·8	0·672		+288			+1·00	+0·0000
		− 328					
+1·0	−1·000						

Using Bessel's formula, we wish to find $h^2\lambda = 0{\cdot}4+0{\cdot}2n$, where n is given by

$$0{\cdot}896-0{\cdot}952n+0{\cdot}720B_2(n)-0{\cdot}048B_3(n) = 0, \tag{49}$$

where $B_2(n)$ and $B_3(n)$ are respectively coefficients of the double second difference and third difference. The usual method is to solve this by an iterative process, according to the scheme

$$0{\cdot}952n_r = 0{\cdot}896+0{\cdot}720B_2(n_{r-1})-0{\cdot}048B_3(n_{r-1}). \tag{50}$$

Using tables of Bessel coefficients the whole calculation can be performed on two calculating machines without intermediate recording, and we find $n = 0{\cdot}929$. The required value of $h^2\lambda$ is then 0·5858. We could now get the required eigenfunction by direct interpolation for this value of $h^2\lambda$, at each pivotal value of x, but it is quicker to repeat the recurrence process, as shown in Table 16. This result agrees with that obtained by a different method in Chapter 7.

24. This method is satisfactory, but may involve a large number of trial solutions. In particular we need at least two solutions before we can decide even the sign of the required modification of a guessed λ. We therefore look for a method which provides automatic and fairly rapid convergence.

We have already remarked, in the example of the last section, that $y(1)$ is a function of λ. If we can decide what change in $y(1)$ is caused by a small change in λ, we have some idea of the modification necessary in λ for the satisfaction of the boundary condition at $x = 1$. For we want λ such that

$$y = 0 \text{ at } x = 1, \tag{51}$$

and Newton's rule suggests that the necessary correction to an approximate λ_1 is given by

$$\delta\lambda = -\frac{y_1(1)}{z_1(1)}, \tag{52}$$

where $y_1(1)$ and $z_1(1)$ are evaluated for $\lambda = \lambda_1$, and

$$z = \partial y/\partial\lambda. \tag{53}$$

From the differential equation (47), we have, differentiating with respect to λ, the equation

$$z'' + \lambda z + y = 0. \tag{54}$$

The function y is known for any fixed λ, so that (54) can be regarded as an ordinary differential equation for $z = \partial y/\partial\lambda$. For initial conditions we have $z_0 = z_1 = 0$, for in our solution the values of y are fixed at the first two points, and are independent of λ.

Equations (47) and (54) can be solved successively, with effectively the same finite-difference equations except that (54) is not homogeneous, its finite-difference equation being given by

$$z_1 + z_{-1} - (2 - h^2\lambda)z_0 + h^2 y_0 = 0. \tag{55}$$

Starting with $h^2\lambda = 0$ as in the first solution of Table 15, we produce the corresponding first solutions y and z given in Table 17. From $z(1)$ and $y(1)$ we find, from (52), $\delta\lambda = +6{\cdot}4$, $h^2(\lambda+\delta\lambda) = 0{\cdot}40$. This value of λ gives second solutions y and z shown in Table 17, and the other solutions are obtained in a similar manner.

TABLE 17

	$h^2\lambda = 0$		$h^2\lambda = 0{\cdot}40$		$h^2\lambda = 0{\cdot}5577$		$h^2\lambda = 0{\cdot}5850$		$h^2\lambda = 0{\cdot}5858$	
x	y	z	y	z	y	z	y	z	y	z
+0·00	+0·0000	−0·0000	+0·0000	−0·0000	+0·0000	−0·0000	+0·0000	−0·0000	+0·0000	−0·0000
·25	1·0000	·0000	1·0000	·0000	1·0000	·0000	1·0000	·0000	1·0000	·0000
·50	2·0000	·0625	1·6000	·0625	1·4423	·0625	1·4150	·0625	1·4142	·0625
0·75	3·0000	·2500	1·5600	·2000	1·0802	·1803	1·0022	·1769	1·0000	·1768
+1·00	+4·0000	−0·6250	+0·8960	−0·3550	+0·1157	−0·2651	+0·0031	−0·2505	+0·0000	−0·2500
	$\delta\lambda = 6{\cdot}4$		$\delta\lambda = 2{\cdot}5239$		$\delta\lambda = 0{\cdot}4364$		$\delta\lambda = 0{\cdot}0124$			

The convergence of the Newton method is remarkably quick, considering the poorness of the first guess. In fact this iterative method is of second-order convergence, the error being ultimately squared at each step.

The difference correction

25. With the method of § 23, our final solutions y and λ have satisfied the relevant boundary conditions and the approximate finite-difference equations (48). We require solutions of the equations which include the difference corrections on the left of (48), given here by

$$Cy = (-\tfrac{1}{12}\delta^4+\tfrac{1}{90}\delta^6-\tfrac{1}{560}\delta^8+...)y. \tag{56}$$

Our first approximations to both y and λ will in general need modification when Cy is included.

We could of course improve each of the trial solutions in Table 15 by incorporating the relevant Cy in each, though this involves a large amount of work. Alternatively, we can examine an equation for the necessary corrections to y and λ. If the corrections are η and $\delta\lambda$, we have, from (48) and (56), the equation

$$\eta_1+\eta_{-1}-(2-h^2\lambda)\eta_0+h^2\delta\lambda y_0+Cy_0 = 0, \tag{57}$$

so that $y+\eta$, $\lambda+\delta\lambda$ satisfy the differential equation provided that $C\eta$ and $h^2\eta\delta\lambda$ are negligible.

In (57) η has its first two values zero, and we find $\delta\lambda$ and η at other points by using a recurrence process, as previously, taking various values of $\delta\lambda$ and choosing that for which η is zero at the other end of the range. We have seen in Chapter 7 that $\delta\lambda$ is often larger than η, so that a reasonable guess at $\delta\lambda$ is provided from (57) by assuming η to be zero everywhere, and calculating $h^2\delta\lambda = -(Cy/y)$, choosing for this purpose the largest pivotal value in the range.

The difference corrections given by (56) for the y of Table 16 have the values $-0{\cdot}0311$ at $x = 0{\cdot}25$ and $0{\cdot}75$, and $-0{\cdot}0439$ at $x = 0{\cdot}50$. If we take $h^2\delta\lambda = +0{\cdot}02$, $+0{\cdot}03$, $+0{\cdot}04$ we find the respective η, given by (57), to have the values shown in Table 18.

TABLE 18

x	$(h^2\delta\lambda = +0{\cdot}02)$	$(h^2\delta\lambda = +0{\cdot}03)$	$(h^2\delta\lambda = +0{\cdot}04)$	$(h^2\delta\lambda = +0{\cdot}03105)$
+0·00	+0·0000	+0·0000	−0·0000	+0·00000
·25	·0000	·0000	·0000	·00000
·50	·0111	·0011	·0089	·00005
0·75	·0313	·0030	·0253	·00006
+1·00	+0·0443	+0·0042	−0·0358	+0·00008

Interpolating inversely for $h^2\delta\lambda$ such that $\eta(1{\cdot}0) = 0$, we find $h^2\delta\lambda = +0{\cdot}03105$, and the corresponding η is given in the last column of Table 18. For this particular example η should really be zero everywhere, as we saw in Chapter 7, and the error, a maximum of 1 unit in the fourth decimal, is caused by the fact that the difference correction was recorded to only four decimals. The final λ is correct in its fifth decimal.

26. With the method of § 24 we proceed to incorporate the difference correction in the first approximation y, given in the last column of Table 17, finding a solution, given on the left of Table 19, which no longer satisfies the boundary condition at $x = 1$.

TABLE 19

x	$(h^2\lambda = 0{\cdot}5858)$	$(h^2\lambda = 0{\cdot}6169)$
+0·00	+0·0000	+0·0000
·25	1·0000	1·0000
·50	1·4453	1·4142
0·75	1·0878	+0·9999
+1·00	+0·1242	−0·0001

Using the previous value of z, we find a correction $\delta\lambda = \dfrac{0{\cdot}1242}{0{\cdot}2500}$, or $h^2\delta\lambda = 0{\cdot}0311$, and, with the new λ and the inclusion of the difference correction of the previous y, we find the new y shown on the right of Table 19.

The continuation of this process, if necessary due to further changes in Cy, involves the solution of the equations

$$\left.\begin{aligned} y_1^{(n)}+y_{-1}^{(n)}-\{2-h^2\lambda^{(n)}\}y_0^{(n)}+Cy_0^{(n-1)} &= 0 \\ z_1^{(n)}+z_{-1}^{(n)}-\{2-h^2\lambda^{(n)}\}z_0^{(n)}+Cz_0^{(n-1)}+h^2y_0^{(n)} &= 0 \\ \lambda^{(n)} = \lambda^{(n-1)}-y^{(n-1)}(1)/z^{(n-1)}(1) & \end{aligned}\right\}. \tag{58}$$

In the calculation of Table 19 we ignored not only Cz_0 but also any change in z caused by the change in y. This is generally satisfactory, since it is only necessary for $y(1)$ to vanish, $z(1)$ being unwanted, and the latter is probably sufficiently accurate at this stage to ensure convergence. Nevertheless the full advantage of Newton's method will be lost if several more steps are needed, convergence no longer being quadratic.

Fourth-order eigenvalue problems

27. In the solution of second-order eigenvalue problems the first two pivotal values could be fixed in advance, to satisfy the first boundary condition, and λ could be varied to satisfy the second. In the case of fourth-order equations, with a pair of conditions given at each boundary, we can fix the first three pivotal values, to satisfy two initial conditions, but the fourth value is arbitrary together with λ, and these two values must be altered until two boundary conditions are satisfied at the other end.

Consider, for example, the problem treated in Chapter 7, defined by the equations

$$y^{\text{iv}}-\lambda y = 0, \qquad y(0) = y'(0) = y(1) = y'(1) = 0. \tag{59}$$

The general approximate finite-difference equation is given by

$$y_{r+2}-4y_{r+1}+(6-h^4\lambda)y_r-4y_{r-1}+y_{r-2} = 0. \tag{60}$$

If we use the simplest central-difference expression for the derivative at the origin we have $y(-h) = y(h)$, and with $y(0) = 0$ we can also take $y(h) = y(-h) = 1$, thus fixing the 'size' of our solution. For the calculation of successive values $y(rh)$, however, both λ and $y(2h)$ have to be given numerical values.

The values of $y(1)$, $y'(1)$ are clearly functions of both $y(2h)$ and λ. If we denote $y(2h)$ by p, $\partial y/\partial\lambda$ by z, and $\partial y/\partial p$ by q, the corrections δp, $\delta\lambda$, designed to make $y(1) = y'(1) = 0$, are given by a pair of equations, analogous to the single equation (52) for the second-order problem, of the form

$$\left.\begin{aligned} y(1)+z(1)\delta\lambda+q(1)\delta p &= 0 \\ y'(1)+z'(1)\delta\lambda+q'(1)\delta p &= 0 \end{aligned}\right\}. \tag{61}$$

The functions z and q satisfy the equations

$$\left.\begin{aligned} z^{\text{iv}}-\lambda z-y &= 0 \\ q^{\text{iv}}-\lambda q \quad &= 0 \end{aligned}\right\}, \tag{62}$$

and their initial conditions are given by

$$z(-h) = z(0) = z(h) = z(2h) = 0,$$

$$q(-h) = q(0) = q(h) = 0, \qquad q(2h) = 1.$$

For our particular problem, we first calculate y with λ and $y(2h)$ assumed to be zero, obtaining the first solution of Table 20. Corresponding values of z and q are also given on the left of Table 20.

TABLE 20

		$h^4\lambda = 0$			$h^4\lambda = -0{\cdot}2830$	
x	y	z	q	y	z	q
−0·2	+ 1·0000	+0·0000	+ 0·0000	+0·3581	+0·0000	+ 0·0000
+ ·0	0·0000	·0000	0·0000	0·0000	·0000	0·0000
·2	1·0000	·0000	0·0000	0·3581	·0000	0·0000
·4	+ 0·0000	·0000	1·0000	1·0000	·0000	1·0000
·6	− 7·0000	·0016	4·0000	1·3920	·0006	4·0000
0·8	24·0000	·0064	10·0000	+0·7174	·0040	9·7170
1·0	55·0000	+ ·0048	20·0000	−2·2344	·0145	17·7360
+1·2	−104·0000	−0·0512	+35·0000	−8·8770	+0·0364	+24·8921

Using the first term in the central-difference expression for the derivative condition at $x = 1$, we find

$$\left.\begin{aligned} -55{\cdot}0000+0{\cdot}0048\delta\lambda+20{\cdot}0000\delta p &= 0 \\ -80{\cdot}0000-0{\cdot}0576\delta\lambda+25{\cdot}0000\delta p &= 0 \end{aligned}\right\}, \tag{63}$$

giving $\delta p = 2{\cdot}7925$, $\delta\lambda = -176{\cdot}89$, $h^4\delta\lambda = -0{\cdot}2830$.

To prevent the occurrence of very large values, in our new y we take $p = 1$, reducing $y(h)$ and $y(-h)$ to $(2{\cdot}7925)^{-1}$. New values of y, z, and q are shown on the right of Table 20. At this stage we find new corrections $\delta p = -0{\cdot}1881$, $\delta(h^4\lambda) = +0{\cdot}6148$, giving a current $h^4\lambda$ of $+0{\cdot}3318$, and repeat the process as shown in the first part of Table 21.

TABLE 21

	$h^4\lambda = 0{\cdot}3318$			$h^4\lambda = 0{\cdot}5767$			0·5949	0·5949
x	y	z	q	y	z	q	y	y
−0·2	+0·3581	+0·0000	+0·0000	+0·3581	+0·0000	+ 0·0000	+0·3581	+0·3581
+ ·0	·0000	·0000	0·0000	·0000	·0000	0·0000	·0000	·0000
·2	·3581	·0000	0·0000	·3581	·0000	0·0000	·3581	·3581
·4	·8119	·0000	1·0000	·7654	·0000	1·0000	·7645	·7646
·6	·8597	·0006	4·0000	·7614	·0006	4·0000	·7643	·7647
0·8	+ ·2692	·0037	10·3318	+ ·3270	·0036	10·5767	+ ·3574	·3585
1·0	−0·9067	·0128	22·6544	− ·1178	·0124	24·6136	− ·0016	·0010
+1·2	−2·5258	+0·0331	+47·0549	+0·0356	+0·0330	+56·0938	+0·3545	+0·3605
	$h^4\delta\lambda = +0{\cdot}2449$			$h^4\delta\lambda = +0{\cdot}0182$			$h^4\delta\lambda = 0$	
	$\delta p = -0{\cdot}0465$			$\delta p = -0{\cdot}0009$			$\delta p = 0{\cdot}0001$	

In the middle of Table 21 we give another approximation, and in the last part of the table two further approximations to y, the last being obtained without necessity for further change in z and q. Better accuracy cannot be achieved without the retention of more figures in the recurrence process, or some equivalent refinement.

Difference corrections can be incorporated in y, and ignored in z and q, as in previous applications.

28. If we use the Taylor-series method to avoid any difference correction at the boundaries we shall need a slight change in technique. With $y(0)$ and $y'(0)$ equated to zero in the last example, we find

$$y(h) = \left(\frac{h^2}{2!}+\frac{h^6}{6!}\lambda+\frac{h^{10}}{10!}\lambda^2+\ldots\right)y''(0)+\left(\frac{h^3}{3!}+\frac{h^7}{7!}\lambda+\frac{h^{11}}{11!}\lambda^2+\ldots\right)y'''(0), \tag{64}$$

and by eliminating $y''(0)$ and $y'''(0)$ we can find a linear relation of the form

$$a_0\,y(-h)+b_0\,y(h)+c_0\,y(2h) = 0, \tag{65}$$

where a_0, b_0, and c_0 are functions of λ as well as of h. If we fix $y(-h)$ and give $y(2h)$ the arbitrary value p as before, then $y(h)$ is uniquely determined by (65), and with $y(0)$ equal to zero we can recur forwards as before. Any subsequent change in λ and p will cause $y(h)$ to change in accordance with (65), and in the initial conditions for q we must satisfy (65) with $y(-h) = 0$, so that $y(h) = -c_0 p/b_0$.

At the other end of the range we can find a linear relation similar to (65), which has to be satisfied together with the condition $y(1) = 0$. If the linear relation is

$$a_1 y(1-h)+b_1 y(1+h)+c_1 y(1+2h) = 0, \tag{66}$$

we must remember that a_1, b_1, and c_1 are also functions of λ and allow for this in equations like (61). The method then becomes rather tedious.

Non-linear equations

29. Non-linear equations present a more formidable problem. For example, as we noticed in earlier chapters, there may be more than one solution and the acceptable one is then determined by physical considerations. By a combination of methods already outlined, however, we can usually get some solution to a non-linear equation.

There is an obvious simple method, analogous to those of the early part of this chapter. For example, a second-order equation of the form

$$y''+f(x,y,y') = 0, \tag{67}$$

with boundary conditions

$$g_0(y,y') = 0 \text{ at } x = x_0, \qquad g_n(y,y') = 0 \text{ at } x = x_n, \tag{68}$$

might be solved by step-by-step methods. We can always find a relation between $y(x_0)$ and $y(x_0+h)$ which satisfies the initial condition accurately or approximately and, with one of these values fixed arbitrarily, we can integrate to $x = x_n$ and calculate the corresponding value of $g_n(y,y')$ at that boundary. From a set of such solutions, with different values of the arbitrary pivotal value, we can choose by inverse interpolation that solution for which $g_n = 0$, and our problem is solved, apart from any truncation errors.

30. Consider the problem defined by the equations

$$y''-2y^3 = 0, \qquad y(1) = 1, \qquad y'(2)+y^2(2) = 0, \tag{69}$$

first solved in Chapter 4. The truncated finite-difference representation of the differential equation is given by

$$y_{r+1} = 2y_r+2h^2y_r^3-y_{r-1}. \tag{70}$$

Starting with $y(1) = 1$ and $y(1{\cdot}2) = 0, 1$, we find the first two trial solutions $y^{(1)}$ and $y^{(2)}$ shown on the left of Table 22.

TABLE 22

x	$y^{(1)}$	$y^{(2)}$	$y^{(3)}$	$y^{(4)}$	$y^{(5)}$	$y^{(6)}$
+1·0	+ 1·0000	+1·0000	+1·0000	+1·0000	+1·0000	+1·0000
1·2	− 0·0000	1·0000	0·9000	0·8500	0·8400	0·8300
1·4	1·0000	1·0800	0·8583	·7491	·7274	·7057
1·6	2·0800	1·2608	+0·8671	·6818	·6456	·6095
1·8	3·8799	1·6019		·6399	·5853	·5314
2·0	10·3523	2·2718		·6190	·5410	·4653
+2·2	−105·5813	+3·8797		+0·6171	+0·5094	+0·4073

For the second boundary condition we use the simple formula

$$y(2{\cdot}2)-y(1{\cdot}8)+0{\cdot}4y^2(2{\cdot}0) = 0. \tag{71}$$

For $y^{(1)}$ and $y^{(2)}$ the left-hand sides of (71) are respectively $-58{\cdot}8$ and $+4{\cdot}3$, so that we expect that the true solution has $y(1{\cdot}2)$ somewhere between those selected, and by rough linear inverse interpolation we find $y^{(3)}(1{\cdot}2) = 0{\cdot}9$. The start of the solution for $y^{(3)}$ is given in Table 22, but it soon becomes clear that $y^{(3)}$ is going to increase steadily, so that (71) cannot be satisfied. Clearly $y(1{\cdot}2)$ must be less than $0{\cdot}9$, and we start with $0{\cdot}85$, as shown in $y^{(4)}$ of Table 22. For the left-hand side of (71) we find for $y^{(4)}$ the value $0{\cdot}1305$, so that $y^{(4)}$ is close to a solution though $y^{(4)}(1{\cdot}2)$ is still too large. For two further solutions the values of (71) are $+0{\cdot}0412$ and $-0{\cdot}0375$, so that $y(1{\cdot}2)$ is certainly between $0{\cdot}83$ and $0{\cdot}84$, and an accurate value can be found by interpolation.

31. This method of converging to a solution is not very systematic, and we may do better by using Newton's method in a manner analogous to the 'η' process of Chapter 4 and the method of this chapter for eigenvalue problems. If y is an approximate solution of (67), then $y+\eta$ is a better solution, where η is given by the equation

$$\eta''+\eta\frac{\partial f}{\partial y}+\eta'\frac{\partial f}{\partial y'}+y''+f(x,y,y') = 0, \tag{72}$$

and boundary conditions

$$\left.\begin{aligned} \eta'\frac{\partial g_0}{\partial y'}+\eta\frac{\partial g_0}{\partial y}+\ g_0(y,y') &= 0 \text{ at } x = x_0 \\ \eta'\frac{\partial g_n}{\partial y'}+\eta\frac{\partial g_n}{\partial y}+g_n(y,y') &= 0 \text{ at } x = x_n \end{aligned}\right\}. \tag{73}$$

The equations for η are linear, and can be solved by the methods of this chapter. The approximation $y+\eta$ is then a more accurate solution to our problem, and can be used in place of the original y to give equations for a second correction of η type.

Convergence of this method is not certain, and it is advisable to do a little work by the previous method to get a reasonably accurate first approximation. We therefore start with the second solution of Table 22, here denoted by $y^{(1)}$ for future convenience.

In general η will be evaluated in the form $\eta^{(1)}+\alpha\eta^{(2)}$, where $\eta^{(1)}$ is a particular solution of (72) satisfying the first condition (73) at x_0, $\eta^{(2)}$ a solution satisfying the corresponding homogeneous equations, and α determined to satisfy the second of (73). For our first correction, however, since $y^{(1)}$ already satisfies the finite-difference equivalent of (69) and the initial condition, (72) and the first of (73) for η are already homogeneous, so that $\eta^{(1)}$ is zero and we need only one trial solution $\eta^{(2)}$. In our problem equations (72) and (73) are given by

$$\left.\begin{array}{r} \eta''-6y^2\eta+y''-2y^3 = 0, \qquad \eta = 0 \text{ at } x = 1 \\ \eta'+2y\eta+y'+y^2 = 0 \text{ at } x = 2 \end{array}\right\}, \tag{74}$$

and the simple finite-difference equation for η is

$$\eta_1+\eta_{-1}-(2+6h^2y_0^2)\eta_0+\delta^2y_0-2h^2y_0^3 = 0. \tag{75}$$

The trial solution $\eta^{(2)}$ is shown in Table 23. To satisfy the simplest finite-difference form of the third of (74) we find $\alpha = -0{\cdot}06796$, and $y^{(2)} = y^{(1)}+\alpha\eta^{(2)}$ is also shown in Table 23.

The function $y^{(2)}$ no longer satisfies exactly the approximate finite-difference equations so that our next η correction has two parts, $\eta^{(1)}$ and $\eta^{(2)}$, and the next approximation $y^{(3)}$ is $y^{(2)}+\eta^{(1)}+\alpha\eta^{(2)}$ where α, chosen so that $\eta^{(1)}+\alpha\eta^{(2)}$ satisfies the finite-difference approximation to the last of (74), is found to be $-0{\cdot}06324$, and the solution $y^{(3)}$ is assembled in Table 23. The cycle of operations is repeated as shown in Table 24.

32. At the end of Table 24 the approximation $y^{(5)}$ satisfies the approximate finite-difference equations almost exactly, for both differential equation and boundary condition at $x = 2$, and we must now correct it by incorporating the difference corrections. As in Chapter 4, however, we shall find it more convenient to assess the error of our solution by evaluating accurately $y''-2y^3$ at every point and $y'+y^2$ at $x = 2$, using these quantities in the first and last of (74) respectively, from which the next correction η can be derived. Where derivatives of η occur in these equations, of course, we use the simple derivative formulae as before.

The differences, the values of $y''-2y^3$ and of $y'(2)+y^2(2)$ are calculated for the function $y^{(5)}$ of Table 24, with external values of $y^{(5)}$ obtained by recurrence in the usual way. The quantities $y''-2y^3$ are used, of course,

TABLE 23

x	$\eta^{(2)}$	$y^{(2)}$		δ^2	$\delta^2 y-2h^2y^3$	$\eta^{(1)}$	$\eta^{(2)}$	$y^{(3)}$		δ^2	$\delta^2 y-2h^2y^3$
+1·0	+ 0·0000	+1·0000				+0·0000	+ 0·0000	+1·0000			
			− 680						−1312		
1·2	1·0000	0·9320		+ 638	−0·0010	·0000	1·0000	0·8688		+515	−0·0010
			− 42						797		
1·4	2·2400	0·9278		581	·0058	·0010	2·2085	·7891		+354	·0039
			+ 539						443		
1·6	4·1070	0·9817		538	·0219	·0080	3·8733	·7448		+208	·0122
			1077						235		
1·8	7·5408	1·0894		+ 132	·0902	·0388	6·4340	·7213		− 13	·0313
			+1209						248		
2·0	15·6190	1·2103		−3768	−0·5186	·1709	10·8271	·6965		−389	−0·0659
			−2559						− 637		
+2·2	+43·0445	+0·9544				+0·8817	+19·0270	+0·6328			

TABLE 24

x	$\eta^{(1)}$	$\eta^{(2)}$	$y^{(4)}$		δ^2	$\delta^2 y - 2h^2y^3$	$\eta^{(1)}$	$\eta^{(2)}$	$y^{(5)}$		δ^2	$\delta^2 y - 2h^2y^3$
+1·0	+0·0000	+ 0·0000	+1·0000				+0·0000	+ 0·0000	+1·0000			
				−1608						−1650		
1·2	·0000	1·0000	0·8392		+472	−0·0001	·0000	1·0000	0·8350		+465	−0·0001
				1136						1185		
1·4	·0010	2·1812	·7256		297	·0009	·0001	2·1690	·7165		295	+ ·0001
				839						890		
1·6	·0060	3·6883	·6417		193	·0018	·0011	3·6122	·6275		197	− ·0001
				646						693		
1·8	·0240	5·6863	·5771		120	·0034	·0040	5·4123	·5582		138	− ·0001
				526						555		
2·0	·0763	8·3945	·5245		+ 70	−0·0045	·0106	7·6448	·5027		+101	−0·0001
				− 456						− 454		
+2·2	+0·2034	+12·0798	+0·4789				+0·0224	+10·3819	+0·4573			

$\alpha = -0{\cdot}02958$ $\alpha = -0{\cdot}00424$

only in the solution $\eta^{(1)}$. All the calculations are shown in Table 25, the final solution being $y^{(5)}+\eta^{(1)}+\alpha\eta^{(2)}$, where α is determined so that, with $\eta = \eta^{(1)}+\alpha\eta^{(2)}$,

$$\eta_{2\cdot2}-\eta_{1\cdot8}+0\cdot4(2y\eta+y'+y^2)_{2\cdot0} = 0. \tag{76}$$

TABLE 25

x	$y^{(5)}$		δ^2		δ^4		δ^6	$h^2(y''-2y^3)$	$\eta^{(1)}$	$\eta^{(2)}$	y
+0·6	+1·6444										
		−3994									
0·8	1·2450		+1544								
		2450		−744							
1·0	1·0000		800		+409				+0·0000	+ 0·0000	+1·0000
		1650		335		−244					
1·2	0·8350		465		165		+151	−0·0013	·0000	1·0000	0·8334
		1185		170		93					
1·4	·7165		295		72		60	·0005	·0013	2·1673	·7143
		890		98		33					
1·6	·6275		197		39		16	·0004	·0033	3·6016	·6250
		693		59		17					
1·8	·5582		138		22		7	·0003	·0060	5·3763	·5555
		555		37		10					
2·0	·5027		101		12		+ 6	−0·0002	·0094	7·5531	·4999
		454		25		− 4					
2·2	·4573		76		+ 8				+0·0136	+10·1876	+0·4545
		378		− 17							
2·4	·4195		+ 59								
		− 319									
+2·6	+0·3876										

$0\cdot4\{y'(2)+y^2(2)\} = +0\cdot0012$, $\alpha = -0\cdot00161$

The final y differs from the analytical solution x^{-1} by a maximum of 1 unit in the last place retained.

Simultaneous differential equations

33. Simultaneous differential equations provide no inherent extra difficulty. Suppose the equations are linear, of second order, and given by

$$\left.\begin{aligned} y''+a_1y'+b_1y+c_1z'+d_1z+e_1 &= 0\\ z''+a_2y'+b_2y+c_2z'+d_2z+e_2 &= 0\end{aligned}\right\}, \tag{77}$$

where the coefficients are functions of the independent variable x, with boundary conditions

$$\left.\begin{aligned} &\left.\begin{aligned} y'+p_1y+q_1z+r_1 &= 0\\ z'+p_2y+q_2z+r_2 &= 0\end{aligned}\right\}\text{at } x = x_0\\ &\left.\begin{aligned} y'+l_1y+m_1z+n_1 &= 0\\ z'+l_2y+m_2z+n_2 &= 0\end{aligned}\right\}\text{at } x = x_n\end{aligned}\right\}. \tag{78}$$

Our standard methods would produce solutions

$$\left.\begin{aligned} y &= Ay^{(1)}+By^{(2)}+y^{(3)}\\ z &= Az^{(1)}+Bz^{(2)}+z^{(3)}\end{aligned}\right\}, \tag{79}$$

where A and B are chosen so that the last two equations of (78) are satisfied, and the trial solutions are chosen as follows.

$y^{(3)}$ and $z^{(3)}$ satisfy (77) and the first pair of (78), which means that one of the two necessary starting values of both $y^{(3)}$ and $z^{(3)}$ is arbitrary. $y^{(1)}$ and $z^{(1)}$, $y^{(2)}$ and $z^{(2)}$, satisfy the homogeneous forms of (77) and the first pair of (78), and are independent pairs of functions, so that their necessary arbitrary values are fixed at different points. If, in particular, the first pair of (78) specified y and z at $x = x_0$, we should have, for initial conditions of our trial solutions, the values

$$\left.\begin{array}{llllll} y_0^{(3)} = y_0, & z_0^{(3)} = z_0; & y_0^{(1)} = 0, & z_0^{(1)} = 0; & y_0^{(2)} = 0, & z_0^{(2)} = 0 \\ y_1^{(3)} = 0, & z_1^{(3)} = 0; & y_1^{(1)} = 1, & z_1^{(1)} = 0, & y_1^{(2)} = 0, & z_1^{(2)} = 1 \end{array}\right\}. \tag{80}$$

The process is quite straightforward, and its extension to other types of linear simultaneous equations will be obvious.

34. If the simultaneous equations come from an eigenvalue problem the situation is not so favourable. Consider, for example, the equations

$$\left.\begin{array}{l} y''+f(y',z',y,z,x,\lambda) = 0 \\ z''+g(y',z',y,z,x,\lambda) = 0 \end{array}\right\}, \tag{81}$$

where f and g are linear functions of y', z', y, z, and λ and are homogeneous. The boundary conditions may have the form

$$y = z = 0 \text{ at } x = x_0, \qquad y = z = 0 \text{ at } x = x_n. \tag{82}$$

With a single second-order equation we were able to fix both y_0 and y_1, so that y_n was a function only of λ, and by solving a variational equation for $\partial y/\partial\lambda$ we were able to alter λ systematically until a solution was obtained. With more than one equation only one function can be fixed at x_1, this determining the 'size' of all the functions. If y_1 is fixed, the values of y_2 and z_2 obtained from (81) are functions not only of λ but also of z_1, and the situation is similar to that of a single fourth-order equation. We then need to solve variational equations for $\partial y/\partial q$, $\partial z/\partial q$ as well as for $\partial y/\partial\lambda$, $\partial z/\partial\lambda$, where q is z_1.

35. As an example we consider the equations

$$\left.\begin{array}{l} y''+\ z+\lambda y = 0 \\ z''+2y+\lambda z = 0 \end{array}\right\}, \tag{83}$$

with boundary conditions

$$y(0) = y(1) = z(0) = z(1) = 0. \tag{84}$$

It is rather unsymmetrical to fix $y(h)$ and vary $z(h)$, so that in the analysis we shall allow both to vary, with notation $y(h) = p$, $z(h) = q$, and if

$\partial y/\partial\lambda = \eta$, $\partial z/\partial\lambda = \zeta$, we have the following equations and associated boundary conditions:

$$\left.\begin{aligned} y''+z+\lambda y &= 0 \\ z''+2y+\lambda z &= 0 \end{aligned}\right\} \text{ with } y(0) = z(0) = 0,\ y(h) = p,\ z(h) = q, \tag{85}$$

$$\left.\begin{aligned} \left(\frac{\partial y}{\partial p}\right)''+\ \frac{\partial z}{\partial p}+\lambda\frac{\partial y}{\partial p} &= 0 \\ \left(\frac{\partial z}{\partial p}\right)''+2\frac{\partial y}{\partial p}+\lambda\frac{\partial z}{\partial p} &= 0 \end{aligned}\right\} \text{with } \frac{\partial y(0)}{\partial p} = 0,\ \frac{\partial z(0)}{\partial p} = 0,\ \frac{\partial y(h)}{\partial p} = 1,\ \frac{\partial z(h)}{\partial p} = 0, \tag{86}$$

$$\left.\begin{aligned} \left(\frac{\partial y}{\partial q}\right)''+\ \frac{\partial z}{\partial q}+\lambda\frac{\partial y}{\partial q} &= 0 \\ \left(\frac{\partial z}{\partial q}\right)''+2\frac{\partial y}{\partial q}+\lambda\frac{\partial z}{\partial q} &= 0 \end{aligned}\right\} \text{with } \frac{\partial y(0)}{\partial q} = 0,\ \frac{\partial z(0)}{\partial q} = 0,\ \frac{\partial y(h)}{\partial q} = 0,\ \frac{\partial z(h)}{\partial q} = 1, \tag{87}$$

$$\left.\begin{aligned} \eta''+\zeta+\lambda\eta+y &= 0 \\ \zeta''+2\eta+\lambda\zeta+z &= 0 \end{aligned}\right\} \text{ with } \eta(0) = \zeta(0) = \eta(h) = \zeta(h) = 0. \tag{88}$$

When these equations have been solved, with arbitrary starting values of λ, p, and q, we find corrections to these quantities from the equations

$$\left.\begin{aligned} y(1)+\eta(1)\delta\lambda+\frac{\partial y(1)}{\partial p}\delta p+\frac{\partial y(1)}{\partial q}\delta q &= 0 \\ z(1)+\zeta(1)\delta\lambda+\frac{\partial z(1)}{\partial p}\delta p+\frac{\partial z(1)}{\partial q}\delta q &= 0 \end{aligned}\right\}, \tag{89}$$

remembering that one of δp, δq is arbitrary; at this stage we can decide the one to be fixed from considerations of the size of the numbers involved.

Using the simplest finite-difference equations, with the interval $h = 0{\cdot}2$, we find our first results, with $\lambda = 0$, $p = q = 1$ initially, as shown in Table 26.

TABLE 26

x	y	z	η	ζ	$\frac{\partial y}{\partial p}$	$\frac{\partial z}{\partial p}$	$\frac{\partial y}{\partial q}$	$\frac{\partial z}{\partial q}$
+0·0	+0·0000	+0·0000	−0·0000	−0·0000	+0·0000	−0·0000	−0·0000	+0·0000
·2	1·0000	1·0000	·0000	·0000	1·0000	·0000	·0000	1·0000
·4	1·9600	1·9200	·0400	·0400	2·0000	·0800	·0400	2·0000
·6	2·8432	2·6832	·1568	·1536	3·0032	·3200	·1600	3·0032
0·8	3·6191	3·2189	·3812	·3620	4·0192	0·8003	·4001	4·0192
+1·0	+4·2662	+3·4651	−0·7359	−0·6687	+5·0672	−1·6021	−0·8010	+5·0672

We then have

$$\left.\begin{aligned} 4{\cdot}2662-0{\cdot}7359\delta\lambda+5{\cdot}0672\delta p-0{\cdot}8010\delta q &= 0 \\ 3{\cdot}4651-0{\cdot}6687\delta\lambda-1{\cdot}6021\delta p+5{\cdot}0672\delta q &= 0 \end{aligned}\right\}, \tag{90}$$

and, with no apparent reason for discrimination, we fix $\delta q = 0$, so that $\delta p = -0{\cdot}0663$, $h^2\delta\lambda = +0{\cdot}2136$. Table 27 gives the next set of results, and we have decided to make no change in q so that we need no longer solve equations (87).

TABLE 27

x	y	z	η	ζ	$\partial y/\partial p$	$\partial z/\partial p$
+0·0	+0·0000	+0·0000	−0·0000	−0·0000	+0·0000	−0·0000
·2	0·9337	1·0000	·0000	·0000	1·0000	·0000
·4	1·6280	1·7117	·0373	·0400	1·7864	·0800
·6	1·9061	1·9275	·1302	·1369	2·1944	·2858
0·8	1·7000	1·5791	·2661	·2712	2·1451	·6061
+1·0	+1·0676	+0·7574	−0·4023	−0·3894	+1·6619	−0·9685

Further changes in p and λ are now obtained from the equations

$$\left.\begin{aligned} 1{\cdot}0676 - 0{\cdot}4023\delta\lambda + 1{\cdot}6619\delta p &= 0 \\ 0{\cdot}7574 - 0{\cdot}3894\delta\lambda - 0{\cdot}9685\delta p &= 0 \end{aligned}\right\}, \tag{91}$$

so that $\delta p = -0{\cdot}1071$, $h^2\delta\lambda = +0{\cdot}0885$. Further approximations are summarized in Table 28. Convergence to the final solution is quite rapid.

TABLE 28

$y(1)$	$z(1)$	$\eta(1)$	$\zeta(1)$	$\dfrac{\partial y(1)}{\partial p}$	$\dfrac{\partial z(1)}{\partial p}$	$h^2\delta\lambda$	δp
0·2088	0·0857	−0·2668	−0·3025	0·7082	−0·7532	+0·0210	−0·0971
0·0232	0·0009	−0·2142	−0·2899	0·5157	−0·7057	+0·0022	−0·0217
0·0010	0·0001	−0·2050	−0·2894	0·4962	−0·7009	+0·0001	−0·0009

The solution can be improved in the usual way by including difference corrections in solutions for y and z only.

36. With n simultaneous second-order equations in n dependent variables we have to solve, at each stage, $n+1$ sets of n simultaneous differential equations and one set of n algebraic equations to give the starting values and λ for the next step. If the variables are $y_1,\ldots, y_n$ there is one set of n differential equations to give solutions analogous to y and z of the previous section, one set of n variational equations for λ corresponding to η and ζ, and $(n-1)$ sets of variational equations, assuming one of the variables $y(h)$ is held fixed, for changes in $y_r(h)$, corresponding to $\partial y/\partial p$, $\partial z/\partial p$. The starting values for the latter are given by

$$\frac{\partial y_r(0)}{\partial p_r} = 0 \text{ for all } r, \quad \frac{\partial y_r(h)}{\partial p_s} = \begin{cases} 0 & (r \neq s) \\ 1 & (r = s). \end{cases} \tag{92}$$

This is clearly a formidable undertaking. If no *a priori* knowledge is available of the form of the required solutions and corresponding eigenvalue, the Newton process, though ultimately securing powerful convergence, may cause large and apparently random variations in the early stages. Even for a single equation, where the variational function depends on λ alone, we shall have trouble if the original estimate of λ is near a turning-point of $y_n(\lambda)$.

For example, if in the problem and method of § 24 we take the first estimate of $h^2\lambda$ to be 1·2, we find successive corrections of $+13{\cdot}6$, $-4{\cdot}23$, $-2{\cdot}80$, $-1{\cdot}84$,..., and, after the first rather unforeseen development, convergence is slow at first but sure. The solution obtained will correspond, however, to the third eigensolution, for which $h^2\lambda = 3{\cdot}41$, even though our initial guess was somewhere between the first ($h^2\lambda = 0{\cdot}59$) and the second ($h^2\lambda = 2{\cdot}0$).

We cannot therefore ensure, with no initial reasonable knowledge, that our method will converge to any required solution, or that the convergence will be very fast. With several equations these difficulties increase. If there is no real solution, of course, the process will never converge, but we may have to perform many steps before we can be certain that the lack of convergence is real and not caused by initial vagaries of the Newton process.

Difficulties and inaccuracies

37. There are two main sources of inaccuracy in the general methods outlined for linear equations, and we can illustrate them by considering again the simplest problem, that of solving equation (1) subject to given boundary values of y, using the method of § 5. We need consider only the finite-difference equations (2) and ignore all difference corrections.

The first difficulty arises in connexion with the trial solution $y^{(2)}$. If in Table 3 the function $y^{(2)}$ had vanished at $x = 0$ we could not find a finite α for which $y^{(1)}+\alpha y^{(2)}$ has the required boundary value at $x = 0$. In this case there are solutions only for certain combinations of boundary values, and the solutions are not then unique, for we can add any multiple of $y^{(2)}$ without disturbing the differential equation or boundary conditions. In other words, there is a non-trivial solution of the homogeneous equation with homogeneous boundary conditions.

Suppose, for example, we try to solve the finite-difference equations typified by

$$y_{r+1}+y_{r-1}-1{\cdot}4142y_r = 0{\cdot}25, \tag{93}$$

with the conditions $y_0 = 1$, $y_4 = 2$. Our solution of type $y^{(1)}$ has the values

$$y_0 = 1, \quad y_1 = 0, \quad y_2 = -0{\cdot}7500, \quad y_3 = -0{\cdot}8106, \quad y_4 = -0{\cdot}1464,$$

our solution of type $y^{(2)}$ has the values

$$y_0 = 0, \quad y_1 = 1, \quad y_2 = 1{\cdot}4142, \quad y_3 = 1{\cdot}0000, \quad y_4 = 0{\cdot}0000,$$

and we cannot satisfy the second boundary condition.

No assistance is forthcoming from the more general method of § 8. Corresponding to Table 5 we would have the following type solutions:

$y^{(3)}$	0·0000	0·0000	0·2500	0·6035	0·8535
$y^{(1)}$	0·0000	1·0000	1·4142	1·0000	0·0000
$y^{(2)}$	1·0000	0·0000	−1·0000	−1·4142	−1·0000

and equations (16) applied at the boundaries merely give two equations for the constant B, compatible only if $y_0+y_4 = 0{\cdot}8535$. Any multiple of $y^{(1)}$ can be added.

This situation is of course connected with the existence of an eigensolution. Normally the boundary positions are fixed, and a parameter in the differential equation has discrete values for which there are non-trivial solutions. Alternatively we can think of the parameter being held fixed, allowing one boundary position x_n to vary. For one or more values of x_n there may be eigensolutions.

†**38.** In the analytical solution of differential equations this phenomenon causes no great worry, but in the numerical solution we shall have difficulties in obtaining accuracy and there are three main problems. First, the approximate finite-difference equations may have an eigensolution whereas the differential equation has no such solution and, second, there is the converse possibility. Finally, neither may have an eigensolution but such a solution may exist for a closely allied problem, for example for small changes in coefficients or boundary positions. The only way to treat the first two of these difficulties is to test whether the differential equation has an eigensolution by including the difference correction in the 'homogeneous' trial solution. The third difficulty can be overcome only by the retention of more guarding figures. With the method of § 5, an estimate of the precision obtainable is already provided by the size of the value of $y_n^{(2)}$, the coefficient of the required α in the case when y_n is specified.

39. This value of $y_n^{(2)}$ is of course proportional to the value of the determinant of the matrix of the coefficients of the finite-difference equations, for the components of $y^{(2)}$ satisfy the equations

$$\left.\begin{aligned} b_1 y_1 + c_1 y_2 \qquad\qquad\qquad\qquad &= 0 \\ a_2 y_1 + b_2 y_2 + c_2 y_3 \qquad\qquad &= 0 \\ a_3 y_2 + b_3 y_3 + c_3 y_4 \qquad &= 0 \\ \cdots\cdots\cdots\cdots\cdots & \\ a_{n-1} y_{n-2} + b_{n-1} y_{n-1} + c_{n-1} y_n &= 0 \end{aligned}\right\}, \tag{94}$$

and their solution is given by

$$\frac{y_1}{\Delta_1} = -\frac{y_2}{\Delta_2} = \dots = \frac{(-1)^{n+1}y_n}{\Delta_n}, \tag{95}$$

where Δ_r is the determinant obtained from (94) by omitting the rth column. Clearly $\Delta_1 = c_1 c_2 c_3 \dots c_{n-1}$, and $\Delta_n = |A|$, where the matrix A excludes the term in y_n in (94), so that the result is proved, the y_r being obtained successively from (94) with $y_1 = 1$.

†**40.** The second difficulty of a practical nature will arise if $y_n^{(2)}$ is very large, which will happen if one of the complementary functions increases rapidly. To obtain n-decimal accuracy in our final result we shall have to record every component of both $y^{(1)}$ and $y^{(2)}$ to n decimals, which may involve the recording and multiplication of very large numbers.

Consider, for example, the six-decimal solution of the equations typified by

$$y_{r+1}+y_{r-1}-50{\cdot}0625y_r = 0{\cdot}015625, \tag{96}$$

with $y_0 = y_5 = 1$. The trial solutions are shown in Table 29.

TABLE 29

x	$y^{(1)}$	$y^{(2)}$	y
+0	+1·000000	0·000000	+1·000000
1	0·000000	+1·000000	0·019665
2	−0·984375	+50·062500	0·000082
3	−49·264648	+2505·253906	0·000082
4	−2465·311440	+125369·211169	0·019665
+5	−123370·373692	+6273790·880242	+1·000000

We now calculate α from the condition $y^{(1)}(5)+\alpha y^{(2)}(5) = 1$, finding the value $\alpha = 0{\cdot}019664565818$, and we must take as many figures as are accurate, so that $\alpha y^{(2)}$ has nowhere rounding errors in the sixth decimal. We then find y as shown in Table 29.

It might be thought that this method could not give good results because of the growth of rounding errors in both $y^{(2)}$ and $y^{(1)}$. It is clear, however, that both $y^{(2)}$ and $y^{(1)}$ satisfy their respective equations, in the sense that the residuals are not greater than 5 units in the seventh decimal, so that this is true also of $y^{(1)}+\alpha y^{(2)}$ for any value of α, provided $y^{(1)}+\alpha y^{(2)}$ is correctly evaluated to six decimals. For this to be possible we need to be able to calculate α from its equation at the boundary with such precision that $\alpha y^{(2)}$ has everywhere the required precision. In the present case this is obviously possible.

Another point of view would regard the building-up error as systematic and involving only one of the complementary functions. If $y^{(1)}+\alpha y^{(2)}$ eliminates this error at the end point it eliminates it everywhere.

41. Though this case is clearly more favourable than the oscillatory case as regards the accuracy obtainable, the work may be very laborious due to manipulations with large numbers. We might in the last example retain only six significant figures in $y^{(1)}$ and $y^{(2)}$, knowing that α is then calculable to six significant figures. We find $\alpha = +0{\cdot}0196645$, and are then certain that the calculated $y = y^{(1)}+\alpha y^{(2)}$ is correct at x_1. For x_2, however, α is multiplied by 50, so that y may be in error in the sixth decimal.

In this case we may start our problem afresh with the knowledge that $y = 0{\cdot}019665$ at $x = 1$, $1{\cdot}000000$ at $x = 5$, constructing new solutions of types $y^{(1)}$ and $y^{(2)}$ in which, for example, $y^{(2)} = 0$ at $x = 1$, 1 at $x = 2$. The method becomes laborious, though in solving differential equations the diagonal terms are not often so dominant as in the present example.

Matching in the middle

†42. Very large numbers might be avoided by recurring from each end and matching in the middle. Suppose that we take the trial solutions $y^{(1)}$ and $y^{(2)}$ in Table 29 only as far as $x = 3$, and find similar solutions $y^{(3)}$ and $y^{(4)}$ by recurrence from $x = 5$ to $x = 2$ with the conditions $y^{(4)}(5) = 0$, $y^{(4)}(4) = 1$, $y^{(3)}(5) = 1$, $y^{(3)}(4) = 0$. Then $y^{(1)}+\alpha y^{(2)}$ satisfies the differential equation and one boundary condition, $y^{(3)}+\beta y^{(4)}$ satisfies the differential equation and the other boundary condition, and if these are identical functions they are the solution of our problem. They are identical if they agree at two points, say $x = 2$ and $x = 3$. In this case the symmetry of the equations means that $y^{(2)}(x) = y^{(4)}(5-x)$, $y^{(1)}(x) = y^{(3)}(5-x)$, and the simultaneous equations for α and β are given by

$$\left.\begin{aligned} -\ 0{\cdot}984375+\ \ 50{\cdot}0625\alpha &= -49{\cdot}264648+2505{\cdot}253906\beta \\ -49{\cdot}264648+2505{\cdot}253906\alpha &= -\ 0{\cdot}984375+\ \ 50{\cdot}0625\beta \end{aligned}\right\}, \tag{97}$$

from which we find $\alpha = \beta = 0{\cdot}0196645666$. No greater precision is obtainable, but none is required, since we never multiply α or β by a number greater than 2500, so that six-decimal accuracy is assured.

Even if there is no rapid increase in the complementary functions it may be desirable to match in the middle of the range, particularly if the range is large. For now accumulation of rounding error, although much smaller than before, is not confined to one of the complementary functions, and its overall effect is likely to be smaller with the use of twice as many recurrence processes taken over only half the range.

This technique is useful if one of the boundaries is at infinity, when an asymptotic solution might be known. For some eigenvalue problems also we may find it desirable, but discussion of difficult eigenvalue problems is deferred until Chapter 10.

Connexion with matrix methods and pure mathematics

43. We end this chapter with a short discussion of the connexion between the combination of step-by-step methods used in this chapter, the methods using triangular matrices developed in earlier chapters, and analytical methods.

The simplest case is that of the second-order equation

$$y''+f(x)y'+g(x)y = k(x), \tag{98}$$

with specified boundary values. In § 5 we solved the approximate finite-difference form of this problem by considering two trial solutions, $y^{(1)}$ and $y^{(2)}$, of which the former satisfied the inhomogeneous finite-difference equations and the correct initial condition, and $y^{(2)}$ satisfied the homogeneous forms of the finite-difference equations and had initial value zero. The final approximate solution was then $y^{(1)}+\alpha y^{(2)}$, with α chosen to satisfy the second boundary value.

Here $y^{(2)}$ corresponds to one complementary solution of the differential equation, but $y^{(1)}$ is a combination of the second complementary solution and a particular integral. To avoid further complications in notation we take $y^{(2)}$ and $y^{(3)}$ to be the two independent complementary solutions of the difference equations, with zero and non-zero values as respective initial conditions. For the differential equation we take y_2 and y_3 to be the corresponding complementary functions.

†44. With y_2 and y_3, independent solutions of the homogeneous form of (98), there is an associated quantity, the Wronskian, given by

$$W = y_2' y_3 - y_3' y_2 = \exp\left(-\int f\,dx\right). \tag{99}$$

We can use (99) to determine y_3 in terms of y_2 and W, finding

$$y_3 = -y_2 \int \frac{W}{y_2^2}\,dx. \tag{100}$$

It is known further that if the independent complementary functions are taken in the form

$$y_2 = t_2, \qquad y_3 = t_2 \int t_3\,dx, \tag{101}$$

then the differential equation (98) can be expressed in the form

$$-\frac{W}{y_2}\frac{d}{dx}\left(\frac{1}{t_3}\frac{d}{dx}\frac{1}{t_2}\right)y = k, \tag{102}$$

so that a particular integral of (98) is given by

$$y = y_2 \int \frac{W}{y_2^2} \left(\int k \frac{y_2}{W} \, dx \right) dx. \tag{103}$$

45. There is a direct analogy, which we proceed to develop, between this result and that obtained in the matrix operations of our standard boundary-value techniques. Here we express (98) as an approximate difference equation of the form

$$a_r y_{r-1} + b_r y_r + c_r y_{r+1} = d_r, \tag{104}$$

and, with the values y_0 and y_n known, we consider the matrix

$$\begin{bmatrix} b_1 & c_1 & & & & \\ a_2 & b_2 & c_2 & & & \\ & a_3 & b_3 & c_3 & & \\ & \cdot & \cdot & \cdot & \cdot & \cdot & \cdot \\ & & & a_{n-2} & b_{n-2} & c_{n-2} \\ & & & & a_{n-1} & b_{n-1} \end{bmatrix}. \tag{105}$$

The first step is to express (105) as a product LU of triangular matrices, in the form

$$\overset{L}{\begin{bmatrix} 1 & & & & & \\ l_2 & 1 & & & & \\ & l_3 & 1 & & & \\ & \cdot & \cdot & \cdot & \cdot & \\ & & l_{n-2} & 1 & & \\ & & & l_{n-1} & 1 \end{bmatrix}} \overset{U}{\begin{bmatrix} u_1 & v_1 & & & & \\ & u_2 & v_2 & & & \\ & & u_3 & v_3 & & \\ & & \cdot & \cdot & \cdot & \cdot & \cdot & \cdot \\ & & & & u_{n-2} & v_{n-2} \\ & & & & & u_{n-1} \end{bmatrix}}. \tag{106}$$

The general equations for the determination of l_r, u_r, v_r are given by

$$\left. \begin{aligned} l_r u_{r-1} &= a_r \\ l_r v_{r-1} + u_r &= b_r \\ v_r &= c_r \end{aligned} \right\}, \tag{107}$$

and there is one particular equation

$$u_1 = b_1. \tag{108}$$

The coefficients l_r and u_r can be expressed as continued fractions, for by eliminating u we can find

$$l_{r+1} = \frac{a_{r+1}}{b_r - c_{r-1} l_r}, \qquad l_r = \frac{b_r}{c_{r-1}} - \frac{a_{r+1}}{c_{r-1} l_{r+1}}, \tag{109}$$

and by eliminating l we have

$$u_r = b_r - \frac{a_r c_{r-1}}{u_{r-1}}, \qquad u_{r-1} = \frac{a_r c_{r-1}}{b_r - u_r}. \tag{110}$$

We can also write the second of (110) in the form

$$\frac{u_{r-1}}{c_{r-1}} = \frac{a_r/c_r}{b_r/c_r - u_r/c_r}. \tag{111}$$

Now if y is any solution of the homogeneous form of (104), we have

$$-\frac{y_r}{y_{r-1}} = \frac{a_r/c_r}{b_r/c_r + y_{r+1}/y_r}. \tag{112}$$

Comparing (112) with (111), and using the initial condition (108), we see that

$$\frac{u_r}{c_r} = -\frac{y_{r+1}^{(2)}}{y_r^{(2)}}, \tag{113}$$

so that u_r/c_r is equal to the ratio of successive elements of that solution of the homogeneous difference equation for which $y_0^{(2)} = 0$. This $y^{(2)}$ is identical with our trial solution of the same notation used in § 5.

We found subsequently that the last pivotal value $y_n^{(2)}$ of this solution is proportional to the determinant of the matrix (105). This is easily verified, for the determinant is equal to the product of all the u_r, and from (113) we find

$$\prod_{r=1}^{n-1} u_r = (-1)^{n-1} \frac{y_n^{(2)}}{y_1^{(2)}} \prod_{r=1}^{n-1} c_r. \tag{114}$$

46. If we write (104) in the form

$$(a_r + b_r E + c_r E^2) y_{r-1} = d_r, \tag{115}$$

it becomes clear that the triangular matrix-resolution process is equivalent to decomposition of (115) in the form

$$(l_r + E)(u_{r-1} + v_{r-1} E) y_{r-1} = d_r. \tag{116}$$

This equation is equivalent to the pair of equations

$$(l_r + E) z_{r-1} = d_r, \qquad (u_{r-1} + v_{r-1} E) y_{r-1} = z_{r-1}, \tag{117}$$

which correspond to the respective matrix equations

$$Lz = d, \qquad Uy = z. \tag{118}$$

We consider first the solution for y, with z assumed known, of the second of (117), written in its expanded form

$$u_{r-1} y_{r-1} + v_{r-1} y_r = z_{r-1}. \tag{119}$$

Since $v_r = c_r$, it follows from (113) that a complementary solution of (119) is $y_r^{(2)}$, and with the method of variation of parameters we find a complete solution of (119) in the form

$$y_r = p_r y_r^{(2)}. \tag{120}$$

The function p_r satisfies the equation

$$v_{r-1}y_r^{(2)}\Delta p_{r-1} = z_{r-1}, \tag{121}$$

so that
$$p_{r-1} = \Delta^{-1}\left\{\frac{z_{r-1}}{v_{r-1}y_r^{(2)}}\right\}, \qquad y_r = y_r^{(2)}\Delta^{-1}\left\{\frac{z_r}{c_r y_{r+1}^{(2)}}\right\}, \tag{122}$$

where Δ^{-1} is the operation of 'summing', analogous to that of integrating.

47. Turning now to the first of (117), written in its expanded form

$$l_r z_{r-1}+z_r = d_r, \tag{123}$$

it is not difficult to show, with the help of (107), that a complementary solution of (123) is given by

$$z_r = c_r y_{r+1}^{(3)}+u_r y_r^{(3)}, \tag{124}$$

where $y_r^{(3)}$ is a second complementary solution, independent of $y_r^{(2)}$, of the original equation (104).

The complementary solutions $y_r^{(2)}$ and $y_r^{(3)}$ are connected by the relation

$$a_r\{y_{r-1}^{(2)}y_r^{(3)}-y_{r-1}^{(3)}y_r^{(2)}\} = c_r\{y_r^{(2)}y_{r+1}^{(3)}-y_r^{(3)}y_{r+1}^{(2)}\}. \tag{125}$$

The expression $y_{r-1}^{(2)}y_r^{(3)}-y_{r-1}^{(3)}y_r^{(2)}$, which we shall denote by W_r, is clearly analogous to the Wronskian in differential equations, and from (125) it satisfies the equation

$$a_r W_r = c_r W_{r+1}. \tag{126}$$

We can write this in the form

$$\Delta \log W_r = \log(a_r/c_r), \tag{127}$$

so that
$$W_r = \exp\{\Delta^{-1}\log(a_r/c_r)\}, \tag{128}$$

in analogy with (99) for the differential equation. This analogy is verified quantitatively by noting that, to the first order in h,

$$a_r = 1-\tfrac{1}{2}hf_r, \qquad c_r = 1+\tfrac{1}{2}hf_r, \tag{129}$$

so that
$$\log(a_r/c_r) = -hf_r, \qquad W_r = \exp\{\Delta^{-1}(-hf_r)\}, \tag{130}$$

correct to the first order in h.

48. As in the case of the differential equation, so here we can express $y_r^{(3)}$ in terms of W_r. For

$$W_r = y_{r-1}^{(2)}y_r^{(2)}\left\{\frac{y_r^{(3)}}{y_r^{(2)}}-\frac{y_{r-1}^{(3)}}{y_{r-1}^{(2)}}\right\} = y_{r-1}^{(2)}y_r^{(2)}\Delta\left\{\frac{y_{r-1}^{(3)}}{y_{r-1}^{(2)}}\right\}, \tag{131}$$

so that
$$y_r^{(3)} = y_r^{(2)}\Delta^{-1}\left\{\frac{W_{r+1}}{y_r^{(2)}y_{r+1}^{(2)}}\right\}, \tag{132}$$

in analogy with (100).

Now, using (113) and (126), we can write (124) in the form

$$z_r = c_r y_{r+1}^{(3)} + u_r y_r^{(3)} = c_r \left\{ y_{r+1}^{(3)} - \frac{y_{r+1}^{(2)}}{y_r^{(2)}} y_r^{(3)} \right\} = \frac{a_r W_r}{y_r^{(2)}}. \tag{133}$$

This is the solution of the homogeneous form of (123), and the solution of the complete equation is given by

$$z_r = \frac{a_r W_r}{y_r^{(2)}} q_r. \tag{134}$$

We find that q satisfies the equation

$$\frac{a_r}{y_r^{(2)}} W_r (q_r - q_{r-1}) = d_r, \tag{135}$$

so that $$q_{r-1} = \Delta^{-1} \left\{ \frac{d_r y_r^{(2)}}{a_r W_r} \right\}, \qquad z_r = \frac{a_r W_r}{y_r^{(2)}} \Delta^{-1} \left\{ \frac{d_{r+1} y_{r+1}^{(2)}}{a_{r+1} W_{r+1}} \right\}. \tag{136}$$

Finally, from (122) and (126), we have the required solution

$$y_r = y_r^{(2)} \Delta^{-1} \left[\frac{W_{r+1}}{y_r^{(2)} y_{r+1}^{(2)}} \Delta^{-1} \left\{ \frac{d_{r+1} y_{r+1}^{(2)}}{a_{r+1} W_{r+1}} \right\} \right], \tag{137}$$

or $$y_r = y_r^{(2)} \Delta^{-1} \left[\frac{W_{r+1}}{y_r^{(2)} y_{r+1}^{(2)}} \Delta^{-1} \left\{ \frac{d_{r+1} y_{r+1}^{(2)}}{c_{r+1} W_{r+2}} \right\} \right], \tag{138}$$

in complete analogy with the solution (103) of the differential equation. Remembering the result (132) for $y_r^{(3)}$, we note further the existence of a theorem analogous to that of (102).

49. We have shown that the $y_r^{(2)}$ of our step-by-step method of § 5 is closely connected, by equation (113), with the quantity u_r of the matrix method. We shall now extend the relationship between the two methods by connecting the auxiliary quantities z_r with the trial solutions $y^{(1)}$ and $y^{(2)}$ of § 5.

In the set of equations (123) for the quantities z_r, the first equation is special, involves the initial condition, and is given by

$$z_1 = d_1 - a_1 y_0. \tag{139}$$

The function $y_r^{(1)}$, which takes initially the values $y_0^{(1)} = y_0$, $y_1^{(1)} = 0$, satisfies initially the similar equation

$$c_1 y_2^{(1)} = d_1 - a_1 y_0, \tag{140}$$

so that $c_1 y_2^{(1)} = z_1$. Since any solution of the difference equations satisfies (119), it follows that z_r is given precisely by the equation

$$z_{r-1} = u_{r-1} y_{r-1}^{(1)} + v_{r-1} y_r^{(1)}, \tag{141}$$

for values of r up to and including $n-1$. Using previous results, we can write (141) in the form

$$z_{r-1} = c_{r-1}\left\{\frac{u_{r-1}}{c_{r-1}} y_{r-1}^{(1)} + y_r^{(1)}\right\} = c_{r-1}\left\{y_r^{(1)} - \frac{y_r^{(2)}}{y_{r-1}^{(2)}} y_{r-1}^{(1)}\right\}, \tag{142}$$

so that the z_r of the matrix method is expressible completely in terms of the two trial solutions of the step-by-step method.

Finally, the value of α in the combined solution $y = y^{(1)} + \alpha y^{(2)}$ comes from the satisfaction of the second boundary value. The inclusion of the latter in the last of equations (104) gives (123) a second special equation, given by

$$l_{n-1} z_{n-2} + z_{n-1} = d_{n-1} - c_{n-1} y_n, \tag{143}$$

and there is a corresponding special equation,

$$u_{n-1} y_{n-1} = z_{n-1}, \tag{144}$$

belonging to the set (119). Substituting in (143) for z_{n-2} from (141) and z_{n-1} from (144), replacing y_{n-1} by $y_{n-1}^{(1)} + \alpha y_{n-1}^{(2)}$, and remembering that $y^{(1)}$ is a solution of (104), we find that

$$\alpha = \frac{c_{n-1}\{y_n^{(1)} - y_n\}}{u_{n-1} y_{n-1}^{(2)}} = \frac{y_n - y_n^{(1)}}{y_n^{(2)}}, \tag{145}$$

in accordance with the expected result.

50. To illustrate these results, we consider the solution by matrix methods of the example of § 40. Working throughout with six decimals, we find

$$\begin{bmatrix} -50\cdot0625 & 1 & & \\ 1 & -50\cdot0625 & 1 & \\ & 1 & -50\cdot0625 & 1 \\ & & 1 & -50\cdot0625 \end{bmatrix}$$

$$= \begin{bmatrix} 1 & & & \\ -0\cdot019975 & 1 & & \\ & -0\cdot019983 & 1 & \\ & & -0\cdot019983 & 1 \end{bmatrix} \times$$

$$\times \begin{bmatrix} -50\cdot062500 & 1 & & \\ & -50\cdot042525 & 1 & \\ & & -50\cdot042517 & 1 \\ & & & -50\cdot042517 \end{bmatrix}.$$

The right-hand sides of the original equations are $-0\cdot984375$, $+0\cdot015625$, $+0\cdot015625$, and $-0\cdot984375$, and we find

$$z_1 = -0\cdot984375, \qquad z_2 = -0\cdot004038,$$
$$z_3 = +0\cdot015544, \qquad z_4 = -0\cdot984064.$$

The ratios of $y_{r+1}^{(2)}/y_r^{(2)}$ are found, from $y^{(2)}$ of Table 29, to have values agreeing exactly, with this precision, to the diagonal terms of the U matrix with their signs changed. This is in accordance with (113), since the elements c_r are here all unity. The values of $y_r^{(1)}-y_{r-1}^{(1)}(y_r^{(2)}/y_{r-1}^{(2)})$, calculated for $r = 2, 3, 4$ from Table 29, are found to be $-0{\cdot}984375$, $-0{\cdot}004037$, $+0{\cdot}015545$, agreeing closely with the first three values of z_r in accordance with (142).

51. In the foregoing analysis we made little reference to boundary conditions, and imposed no limits on the integrals in equation (103). If $y(x_0)$ and $y(x_n)$ are zero, it is easy to see that (103) becomes

$$y = y_2 \int\limits_x^{x_n} \frac{W}{y_2^2}\left(\int\limits_{x_0}^{x} k\frac{y_2}{W}\,dx\right)dx. \tag{146}$$

In the difference equations we satisfy the boundary values by incorporating them in the right-hand sides d_r, of which the first is $d_1-a_1 y_0$ and the last is $d_{n-1}-c_{n-1}y_n$. Then z_1 is chosen from (139) to satisfy the initial condition, and the use of (143) and (144) satisfies the other condition.

Thus, in analogy with (146), we see that the calculation of l_r, u_r, and v_r is equivalent to the determination of the function $y^{(2)}$ of the step-by-step process, that of z to forward integration using one factor of the operator, and that of the final y to backward integration using the second part.

Similarly, for equations of fourth order, we found in Chapter 3 that the triangles L and U had in general three non-zero elements in each row, which corresponds to splitting the fourth-order differential operator into two second-order operators. For equations of the third order one of the matrices corresponds to a second-order operator and one to a first-order operator, depending on the positions of the boundary conditions.

Additional notes and bibliography

Similar methods for solving boundary-value problems by initial-value techniques have been described by

GARWICK, J. V., 1955, The solution of boundary-value problems by step-by-step methods. *Arch. Math. Naturv.* **52**, 1–67.

§ 6. If the trial solutions are extended still further beyond $x = -\frac{1}{4}$, we can use the accurate central-difference formula

$$hy_0' = (\mu\delta-\tfrac{1}{6}\mu\delta^3+\tfrac{1}{30}\mu\delta^5-\tfrac{1}{140}\mu\delta^7+\dots)y_0$$

for the evaluation of the derivatives of $y_0^{(1)}$ and $y_0^{(2)}$, calculating α from the equation

$$y_0^{(1)\prime}+y_0^{(1)}+\alpha\{y_0^{(2)\prime}+y_0^{(2)}\} = 1.$$

The difference correction is then not ignored in the boundary condition.

On the other hand, as explained in § 10, the trial solutions may be 'badly behaved' in comparison with the final solution in which case the labour of accurate calculation of $y_0^{(1)\prime}$ and $y_0^{(2)\prime}$ may be both heavy and unnecessary.

§§ 7, 8. When the derivative is involved in both boundary conditions the method of the previous note cannot be applied at the starting-point, with both convenience and full accuracy.

If three trial solutions are used, as in § 8, none of them necessarily satisfy either boundary condition, and by extending the solutions at both ends of the range we can use central differences for the derivatives and satisfy both boundary conditions accurately. Again, however, this may be wasteful.

§ 18. The 'second method' of this section could obviously be used for any type of linear boundary conditions, and in particular in the case when one condition is given at each of three distinct points.

§ 20. We note in Table 13 the rapid increase of both trial solutions $y^{(1)}$ and $y^{(3)}$, compared with their combination which gives the approximate y. The use of accurate finite-difference formulae to satisfy the derivative condition at $x = 1$ would here be difficult and certainly uneconomic.

§ 38. With second-order equations eigensolutions are more likely to occur if the solutions are of oscillatory type. We find in the next chapter that there is a similar lack of precision in matrix methods when the solution oscillates.

§§ 40, 41. The boundary-value techniques, involving relaxation or direct solution of the finite-difference equations, have no such difficulties, and are obviously more convenient than a combination of step-by-step solutions whenever the complementary function is of other than oscillatory type. In the latter case, however, boundary-value techniques, and in particular the use of triangular matrices, may involve special methods when the complementary function has several zeros in the range of integration. This is discussed in Chapter 9.

D. R. Hartree, in his book *Numerical Analysis* (Oxford, 1952), remarks that most practical problems of boundary-value type are of a kind in which a complementary function increases rapidly, so that step-by-step methods are laborious.

§ 42. A technique of 'matching in the middle', for correcting an approximate eigenvalue, has been described by

Ridley, E. C., 1955, The self-consistent field for Mo^+. *Proc. Camb. Phil. Soc.* **51**, 702–6.

§ 44. The properties of differential equations and their solutions, quoted in this section, can be found in

Ince, E. L., 1944, *Ordinary differential equations.* Dover,

and those of difference equations and their solutions in

Milne-Thomson, L. M., 1933, *The calculus of finite differences.* Macmillan.

9

Accuracy and Precision of Boundary-value Methods

Introduction

†**1.** In estimating the expected error in the results obtained by boundary-value techniques we must consider three main problems. The first is the question of the precision attainable in the solution of the finite-difference equations when the coefficients are not exact numbers but are rounded to a fixed number of decimals or significant figures. The second is the error of the calculated solutions caused by further rounding errors introduced by our inability to retain an infinite number of decimals in the computing process. The third is the possible introduction of further error in the inclusion of difference corrections.

Eigenvalue problems raise further issues which are discussed at the end of the chapter.

Precision of solutions of algebraic equations

2. If the coefficients and right-hand sides have errors of as much as half a unit in the last figure, there is clearly a limit to the precision of the solution, that is to the number of figures in the answers which agree with the true solution of the original exact equations. Consider, for example, the equations

$$\left.\begin{aligned} 1{\cdot}24z_1+1{\cdot}36z_2-0{\cdot}62 &= 0 \\ 0{\cdot}96z_1+2{\cdot}43z_2-0{\cdot}87 &= 0 \end{aligned}\right\}, \tag{1}$$

whose solution, accurate to four decimals, is $z_1 = 0{\cdot}1894$, $z_2 = 0{\cdot}2832$. The equations

$$\left.\begin{aligned} 1{\cdot}2y_1+1{\cdot}4y_2-0{\cdot}6 &= 0 \\ 1{\cdot}0y_1+2{\cdot}4y_2-0{\cdot}9 &= 0 \end{aligned}\right\}, \tag{2}$$

obtained from (1) by rounding the coefficients to one decimal, have the four-decimal solution $y_1 = 0{\cdot}1216$, $y_2 = 0{\cdot}3243$.

The solutions of (2) are here poor approximations to those of (1), our real requirement, with errors of 7 and 4 units in the second place of decimals.

3. More generally, the equations we actually solve can be written in the matrix form

$$Ay = b, \tag{3}$$

while the original accurate equations are given by

$$(A+\delta A)z = b+\delta b, \tag{4}$$

where δA and δb are the truncations effected in (4) to produce (3). The difference between the solutions y and z can be written as

$$\delta y = z-y = A^{-1}(\delta b-\delta Ay), \tag{5}$$

in which second-order terms have been neglected.

If the coefficients of (3) are obtained by rounding those of (4) to $m-1$ decimal places, the coefficients of δA and δb lie between 5.10^{-m} and -5.10^{-m}. It is then not difficult to estimate from (5) the worst possible 'precision-error'. We find, for the rth component of the required solution, the result

$$|\delta y_r| = 5.10^{-m}\Big(1+\sum_1^n |y_s|\Big)\sum_{s=1}^n |\alpha_{rs}|, \tag{6}$$

where n is the order of the matrix and α_{rs} the element of A^{-1} in the rth row and sth column. For equations (2) we find

$$A^{-1} = \begin{bmatrix} 1{\cdot}622 & -0{\cdot}946 \\ -0{\cdot}676 & 0{\cdot}811 \end{bmatrix} \quad \text{and} \quad |\delta y_1| = 0{\cdot}186, \qquad |\delta y_2| = 0{\cdot}108. \tag{7}$$

These maximum limits can be reached, and are approached in the case of the equations

$$\left.\begin{aligned} 1{\cdot}25z_1+1{\cdot}45z_2-0{\cdot}55 &= 0 \\ 0{\cdot}95z_1+2{\cdot}35z_2-0{\cdot}94 &= 0 \end{aligned}\right\}, \tag{8}$$

of which (2) are rounded equivalents. The solutions of (8) are $z_1 = -0{\cdot}0452$, $z_2 = +0{\cdot}4183$, so that the errors caused by replacing (8) by (2) are $|\delta y_1| = 0{\cdot}167$, $|\delta y_2| = 0{\cdot}094$.

4. The maximum error is generally a considerable over-estimation of the expected error. If the rounding errors in A and b are random, and uniformly distributed between 5.10^{-m} and -5.10^{-m}, we can show that the standard deviation of the error δy_r is given by

$$\text{S.D. }(\delta y_r) = \frac{5.10^{-m}}{\sqrt{3}}\Big\{\sum_{s=1}^n (\alpha_{rs})^2\Big(1+\sum_{s=1}^n y_s^2\Big)\Big\}^{\frac{1}{2}}. \tag{9}$$

The probable error is two-thirds of the standard deviation, and it is '95 per cent. certain' that the error is less than twice the standard deviation.

For equations (2) we find the standard deviations of δy_1 and δy_2 to have the respective values 0·057 and 0·032, and the errors are 95 per cent. certain to be less than 0·114 and 0·064. The solutions of (2), regarded as approximations to those of (1), for example, satisfy this criterion.

We note further that the solutions of (2) will not reasonably guarantee the solution of the unrounded equations to better than 11 and 6 units in the second decimal so that, in this particular example, even the first figure of z_1 is uncertain.

5. Equation (9) gives a theoretical assessment of the error, indicating in what circumstances this can be expected to be large or small, and the value of m necessary to obtain results of a given precision. In practice, however, matrix inversion is often inconvenient, and we may assess precision by calculating two solutions, with coefficients rounded to different numbers of decimals. If the solving process introduces no error, we can then assert that the required solution of the original problem (4) is certainly given accurately to the number of figures which our two solutions have in common.

We might, for example, solve separately the equations

$$\left.\begin{aligned} Ay_0 &= b \\ (A+\delta_1 A)y_1 &= b+\delta_1 b \end{aligned}\right\}, \tag{10}$$

both sets of equations being rounded versions of (4), but with the second of (10) having more accurate figures than the first. An economic method of solution is as follows. Solving first for y_0 in (10), we obtain y_1 by an iterative process, taking y_0 as a first approximation, calculating the residuals

$$R_1 = \delta_1 Ay_0 - \delta_1 b, \tag{11}$$

and obtaining a correction δy_0 from the equation

$$A\delta y_0 + R_1 = 0, \tag{12}$$

whose solution is equivalent to that of (5). Further corrections, if significant, can be obtained by further applications of this process, representable by the series

$$y_1 = (I-B+B^2-B^3+\dots)A^{-1}(b+\delta_1 b), \tag{13}$$

where B is the matrix $A^{-1}\delta_1 A$.

If this process converges we have solved the second of (10) fairly trivially, using throughout the same matrix A and several easily calculated sets of right-hand sides. Any operation on the matrix, perhaps involving the triangular resolution process but which in any case is the hardest part of the work, has to be performed once only.

6. The series (13) converges if A is not singular, and if every element of the matrix

$$S_p = I-B+B^2-B^3+\dots+(-)^p B^p \tag{14}$$

converges to a finite limit as $p \to \infty$. A sufficient condition is easily established. For suppose β is the largest coefficient in B: then the largest coefficient in $B^2 \leqslant n\beta^2$, and in $B^r \leqslant n^{r-1}\beta^r$. Thus the largest element in $S_p \leqslant 1+\beta+n\beta^2+...+n^{p-1}\beta^p$, and as $p \to \infty$ this converges to $1+\beta/(1-n\beta)$ if $|n\beta| < 1$. In our case $B = A^{-1}\delta_1 A$ and its largest element cannot exceed $n\alpha\epsilon$, where α is the largest element of A^{-1} and ϵ the largest of $\delta_1 A$. A sufficient condition for convergence is then given by $\alpha < (n^2\epsilon)^{-1}$, and if $\epsilon = 5.10^{-m}$ this becomes

$$\max|\alpha_{rs}| < 10^m/5n^2, \tag{15}$$

where α_{rs} is an element of A^{-1}.

It is likely that the process will converge unless the inverse matrix has very large elements or unless m is small, meaning that we have used few accurate figures in the original matrix. If it converges but gives results of unsatisfactory precision we can extend the process trivially by replacing $\delta_1 A$ by $\delta_2 A$, a closer approximation to the true δA. Convergence is then assured, since the ϵ of $\delta_2 A$ cannot exceed that of $\delta_1 A$; occasionally, however, the inverse A^{-1}, though satisfactory as regards theoretical convergence, may be such a poor approximation to $(A+\delta A)^{-1}$ that a large number of steps is needed in the iterative process. In this case we might use Aitken's acceleration device, or recompute $(A+\delta_2 A)^{-1}$, or carry out any equivalent operations on the more accurate matrix.

7. Using our simple example for illustration, we have for equations (2) the results $\max|\alpha_{rs}| = 1{\cdot}622$, $10^m/5n^2 = 5$, so that the iterative method can be used to solve any equations of which (2) is a rounded equivalent. On the other hand the condition is not satisfied for the equations

$$\left.\begin{aligned} y_1+\ y_2-1 &= 0 \\ y_1+2y_2-1 &= 0 \end{aligned}\right\}, \tag{16}$$

obtained by rounding the coefficients of (2) to one less figure. For now $\max|\alpha_{rs}| = 2$, and exceeds $10^m/5n^2 = 0{\cdot}5$. If the matrix of (16) is used in the first of (10), we find the series (14) divergent, and the successive approximations to the solutions of (2) are found to be (1·0, 0·0), (—0·1, 0·5), (2·3, —0·8),... .

It will be noticed that this iterative method is identical with the relaxation trick (iii) of § 7, Chapter 3, in which we suggested the advantage of using simple numbers for the operations table, but more precise values in calculating the residuals.

The precision of our answers, and the success of our solving processes, clearly depends on the size of the elements of the inverse matrix. We shall later examine some typical matrices, but it is instructive first to consider the question of precision from the point of view of the relaxation method applied to the finite-difference equations arising from ordinary differential equations.

Relaxation and precision

8. This topic had preliminary consideration in Chapter 3. In § 25 of that chapter, for example, we indicated the difficulty of making sure that residuals were negligible, showing that, for simple second-order equations, small residuals may require large increments in y for effective liquidation, and that the boundary conditions as well as the elements of the general finite-difference equation were important factors in this connexion. In § 27 of that chapter we showed that for fourth-order equations the precision was generally poorer than that for second-order equations. We proceed now to amplify these remarks.

We take first the simple second-order equation

$$y'' = f(x), \tag{17}$$

with its typical finite-difference equivalent

$$y_1+y_{-1}-2y_0 = h^2 f_0. \tag{18}$$

In this case the matrix has exact coefficients, so that all the 'precision-error' arises through any roundings performed in $h^2 f$. These rounding errors can be regarded as unliquidated residuals, and we must examine the errors they introduce into the solution y.

If in (18) the boundary values are specified at $x = \pm 1$, and the interval h is chosen so that there are $(2n-1)$ internal pivotal points (so that $nh = 1$), then the increment in y at each point $x = rh$, needed to liquidate a unit residual at the origin, is given by

$$y_r = \tfrac{1}{2}(n-|r|), \tag{19}$$

with a maximum value of $\tfrac{1}{2}n$.

If the derivative vanishes at $x = 1$, a condition represented approximately by the equation $y(1+h) = y(1-h)$, so that the last equation of the set (18) becomes

$$2y(1-h)-2y(1) = h^2 f(1), \tag{20}$$

a unit residual at $x = 1$ causes increments

$$y_r = \tfrac{1}{2}(n+r) \tag{21}$$

with a maximum of n, twice as great as before.

If the derivative vanishes at both ends there is no possibility of liquidating a unit residual by finite increments, and no solution can be obtained unless $f(x)$ satifies certain conditions.

9. For the simplest fourth-order equation

$$y^{\text{iv}} = f(x), \tag{22}$$

with its typical finite-difference equation

$$y_2 - 4y_1 + 6y_0 - 4y_{-1} + y_{-2} = h^4 f_0, \tag{23}$$

we take the boundary conditions $y(\pm 1) = 0$ and $y(\pm 1+h) = y(\pm 1-h)$, corresponding to $y = y' = 0$ at each end of the range. The increments required to liquidate a single residual at the origin, with $(2n-1)$ internal points, are given by

$$y_r = -\tfrac{1}{24}(n^3 + 2n + 2|r|^3 - 3nr^2 - 2|r|). \tag{24}$$

We note that, with conditions as nearly comparable as possible, this error is considerably greater, unless n is very small, than in the case of the second-order equation. Moreover, as the interval decreases and n increases, the errors in all cases increase with n so that greater precision has to be used in $h^2 f$ or $h^4 f$, as the number of pivotal points increases, to give constant precision in the solution of the equations. *Thus the interval should always be as large as possible for best precision.*

With the same differential equation (22), leading to the finite-difference equation (23), but with $y(\pm 1) = 0$, $y(\pm 1+h) = -y(\pm 1-h)$, corresponding to $y = y'' = 0$ at each end, we find, for the increments required to liquidate a single residual at the origin, the results

$$y_r = -\tfrac{1}{12}(2n^3 + n + |r|^3 - 3nr^2 - |r|), \tag{25}$$

whose values are generally greater than those of (24). We return to this point in § 11.

10. These results have their counterparts in the degree of difficulty of liquidating residuals. For equations (18), for example, we observed in Chapter 3 that the algebraic sum of residuals is never altered by changes in y, unless an alteration is made in a pivotal value next to a boundary at which the value is given. The aim is then to 'sweep' residuals towards such boundaries, and the further the distance the bigger will be the necessary increments. In the problem resulting in equation (19) there were two boundary points helpful in this way, but in that of equation (21) there was only one such point, the derivative condition being in this sense useless. The residual at $x = 1$ had thus twice as far to 'travel' as before, and the increments are correspondingly larger. If finally the derivative is specified at both boundary points the algebraic sum of residuals can never be reduced, so that solutions exist only if it is originally zero, a finite-difference equivalent of the mathematical requirement $\int\limits_{-1}^{+1} f(x)\,dx = (b-a)$, when $y'(-1) = a$, $y'(1) = b$. Even then the solution is not unique, any additional arbitrary constant providing an equally valid solution.

The last comment introduces a further important point. Suppose the boundary conditions are of the form

$$ay+by' = c, \tag{26}$$

and consider the boundary point $x = 1$. The simplest finite-difference form of (26), together with (18) at $x = 1$, give, for the approximate equation to be satisfied at that point, the equation

$$2y(1-h)-2\left(1+\frac{a}{b}h\right)y(1) = h^2f(1)-\frac{2c}{b}h. \tag{27}$$

Now if $a = 0$ we have the previous indeterminate case. If a is small compared with b, however, we still have the difficulty that only very small amounts of the total residual are lost at this boundary: we can add a constant everywhere with little effect on the residuals.

11. We see then that for the simplest second-order differential equation we get best precision, in the solution of the corresponding difference equations, with the smallest number of pivotal points, and that, for the general boundary condition (26), precision is best when a is large compared with b, poorest in the converse case.

Similar criteria hold for equations of fourth order. It is now even more important to restrict the number of algebraic equations, and the boundary conditions are 'best' when the order of derivative present is as low as possible, or when the coefficients of higher derivatives are small compared with the rest. For if $s = 0$ in each of two conditions, given at each boundary, of the form

$$py'''+qy''+ry'+sy = 0, \tag{28}$$

then the solution of (22) is indeterminate to the extent of an arbitrary additive constant. If r is also zero we can add a linear term $ax+b$. However, the specification of y does not remove all the difficulty. For if y and y''', for example, are given at each end, we can add with impunity a solution of the form ax^2+bx+c, with one disposable constant.

†12. Turning now to more complicated equations, we find that the same considerations give useful indications of the relative precision of the solution of the second-order equation

$$y''+f(x)y = g(x), \tag{29}$$

with its simplest finite-difference representation

$$y_1+y_{-1}-(2-h^2f_0)y_0 = h^2g_0, \tag{30}$$

according to the sign of f, and assuming boundary values specified.

If f is negative everywhere, then the algebraic sum of the residuals can be decreased by any increment, not only near the boundary, though in general, of course, h^2f is small and the largest effect is at the boundary. Thus *negative f gives better precision than zero f.*

If f is positive everywhere the attempted point-relaxation of a positive residual actually increases the algebraic sum, which is decreased only if the selected point is next to the boundary, and larger increments are necessary than before to liquidate the same residual. Thus *positive f gives poorer precision than zero f.*

From another point of view, the over-relaxation needed for the liquidation of a particular residual is less if f is negative, greater if f is positive, than for the simple case $f = 0$.

This is an important conclusion. It means that, for second-order equations with boundary values specified, boundary-value methods are best when the solutions are of exponential type.

13. The presence of a first derivative rather complicates the issue from the relaxation point of view. Consider, for example, the equation

$$y''+ay'+by = f(x), \tag{31}$$

where a and b are constants, and with boundary values specified. The simple finite-difference equation is given by

$$y_1(1+\tfrac{1}{2}ah)+y_{-1}(1-\tfrac{1}{2}ah)-y_0(2-bh^2) = h^2f_0. \tag{32}$$

If b is negative the situation, as before, is certainly good for relaxation, since some part of the total residual is 'lost' at each step. If b is positive it would appear unsatisfactory, and yet the solutions are of exponential type if $a^2 > 4b$. The explanation is somewhat as follows. Consider first the case of positive a. Here residuals resist being pushed to the right (x increasing) but are prepared to pile up near the left-hand boundary. A positive unit increment next to this boundary results in a total residual decrease of amount $1+\frac{1}{2}ah-bh^2$, and if a is sufficiently larger than b this more than offsets the small residual increases caused by increments in y at a general point. If a is negative we have the same situation at the right-hand end. If $|a|$ is not large enough we are approaching the unsatisfactory oscillatory case $|a| = 0$, $b > 0$ already considered.

If in (31) a and b are functions of x, we can obtain an idea of the 'condition' of the equations by removing the first derivative. Writing $y = Y\exp(\int -\frac{1}{2}a\,dx)$, we find

$$Y'' - (\tfrac{1}{4}a^2 + \tfrac{1}{2}a' - b)Y = \exp\Big(\int \tfrac{1}{2}a\,dx\Big)f(x), \tag{33}$$

and the sign and size of the coefficient of Y is the important consideration. It should be observed, however, that the resulting error in y is $\exp(\int -\frac{1}{2}a\,dx)$ times that in Y.

14. Similar considerations might be applied to fourth-order equations, but the analysis is more difficult, and we are content to observe that, for the equation

$$y^{\mathrm{iv}} + ay = f(x), \tag{34}$$

'condition' is better or worse than that for y^{iv} alone, with comparable boundary conditions, according as a is positive or negative. This criterion corresponds, as in the discussion of (29), to increasing or decreasing the diagonal terms in the finite-difference matrix.

15. Many of these conclusions, of course, cease to be valid for mixed types of boundary conditions. For example, it is quite possible, for second-order equations, to have better precision with oscillatory-type solutions and fixed boundary values than with exponential-type solutions with a boundary condition involving a derivative. For second-order equations with variable coefficients we might have exponential-type solutions in some regions, oscillation in others, and no general conditions for precision can be laid down.

Though, in the last resort, only the elements of the inverse matrix can define the precision, much useful information can be obtained in the direct solving processes, particularly those involving triangular resolution of the matrix. This we now proceed to explore.

Further examination of triangular decomposition of matrices

16. If the original approximate finite-difference equations are expressed in the matrix form

$$Ay = b, \tag{35}$$

we know that A depends only on the homogeneous problem, and b involves any functions of x only which occur in differential equation and boundary conditions. The homogeneous forms of the latter may affect the first few and last few rows of A, the number affected depending on the order of the differential equation. Consider now the homogeneous equation

$$Ay = 0. \tag{36}$$

If we carry out the triangular resolution $A = LU$, we see that, since L is not singular, the elements of U are such that

$$Uy = 0, \tag{37}$$

where y is a solution of the homogeneous equation (36). The first few rows of A, and therefore all the rows of U, depend on the boundary conditions at the first boundary point of the range, but only the last few rows of A and therefore of U depend on the conditions at the second boundary point.

Second-order equations

17. Consider now a second-order equation with constant coefficients, leading to a difference equation whose general homogeneous form is given by

$$a_r y_{r-1} + b_r y_r + c_r y_{r+1} = 0. \tag{38}$$

Here a_r, b_r, and c_r are generally independent of r, except for the influence of the boundary conditions which affect the first and last rows of the corresponding matrix. The general solution of (38) is given by

$$y_r = Ap_1^r + Bp_2^r, \tag{39}$$

where p_1 and p_2 are roots of the algebraic equation

$$a + bp + cp^2 = 0. \tag{40}$$

The roots are real if $b^2 > 4ac$, and in that case positive for meaningful finite-difference equations, giving two solutions of exponential type, and the first boundary condition gives a relation between A and B in (39). Any solution obtained by a step-by-step process, starting with two arbitrary values of y, will ultimately be dominated by the term Ap_1^r, where $p_1 > p_2$. Thus, if the diagonal terms of U are denoted by u_r, the single adjacent line of terms by v_r, we see from (37) that the elements of u_r and v_r will satisfy the equation

$$u_r y_r + v_r y_{r+1} = 0, \tag{41}$$

and that ultimately, provided there is no interference from the second boundary condition, the u and v will attain constant values such that

$$u_r/v_r = -y_{r+1}/y_r = -p_1. \tag{42}$$

At this stage, moreover, it is clear from the discussion starting in § 43 of the last chapter that, if l_r denotes the off-diagonal term of the unit triangular matrix L, we have effectively factorized equation (40), and l_r attains the constant value $-p_2$, so that (40) is equivalent to the equation

$$(l+p_2)(u+p_1 v) = 0. \tag{43}$$

In this case, of course, the effect on U of the initial boundary condition generally gets weaker in successive rows and ultimately disappears. In particular cases, for example when the initial condition is such that A is zero in (39), the elements of L and U have their constant values from the beginning, and $-l$ may then have the larger value p_1.

From (41), and also from the discussion of the previous chapter, we see that u_r/v_r will have a constant sign unless y_r passes through a zero. For exponential-type solutions there can only be one such zero, given by the value of r for which

$$Ap_1^r+Bp_2^r = 0, \tag{44}$$

and only for some types of initial condition will there be any zero at all. If it exists, u_r/v_r will change sign twice on passing through this point, and then settle down, more or less steadily, to its constant value.

18. If the roots of (40) are not real the two complementary functions oscillate, neither is ever dominant, and the elements u, v, and l do not settle down to constant values, but oscillate indefinitely. In particular, the ratio u/v has sudden sign changes, with possibly large values, at points separated by the period of the oscillation.

To illustrate these results we give some triangular matrices arising from typical finite-difference matrices.

19. Consider first the second-order equation

$$y''-y = 0, \tag{45}$$

with initial value of y specified as zero. The general finite-difference equation is

$$y_{r-1}-(2+h^2)y_r+y_{r+1} = 0, \tag{46}$$

and, for $h = \frac{1}{2}$, the finite-difference matrix has the form

$$\begin{array}{cccc} (y_1) & (y_2) & (y_3) & (y_4) \end{array}$$
$$\begin{bmatrix} -2{\cdot}25 & 1 & & \\ 1 & -2{\cdot}25 & 1 & \\ & 1 & -2{\cdot}25 & 1 \\ & & \text{etc.} & \end{bmatrix}. \tag{47}$$

In the triangular resolution the first few four-decimal values of u_r are as follows, v_r having the constant value 1, and $l_r = u_{r-1}^{-1}$:

$-u_r$: 2·2500, 1·8056, 1·6962, 1·6604, 1·6477, 1·6431, 1·6414, 1·6408, 1·6405, 1·6404.

The final constant values of u and l are respectively $-1{\cdot}6404$ and $-0{\cdot}6096$. Here $A = -B$ in (39), and u_r, the ratio of successive values of the function $(1{\cdot}6404)^r-(0{\cdot}6096)^r$, proceeds monotonically to its constant value.

If we choose the initial condition

$$y'+1\cdot 0308y = 0, \tag{48}$$

represented by the finite-difference equation

$$y_1-y_{-1}+1\cdot 0308y_0 = 0, \tag{49}$$

our matrix has the form

$$\begin{matrix} (y_0) & (y_1) & (y_2) & (y_3) \end{matrix}$$
$$\begin{bmatrix} -1\cdot 2192 & 2 & & \\ 1 & -2\cdot 25 & 1 & \\ & 1 & -2\cdot 25 & 1 \\ & & \text{etc.} & \end{bmatrix}, \tag{50}$$

and successive values of u, l, and v are as follows:

$$-u\colon 1\cdot 2192,\ 0\cdot 6096,\ 0\cdot 6096,...,$$
$$-l\colon 0\cdot 8202,\ 1\cdot 6404,\ 1\cdot 6404,...,$$
$$v\colon 2\cdot 0000,\ 1\cdot 0000,\ 1\cdot 0000,... .$$

If the first row of the matrix had been divided by 2 all values of u, l, and v would have been constant, and that of l is here the larger of the two roots p_1 and p_2. This boundary condition ensures that the coefficient A of the term Ap_1^r is zero, so that this term does not appear either in y or U.

Using the finite-difference form of the initial condition

$$y'+1\cdot 2y = 0, \tag{51}$$

we find that the first row of the matrix (50) is replaced by $-1\cdot 05$, 2, and the elements u have the following values:

$$-u\colon 1\cdot 0500,\ 0\cdot 3452,\ -0\cdot 6469,\ 3\cdot 7958,\ 1\cdot 9866,\ 1\cdot 7466,\ 1\cdot 6775,$$
$$1\cdot 6539,\ 1\cdot 6454,\ 1\cdot 6422,\ 1\cdot 6411,\ 1\cdot 6407,\ 1\cdot 6405,... .$$

These are proportional to ratios of successive values of the function $(1\cdot 6404)^r-13\cdot 18(0\cdot 6096)^r$, which has one zero between $r = 2$ and $r = 3$.

20. For the second-order equation

$$y''+y = 0, \tag{52}$$

the general approximate finite-difference equation, at the interval $h = \frac{1}{2}$, is

$$y_{r-1}-1\cdot 75y_r+y_{r+1} = 0. \tag{53}$$

If the initial value is fixed, the resulting matrix has the form (47) with 2·25 replaced by 1·75, and successive values of u are as follows:

$-u$: 1·7500, 1·1786, 0·9015, 0·6407, 0·1892, −3·5354, 2·0329, 1·2581, 0·9552, 0·7031, 0·3277, −1·3016, 2·5183,... .

These are ratios of successive values of the function

$$\sin r\theta \quad (\theta = \tan^{-1}\tfrac{1}{7}\sqrt{15}),$$

and there will be periodic bumps near the points for which $r\theta = n\pi$, that is $r = 6{\cdot}2,\ 12{\cdot}4,...$.

This oscillation in u, and therefore in l, clearly persists indefinitely, and no type of initial boundary condition can prevent it. The period of oscillation, moreover, will always be the same, though the position of the zeros will vary according to the initial condition.

21. Similar phenomena will occur in the case of second-order equations with varying coefficients. In general there will be some regions in which the complementary solutions of the homogeneous equations are oscillatory and some regions where they are of exponential type. In the former the elements u_r will have periodic bumps, in the latter they will be generally monotonic.

Fourth-order equations

22. In the case of fourth-order equations our matrices L and U have the following appearance:

$$\begin{array}{cc} L & U \\ \begin{bmatrix} 1 & & & \\ l_2 & 1 & & \\ m_3 & l_3 & 1 & \\ & m_4 & l_4 & 1 \\ & & \text{etc.} & \end{bmatrix}, & \begin{bmatrix} u_1 & v_1 & w_1 & & & \\ & u_2 & v_2 & w_2 & & \\ & & u_3 & v_3 & w_3 & \\ & & & u_4 & v_4 & w_4 \\ & & & & \text{etc.} & \end{bmatrix}. \end{array} \tag{54}$$

Any solution y_r of the homogeneous equation

$$a_r y_{r-2} + b_r y_{r-1} + c_r y_r + d_r y_{r+1} + e_r y_{r+2} = 0 \tag{55}$$

which satisfies the two given initial conditions also satisfies

$$u_r y_r + v_r y_{r+1} + w_r y_{r+2} = 0, \tag{56}$$

the equivalent of (37).

If the coefficients in (55) are constants, we know that the solution is of the form

$$y_r = A_1 p_1^r + A_2 p_2^r + A_3 p_3^r + A_4 p_4^r, \tag{57}$$

where p_1, p_2, p_3, and p_4 are roots of the quartic equation

$$a+bp+cp^2+dp^3+ep^4 = 0. \tag{58}$$

If the roots are real, and $p_1 > p_2 > p_3 > p_4$, then after some steps the first two terms in (57) are dominant, and we have

$$\left.\begin{aligned} y_r &= A_1 p_1^r + A_2 p_2^r \\ y_{r+1} &= A_1 p_1^{r+1} + A_2 p_2^{r+1} \\ y_{r+2} &= A_1 p_1^{r+2} + A_2 p_2^{r+2} \end{aligned}\right\}. \tag{59}$$

Eliminating the constants A_1 and A_2 we find the relation

$$p_1 p_2 y_r - (p_1+p_2) y_{r+1} + y_{r+2} = 0, \tag{60}$$

so that, comparing with (56), we see that u_r/w_r tends to the product, $-v_r/w_r$ to the sum, of p_1 and p_2, and in fact $u+vp+wp^2$ is ultimately a quadratic factor, involving the two largest roots, of the quartic equation (58). The other roots, of course, belong to the L matrix, and we have decomposed the quartic equation into the form

$$(m+lp+p^2)(u+vp+wp^2) = 0, \tag{61}$$

where p_3 and p_4 are roots of the first factor, p_1 and p_2 of the second.

23. If the roots of (58) are not all real we may not get convergence to quadratic factors. The conditions for convergence are fairly obvious: provided the two roots of largest modulus form a real quadratic expression, then the elements m, l, u, v, and w will settle to constant values. If all roots are complex, therefore, we shall have convergence; if there are two real and two complex roots, we shall have convergence unless the modulus of the complex roots separates those of the two real roots. We exclude the case of equal moduli other than for conjugate complex roots, since this case is unusual and its convergence probably depends on the initial conditions. We have excluded also the case where the initial conditions prohibit the appearance in (57) of either of the two largest terms.

If the coefficients in (55) are not initially constant but tend in the limit to constant values, the criteria of the last paragraph will apply as regards ultimate convergence, but there may initially be periods of oscillation in all cases.

24. To illustrate these results we give triangular decompositions of some finite-difference matrices arising from differential equations of the fourth order. The equation

$$y_{r-2} - 4y_{r-1} + 6{\cdot}0625 y_r - 4y_{r+1} + y_{r+2} = 0 \tag{62}$$

corresponds to the homogeneous differential equation

$$y^{\text{iv}}+y = 0, \tag{63}$$

using the interval $h = 0{\cdot}5$. With initial conditions $y = 0$, $y' = 0$ and the simplest finite-difference form of the latter, our initial matrix can be expressed in the form

$$\begin{bmatrix} 7{\cdot}0625 & -4 & 1 & & & \\ -4 & 6{\cdot}0625 & -4 & 1 & & \\ 1 & -4 & 6{\cdot}0625 & -4 & 1 & \\ & 1 & -4 & 6{\cdot}0625 & -4 & 1 \\ & & & \text{etc.} & & \end{bmatrix}. \tag{64}$$

The elements u and v of the U matrix, and l and m of the L matrix, have the following successive four-decimal values, all the w elements being unity:

u: 7·0625, 3·7969, 2·8159, 2·3960, 2·1958, 2·1017, 2·0618, 2·0479, 2·0443, 2·0436, 2·0436,... .
$-v$: 4·0000, 3·4336, 3·0957, 2·9007, 2·7894, 2·7297, 2·7012, 2·6899, 2·6865, 2·6858, 2·6858,... .
$-l$: 0·5664, 0·9043, 1·0993, 1·2106, 1·2703, 1·2988, 1·3101, 1·3135, 1·3142, 1·3142, 1·3142,... .
m: 0·1416, 0·2634, 0·3551, 0·4174, 0·4554, 0·4758, 0·4850, 0·4883, 0·4892, 0·4893, 0·4893,... .

It is easily verified that

$$(1-4p+6{\cdot}0625p^2-4p^3+p^4)$$
$$= (0{\cdot}4893-1{\cdot}3142p+p^2)(2{\cdot}0436-2{\cdot}6858p+p^2). \tag{65}$$

All the roots are complex, and the second pair in (65) has a larger modulus than the first.

For the equation
$$y^{\text{iv}}-y = 0, \tag{66}$$
with the same interval and initial conditions, the matrix has the form

$$\begin{bmatrix} 6{\cdot}9375 & -4 & 1 & & & \\ -4 & 5{\cdot}9375 & -4 & 1 & & \\ 1 & -4 & 5{\cdot}9375 & -4 & 1 & \\ & 1 & -4 & 5{\cdot}9375 & -4 & 1 \\ & & & \text{etc.} & & \end{bmatrix}, \tag{67}$$

and the elements of U and L have the following values:

u: 6·9375, 3·6311, 2·5655, 2·0190, 1·6414, 1·2952, 0·8419, −0·2682, 12·6923,... .
$-v$: 4·0000, 3·4234, 3·0571, 2·8083, 2·6090, 2·4105, 2·1388, 1·4595, 9·4220,... .
$-l$: 0·5766, 0·9429, 1·1917, 1·3910, 1·5895, 1·8612, 2·5405, −5·4420, 0·7439,... .
m: 0·1441, 0·2754, 0·3898, 0·4953, 0·6092, 0·7721, 1·1878, −3·7286, 0·0788,... .

The roots of the quartic equation

$$1-4p+5{\cdot}9375p^2-4p^3+p^4 = 0 \tag{68}$$

are 1·6404, 0·6096, and a complex pair of unit modulus. In this case the modulus of the complex pair separates the real roots, and periodic oscillations in u, etc., will persist indefinitely.

Effect of the second boundary

25. We have already discussed the general nature of the coefficients of L and U in typical cases, and the next interesting consideration is the influence of the conditions at the second boundary on the last one or two rows.

Consider first the simple second-order problem leading to the matrix (47) in which the solutions are of exponential type. If the second boundary condition specifies the value of y, then the last two rows of the original matrix are

$$\begin{bmatrix} 1 & -2\cdot 25 & 1 \\ & 1 & -2\cdot 25 \end{bmatrix}. \tag{69}$$

Suppose that, before these rows are used, the elements of L and U have settled down to their constant values. With this boundary condition there is no upset, the last two rows of L and U having the values

$$\overset{L}{\begin{bmatrix} -0\cdot 6096 & 1 & \\ & -0\cdot 6096 & 1 \end{bmatrix}}, \quad \overset{U}{\begin{bmatrix} -1\cdot 6404 & 1 \\ & -1\cdot 6404 \end{bmatrix}}. \tag{70}$$

Consider, however, the boundary condition

$$y'+ky = b, \tag{71}$$

represented by the finite-difference equation

$$y_1-y_{-1}+ky_0 = b, \tag{72}$$

where $h = \frac{1}{2}$, and the suffix 0 represents the boundary point. If we eliminate y_1 by use of (46), we find the last two rows of the original matrix to be

$$\begin{bmatrix} 1 & -2\cdot 25 & 1 \\ & 2 & -(2\cdot 25+k) \end{bmatrix}, \tag{73}$$

and the last two rows of the triangular matrices are given by

$$\overset{L}{\begin{bmatrix} -0\cdot 6096 & 1 & \\ & -1\cdot 2192 & 1 \end{bmatrix}}, \quad \overset{U}{\begin{bmatrix} -1\cdot 6404 & 1 \\ & -1\cdot 0308-k \end{bmatrix}}. \tag{74}$$

Now if $k = -1\cdot 0308$ the last diagonal term of U vanishes, the matrix is singular, and we can never find a unique solution. This corresponds to the mathematical problem presented by the differential equation (45), with given initial value of y and a second boundary condition like (71): if $k = -1$ in (71), and this condition is applied at infinity, the homogeneous problem has the solution $A(e^x-e^{-x})$, and the complete solution is not therefore unique.

When k does not have the critical value, but is fairly close to it, precision will be poor, for in the final calculation we use the equations

$$\left.\begin{aligned} Lz &= b \\ Uy &= z \end{aligned}\right\}. \tag{75}$$

The last equation of the second of (75) is given by

$$u_n y_n = z_n, \tag{76}$$

and, with u_n small, y_n is poorly determined. We may note, however, that successive values y_{n-1}, y_{n-2},..., are increasingly of better absolute precision, for in the general equation

$$u_r y_r + y_{r+1} = z_r \tag{77}$$

the coefficients u_r are all greater than unity, except in the particular case when u_r passes near its single possible zero, for an initial condition like (51).

26. In the oscillatory case any second boundary condition may present difficulties. Even if the boundary value is specified the last diagonal element of U may be zero or very small, and a condition involving a derivative may have a similar effect. The effect of small intermediate diagonal terms is less obvious.

These general considerations, therefore, reinforce our previous indications that, in general, precision for second-order equations is best when the solutions are of exponential type, but we now see that this is not true for all types of boundary conditions.

For fourth-order equations there will generally be two conditions at each boundary, and analysis is more complicated.

Some inverse matrices

27. We shall now give the inverses of some typical matrices, to exhibit the effect of various types of boundary conditions and coefficients of the differential equations.

Consider first the differential equation (29), with f constant, y specified at $x = \pm 1$, and with $h = 0{\cdot}2$, giving nine internal pivotal points. The finite-difference matrix is then given by

$$\begin{bmatrix} -(2-h^2f) & 1 & & & & \\ 1 & -(2-h^2f) & 1 & & & \\ & 1 & -(2-h^2f) & 1 & & \\ & & \cdot\;\cdot\;\cdot & \cdot\;\cdot\;\cdot & \cdot\;\cdot\;\cdot & \\ & & 1 & -(2-h^2f) & 1 & \\ & & & 1 & -(2-h^2f) \end{bmatrix}. \tag{78}$$

Tables 1, 2, and 3 give inverse matrices corresponding respectively to the cases $f = 0,\ 1,\ -1$, the corresponding diagonal terms in the original matrices having the values -2, $-1{\cdot}96$, $-2{\cdot}04$. In the second case the diagonal elements of U nowhere pass through a zero in the chosen range.

TABLE 1

$-A^{-1}$ $(f = 0)$ (*boundary values given*)

$$\begin{bmatrix}
0{\cdot}9 & 0{\cdot}8 & 0{\cdot}7 & 0{\cdot}6 & 0{\cdot}5 & 0{\cdot}4 & 0{\cdot}3 & 0{\cdot}2 & 0{\cdot}1 \\
0{\cdot}8 & 1{\cdot}6 & 1{\cdot}4 & 1{\cdot}2 & 1{\cdot}0 & 0{\cdot}8 & 0{\cdot}6 & 0{\cdot}4 & 0{\cdot}2 \\
0{\cdot}7 & 1{\cdot}4 & 2{\cdot}1 & 1{\cdot}8 & 1{\cdot}5 & 1{\cdot}2 & 0{\cdot}9 & 0{\cdot}6 & 0{\cdot}3 \\
0{\cdot}6 & 1{\cdot}2 & 1{\cdot}8 & 2{\cdot}4 & 2{\cdot}0 & 1{\cdot}6 & 1{\cdot}2 & 0{\cdot}8 & 0{\cdot}4 \\
0{\cdot}5 & 1{\cdot}0 & 1{\cdot}5 & 2{\cdot}0 & 2{\cdot}5 & 2{\cdot}0 & 1{\cdot}5 & 1{\cdot}0 & 0{\cdot}5 \\
0{\cdot}4 & 0{\cdot}8 & 1{\cdot}2 & 1{\cdot}6 & 2{\cdot}0 & 2{\cdot}4 & 1{\cdot}8 & 1{\cdot}2 & 0{\cdot}6 \\
0{\cdot}3 & 0{\cdot}6 & 0{\cdot}9 & 1{\cdot}2 & 1{\cdot}5 & 1{\cdot}8 & 2{\cdot}1 & 1{\cdot}4 & 0{\cdot}7 \\
0{\cdot}2 & 0{\cdot}4 & 0{\cdot}6 & 0{\cdot}8 & 1{\cdot}0 & 1{\cdot}2 & 1{\cdot}4 & 1{\cdot}6 & 0{\cdot}8 \\
0{\cdot}1 & 0{\cdot}2 & 0{\cdot}3 & 0{\cdot}4 & 0{\cdot}5 & 0{\cdot}6 & 0{\cdot}7 & 0{\cdot}8 & 0{\cdot}9
\end{bmatrix}$$

TABLE 2

$-A^{-1}$ $(f = 1)$ (*boundary values given*)

$$\begin{bmatrix}
1{\cdot}072 & 1{\cdot}101 & 1{\cdot}086 & 1{\cdot}027 & 0{\cdot}928 & 0{\cdot}791 & 0{\cdot}623 & 0{\cdot}430 & 0{\cdot}219 \\
1{\cdot}101 & 2{\cdot}158 & 2{\cdot}128 & 2{\cdot}014 & 1{\cdot}819 & 1{\cdot}551 & 1{\cdot}221 & 0{\cdot}842 & 0{\cdot}430 \\
1{\cdot}086 & 2{\cdot}128 & 3{\cdot}086 & 2{\cdot}919 & 2{\cdot}637 & 2{\cdot}248 & 1{\cdot}770 & 1{\cdot}221 & 0{\cdot}623 \\
1{\cdot}027 & 2{\cdot}014 & 2{\cdot}919 & 3{\cdot}708 & 3{\cdot}349 & 2{\cdot}856 & 2{\cdot}248 & 1{\cdot}551 & 0{\cdot}791 \\
0{\cdot}928 & 1{\cdot}819 & 2{\cdot}637 & 3{\cdot}349 & 3{\cdot}928 & 3{\cdot}349 & 2{\cdot}637 & 1{\cdot}819 & 0{\cdot}928 \\
0{\cdot}791 & 1{\cdot}551 & 2{\cdot}248 & 2{\cdot}856 & 3{\cdot}349 & 3{\cdot}708 & 2{\cdot}919 & 2{\cdot}014 & 1{\cdot}027 \\
0{\cdot}623 & 1{\cdot}221 & 1{\cdot}770 & 2{\cdot}248 & 2{\cdot}637 & 2{\cdot}919 & 3{\cdot}086 & 2{\cdot}128 & 1{\cdot}086 \\
0{\cdot}430 & 0{\cdot}842 & 1{\cdot}221 & 1{\cdot}551 & 1{\cdot}819 & 2{\cdot}014 & 2{\cdot}128 & 2{\cdot}158 & 1{\cdot}101 \\
0{\cdot}219 & 0{\cdot}430 & 0{\cdot}623 & 0{\cdot}791 & 0{\cdot}928 & 1{\cdot}027 & 1{\cdot}086 & 1{\cdot}101 & 1{\cdot}072
\end{bmatrix}$$

TABLE 3

$-A^{-1}$ $(f = -1)$ (*boundary values given*)

$$\begin{bmatrix}
0{\cdot}811 & 0{\cdot}655 & 0{\cdot}525 & 0{\cdot}417 & 0{\cdot}324 & 0{\cdot}245 & 0{\cdot}176 & 0{\cdot}113 & 0{\cdot}056 \\
0{\cdot}655 & 1{\cdot}337 & 1{\cdot}072 & 0{\cdot}850 & 0{\cdot}662 & 0{\cdot}500 & 0{\cdot}359 & 0{\cdot}231 & 0{\cdot}113 \\
0{\cdot}525 & 1{\cdot}072 & 1{\cdot}661 & 1{\cdot}317 & 1{\cdot}026 & 0{\cdot}775 & 0{\cdot}556 & 0{\cdot}359 & 0{\cdot}176 \\
0{\cdot}417 & 0{\cdot}850 & 1{\cdot}317 & 1{\cdot}837 & 1{\cdot}431 & 1{\cdot}081 & 0{\cdot}775 & 0{\cdot}500 & 0{\cdot}245 \\
0{\cdot}324 & 0{\cdot}662 & 1{\cdot}026 & 1{\cdot}431 & 1{\cdot}893 & 1{\cdot}431 & 1{\cdot}026 & 0{\cdot}662 & 0{\cdot}324 \\
0{\cdot}245 & 0{\cdot}500 & 0{\cdot}775 & 1{\cdot}081 & 1{\cdot}431 & 1{\cdot}837 & 1{\cdot}317 & 0{\cdot}850 & 0{\cdot}417 \\
0{\cdot}176 & 0{\cdot}359 & 0{\cdot}556 & 0{\cdot}775 & 1{\cdot}026 & 1{\cdot}317 & 1{\cdot}661 & 1{\cdot}072 & 0{\cdot}525 \\
0{\cdot}113 & 0{\cdot}231 & 0{\cdot}359 & 0{\cdot}500 & 0{\cdot}662 & 0{\cdot}850 & 1{\cdot}072 & 1{\cdot}337 & 0{\cdot}655 \\
0{\cdot}056 & 0{\cdot}113 & 0{\cdot}176 & 0{\cdot}245 & 0{\cdot}324 & 0{\cdot}417 & 0{\cdot}525 & 0{\cdot}655 & 0{\cdot}811
\end{bmatrix}$$

The different size of the coefficients in the three cases is remarkable in view of the apparently small differences in the three sets of original diagonal terms.

28. From (9) we can find the standard deviation of the errors in our results caused by random rounding errors in the right-hand sides of the equations; the original matrix elements are exact, so that δA in (5) is zero, and we therefore omit the term $\sum y_s^2$ in (9). In each case the centre point suffers most, and we find at this point the respective values

$$\frac{5}{\sqrt{3}}10^{-m}\sqrt{\sum_{s=1}^{n}(\alpha_{rs})^2} = 13.10^{-m},\ 22.10^{-m} \text{ and } 10.10^{-m}$$

for the standard deviation, so that, if the right-hand sides are rounded to a fixed number of decimals, our maximum '95 per cent.' expected error will be about 2·6, 4·4, and 2·0 respectively in units of the last figure.

If, for the same total range, we take the interval $h = 0{\cdot}1$, so that the matrix is of order 19 and the diagonal terms are -2, $-1{\cdot}99$, and $-2{\cdot}01$ respectively, the corresponding error values are 7·5, 12·5, and 5·3 units in the last figure, illustrating the advantage of using as large an interval as possible.

29. We take next the case where, for the oscillatory problem with interval 0·5 and a longer total range, the diagonal terms of U sometimes pass through a zero. If the origin is the first point, we take the second boundary at the various points $x = 2{\cdot}5, 3{\cdot}0, 3{\cdot}5, 4{\cdot}0, 4{\cdot}5, 5{\cdot}0, 5{\cdot}5, 6{\cdot}0$, and record the corresponding inverses for the case 'boundary values specified'. In Table 4 the inverses are denoted by A_n^{-1} according to the order n of the matrix involved.

TABLE 4

A_4^{-1}

$$\begin{bmatrix} -1{\cdot}561 & -1{\cdot}731 & -1{\cdot}469 & -0{\cdot}839 \\ -1{\cdot}731 & -3{\cdot}030 & -2{\cdot}570 & -1{\cdot}469 \\ -1{\cdot}469 & -2{\cdot}570 & -3{\cdot}030 & -1{\cdot}731 \\ -0{\cdot}839 & -1{\cdot}469 & -1{\cdot}731 & -1{\cdot}561 \end{bmatrix}$$

A_5^{-1}

$$\begin{bmatrix} -5{\cdot}281 & -8{\cdot}242 & -9{\cdot}143 & -7{\cdot}758 & -4{\cdot}433 \\ -8{\cdot}242 & -14{\cdot}424 & -16{\cdot}000 & -13{\cdot}576 & -7{\cdot}758 \\ -9{\cdot}143 & -16{\cdot}000 & -18{\cdot}857 & -16{\cdot}000 & -9{\cdot}143 \\ -7{\cdot}758 & -13{\cdot}576 & -16{\cdot}000 & -14{\cdot}424 & -8{\cdot}242 \\ -4{\cdot}433 & -7{\cdot}758 & -9{\cdot}143 & -8{\cdot}242 & -5{\cdot}281 \end{bmatrix}$$

A_6^{-1}

$$\begin{bmatrix} 0{\cdot}283 & 1{\cdot}496 & 2{\cdot}334 & 2{\cdot}589 & 2{\cdot}197 & 1{\cdot}255 \\ 1{\cdot}496 & 2{\cdot}617 & 4{\cdot}085 & 4{\cdot}531 & 3{\cdot}844 & 2{\cdot}197 \\ 2{\cdot}334 & 4{\cdot}085 & 4{\cdot}814 & 5{\cdot}340 & 4{\cdot}531 & 2{\cdot}589 \\ 2{\cdot}589 & 4{\cdot}531 & 5{\cdot}340 & 4{\cdot}814 & 4{\cdot}085 & 2{\cdot}334 \\ 2{\cdot}197 & 3{\cdot}844 & 4{\cdot}531 & 4{\cdot}085 & 2{\cdot}617 & 1{\cdot}496 \\ 1{\cdot}255 & 2{\cdot}197 & 2{\cdot}589 & 2{\cdot}334 & 1{\cdot}496 & 0{\cdot}283 \end{bmatrix}$$

Table 4 (*cont.*)

$$A_7^{-1}$$

$$\begin{bmatrix}
-0.492 & 0.139 & 0.736 & 1.148 & 1.273 & 1.080 & 0.617 \\
0.139 & 0.244 & 1.287 & 2.009 & 2.228 & 1.891 & 1.080 \\
0.736 & 1.287 & 1.517 & 2.368 & 2.626 & 2.228 & 1.273 \\
1.148 & 2.009 & 2.368 & 2.135 & 2.368 & 2.009 & 1.148 \\
1.273 & 2.228 & 2.626 & 2.368 & 1.517 & 1.287 & 0.736 \\
1.080 & 1.891 & 2.228 & 2.009 & 1.287 & 0.244 & 0.139 \\
0.617 & 1.080 & 1.273 & 1.148 & 0.736 & 0.139 & -0.492
\end{bmatrix}$$

$$A_8^{-1}$$

$$\begin{bmatrix}
-0.795 & -0.391 & 0.111 & 0.585 & 0.912 & 1.012 & 0.859 & 0.491 \\
-0.391 & -0.684 & 0.194 & 1.023 & 1.597 & 1.771 & 1.503 & 0.859 \\
0.111 & 0.194 & 0.228 & 1.206 & 1.882 & 2.087 & 1.771 & 1.012 \\
0.585 & 1.023 & 1.206 & 1.087 & 1.697 & 1.882 & 1.597 & 0.912 \\
0.912 & 1.597 & 1.882 & 1.697 & 1.087 & 1.206 & 1.023 & 0.585 \\
1.012 & 1.771 & 2.087 & 1.882 & 1.206 & 0.228 & 0.194 & 0.111 \\
0.859 & 1.503 & 1.771 & 1.597 & 1.023 & 0.194 & -0.684 & -0.391 \\
0.491 & 0.859 & 1.012 & 0.912 & 0.585 & 0.111 & -0.391 & -0.795
\end{bmatrix}$$

$$A_9^{-1}$$

$$\begin{bmatrix}
-1.047 & -0.832 & -0.409 & 0.116 & 0.612 & 0.955 & 1.060 & 0.899 & 0.514 \\
-0.832 & -1.456 & -0.716 & 0.203 & 1.071 & 1.672 & 1.854 & 1.573 & 0.899 \\
-0.409 & -0.716 & -0.844 & 0.239 & 1.262 & 1.970 & 2.185 & 1.854 & 1.060 \\
0.116 & 0.203 & 0.239 & 0.215 & 1.138 & 1.776 & 1.970 & 1.672 & 0.955 \\
0.612 & 1.071 & 1.262 & 1.138 & 0.729 & 1.138 & 1.262 & 1.071 & 0.612 \\
0.955 & 1.672 & 1.970 & 1.776 & 1.138 & 0.215 & 0.239 & 0.203 & 0.116 \\
1.060 & 1.854 & 2.185 & 1.970 & 1.262 & 0.239 & -0.844 & -0.716 & -0.409 \\
0.899 & 1.573 & 1.854 & 1.672 & 1.071 & 0.203 & -0.716 & -1.456 & -0.832 \\
0.514 & 0.899 & 1.060 & 0.955 & 0.612 & 0.116 & -0.409 & -0.832 & -1.047
\end{bmatrix}$$

$$A_{10}^{-1}$$

$$\begin{bmatrix}
-1.422 & -1.489 & -1.183 & -0.582 & 0.165 & 0.871 & 1.359 & 1.507 & 1.279 & 0.731 \\
-1.489 & -2.606 & -2.071 & -1.019 & 0.288 & 1.523 & 2.378 & 2.637 & 2.238 & 1.279 \\
-1.183 & -2.071 & -2.441 & -1.201 & 0.340 & 1.796 & 2.802 & 3.108 & 2.637 & 1.507 \\
-0.582 & -1.019 & -1.201 & -1.082 & 0.306 & 1.619 & 2.526 & 2.802 & 2.378 & 1.359 \\
0.165 & 0.288 & 0.340 & 0.306 & 0.196 & 1.037 & 1.619 & 1.796 & 1.523 & 0.871 \\
0.871 & 1.523 & 1.796 & 1.619 & 1.037 & 0.196 & 0.306 & 0.340 & 0.288 & 0.165 \\
1.359 & 2.378 & 2.802 & 2.526 & 1.619 & 0.306 & -1.082 & -1.201 & -1.019 & -0.582 \\
1.507 & 2.637 & 3.108 & 2.802 & 1.796 & 0.340 & -1.201 & -2.441 & -2.071 & -1.183 \\
1.279 & 2.238 & 2.637 & 2.378 & 1.523 & 0.288 & -1.019 & -2.071 & -2.606 & -1.489 \\
0.731 & 1.279 & 1.507 & 1.359 & 0.871 & 0.165 & -0.582 & -1.183 & -1.489 & -1.422
\end{bmatrix}$$

$$A_{11}^{-1}$$

$$\begin{bmatrix}
-3.052 & -4.340 & -4.544 & -3.612 & -1.776 & 0.503 & 2.657 & 4.146 & 4.599 & 3.902 & 2.230 \\
-4.340 & -7.596 & -7.952 & -6.320 & -3.109 & 0.880 & 4.649 & 7.256 & 8.048 & 6.829 & 3.902 \\
-4.544 & -7.952 & -9.372 & -7.449 & -3.664 & 1.037 & 5.479 & 8.551 & 9.485 & 8.048 & 4.599 \\
-3.612 & -6.320 & -7.449 & -6.715 & -3.303 & 0.935 & 4.940 & 7.709 & 8.551 & 7.256 & 4.146 \\
-1.776 & -3.109 & -3.664 & -3.303 & -2.116 & 0.599 & 3.165 & 4.940 & 5.479 & 4.649 & 2.657 \\
0.503 & 0.880 & 1.037 & 0.935 & 0.599 & 0.113 & 0.599 & 0.935 & 1.037 & 0.880 & 0.503 \\
2.657 & 4.649 & 5.479 & 4.940 & 3.165 & 0.599 & -2.116 & -3.303 & -3.664 & -3.109 & -1.776 \\
4.146 & 7.256 & 8.551 & 7.709 & 4.940 & 0.935 & -3.303 & -6.715 & -7.449 & -6.320 & -3.612 \\
4.599 & 8.048 & 9.485 & 8.551 & 5.479 & 1.037 & -3.664 & -7.449 & -9.372 & -7.952 & -4.544 \\
3.902 & 6.829 & 8.048 & 7.256 & 4.649 & 0.880 & -3.109 & -6.320 & -7.952 & -7.596 & -4.340 \\
2.230 & 3.902 & 4.599 & 4.146 & 2.657 & 0.503 & -1.776 & -3.612 & -4.544 & -4.340 & -3.052
\end{bmatrix}$$

†**30.** These inverse matrices are of considerable interest. If they are computed with the assistance of the L and U triangular matrices, and no advantage is taken of symmetry, the last diagonal elements of U have the successive values 0·6407, 0·1892, $-3{\cdot}5354$,..., given in § 20. By this method of computation, moreover, these terms are the successive diagonal elements, starting at the fourth, of all the U matrices.

It might be expected that the size of these diagonal elements is an important factor determining precision. Certainly for A_5^{-1}, the matrix with the largest elements, the corresponding last U term is the smallest of all, but A_6^{-1}, for which the last diagonal term of U is largest, also has fairly large elements. The main effect on precision is clearly the 'nearness' of the existence of an eigensolution. In this case the differential equation has such solutions with second boundary positions at $x = \pi$, 2π,..., the first two of which correspond most nearly to A_5^{-1} and A_{11}^{-1}, the two largest inverses. Both these matrices have small elements in the last term of their U matrices, as have also those for A_4^{-1} and A_{10}^{-1}, these again being fairly large inverses. A_6^{-1} is also large, and in this case the penultimate diagonal term is small. When a small diagonal term appears in the middle of the range, as in A_9^{-1}, the matrix is much better conditioned.

More simply, we might infer that in this example the largest effect on A^{-1} is the size of the determinant of A, equal to the product of the U diagonal terms. The important quantity is $(-1)^n|A_n|$, and for $n = 4, 5,..., 11$ we find the respective values 1·19, 0·23, $-0{\cdot}80$, $-1{\cdot}62$, $-2{\cdot}04$, $-1{\cdot}95$, $-1{\cdot}37$, $-0{\cdot}45$, and for $n = 12$ the sign changes again with value $+0{\cdot}58$.

The size of the determinant is not, of course, the only consideration affecting nearness to an eigensolution. If we consider the curve of the value of $(-1)^n|A_n|$ plotted against n, we might have a long flat portion of small values, or a steep portion where values are changing rapidly: in the latter case a large value may be nearer to an eigensolution than a small value in the former.

Other interesting points about these inverses are the facts that precision is often poorest for points other than the middle of the range, and that the elements vary in sign. When the latter are all negative, it means that the standard relaxation process of liquidating positive residuals by positive increments in the corresponding pivotal values will converge, but if the inverse elements are positive or mixed in sign the matrix is no longer positive or negative definite, and the iterative process may not converge without considerable refinement.

31. If we repeat the process for the exponential solutions, we find that the elements of the inverses are everywhere negative, their values increase very slowly with the extension of the total range, and in this case do not exceed 1 unit.

There is here, of course, no eigensolution and $(-1)^n|A_n|$ steadily increases in size with n. This is offset by a corresponding increase in the values of the minors of the original determinant, and if A_{rs} is the minor of a_{rs}, a typical element of A, the ratio $A_{rs}/|A|$, which is α_{rs}, remains fairly constant. There is a slow increase in 'precision-error', due here mainly to the larger number of terms which can contribute rounding errors.

†32. For other boundary conditions of the general type $ay+by' = 0$, the oscillatory case has an infinite number of eigen-'situations', that is of separate second-boundary positions, corresponding to given a and b, for which $|A_n| = 0$: the exponential case has only one such position. In particular, if $a = -b$ at the second boundary, the latter will be at infinity for a true eigensolution in the exponential case, but need only be at a reasonably large distance from the first boundary to provide a near eigen-situation to the number of working figures. For example, if y is specified at the first point and $y' = y$ at the second, with five pivotal points in all at interval $h = \frac{1}{2}$, we have the inverse matrix shown in Table 5. We note that its elements are large near the second boundary, but decrease rapidly near the first, a phenomenon not shared by the oscillatory case.

TABLE 5

$$-A^{-1}$$

$$\begin{bmatrix} 0\cdot938 & 1\cdot111 & 1\cdot561 & 2\cdot402 & 1\cdot921 \\ 1\cdot111 & 2\cdot499 & 3\cdot512 & 5\cdot403 & 4\cdot323 \\ 1\cdot561 & 3\cdot512 & 6\cdot342 & 9\cdot756 & 7\cdot805 \\ 2\cdot402 & 5\cdot403 & 9\cdot756 & 16\cdot548 & 13\cdot238 \\ 3\cdot842 & 8\cdot646 & 15\cdot610 & 26\cdot477 & 21\cdot982 \end{bmatrix}$$

33. We take now some simple examples of inverse matrices arising from fourth-order differential equations. For equation (63), with $y = y' = 0$ at both boundaries, and a range $x = 0\cdot0(0\cdot5)5\cdot0$, we have a matrix of order 9 whose first four rows are given in (64) and the remainder are obtained by symmetry considerations. The inverse is given in Table 6, and further calculation shows that as the range is extended the inverse elements gradually increase in value, though their general nature remains unchanged. For no relative positions of the boundary are there eigensolutions to this problem.

TABLE 6

A^{-1}

$$\begin{bmatrix} 0{\cdot}328 & 0{\cdot}432 & 0{\cdot}406 & 0{\cdot}322 & 0{\cdot}224 & 0{\cdot}137 & 0{\cdot}071 & 0{\cdot}028 & 0{\cdot}006 \\ 0{\cdot}432 & 1{\cdot}055 & 1{\cdot}173 & 1{\cdot}022 & 0{\cdot}764 & 0{\cdot}498 & 0{\cdot}276 & 0{\cdot}119 & 0{\cdot}028 \\ 0{\cdot}406 & 1{\cdot}173 & 1{\cdot}825 & 1{\cdot}812 & 1{\cdot}471 & 1{\cdot}025 & 0{\cdot}604 & 0{\cdot}276 & 0{\cdot}071 \\ 0{\cdot}322 & 1{\cdot}022 & 1{\cdot}812 & 2{\cdot}344 & 2{\cdot}152 & 1{\cdot}627 & 1{\cdot}025 & 0{\cdot}498 & 0{\cdot}137 \\ 0{\cdot}224 & 0{\cdot}764 & 1{\cdot}471 & 2{\cdot}152 & 2{\cdot}519 & 2{\cdot}152 & 1{\cdot}471 & 0{\cdot}764 & 0{\cdot}224 \\ 0{\cdot}137 & 0{\cdot}498 & 1{\cdot}025 & 1{\cdot}627 & 2{\cdot}152 & 2{\cdot}344 & 1{\cdot}812 & 1{\cdot}022 & 0{\cdot}322 \\ 0{\cdot}071 & 0{\cdot}276 & 0{\cdot}604 & 1{\cdot}025 & 1{\cdot}471 & 1{\cdot}812 & 1{\cdot}825 & 1{\cdot}173 & 0{\cdot}406 \\ 0{\cdot}028 & 0{\cdot}119 & 0{\cdot}276 & 0{\cdot}498 & 0{\cdot}764 & 1{\cdot}022 & 1{\cdot}173 & 1{\cdot}055 & 0{\cdot}432 \\ 0{\cdot}006 & 0{\cdot}028 & 0{\cdot}071 & 0{\cdot}137 & 0{\cdot}224 & 0{\cdot}322 & 0{\cdot}406 & 0{\cdot}432 & 0{\cdot}328 \end{bmatrix}$$

With the same total range, but with half the interval so that the matrix is now of order 19, the largest element of the inverse increases to 19·476, and the 95 per cent. precision-error level for the middle point goes up from 2·7 to 29·4, a much larger increase than that of the corresponding second-order case. This bears out our previous indications about the advantage of a large interval, and that this is more pronounced for fourth-order equations than for those of second order.

If we take $y = 0$ and the simplest finite-difference form of $y'' = 1{\cdot}2y'$ at the second boundary, the inverse corresponding to Table 6 has the much larger values shown in Table 7. In this case the last element of the U matrix is rather small.

TABLE 7

A^{-1}

$$\begin{bmatrix} 0{\cdot}330 & 0{\cdot}439 & 0{\cdot}425 & 0{\cdot}359 & 0{\cdot}285 & 0{\cdot}225 & 0{\cdot}182 & 0{\cdot}146 & 0{\cdot}096 \\ 0{\cdot}439 & 1{\cdot}092 & 1{\cdot}266 & 1{\cdot}201 & 1{\cdot}057 & 0{\cdot}919 & 0{\cdot}807 & 0{\cdot}683 & 0{\cdot}458 \\ 0{\cdot}425 & 1{\cdot}266 & 2{\cdot}059 & 2{\cdot}265 & 2{\cdot}211 & 2{\cdot}088 & 1{\cdot}945 & 1{\cdot}701 & 1{\cdot}156 \\ 0{\cdot}359 & 1{\cdot}201 & 2{\cdot}265 & 3{\cdot}216 & 3{\cdot}580 & 3{\cdot}679 & 3{\cdot}611 & 3{\cdot}247 & 2{\cdot}230 \\ 0{\cdot}285 & 1{\cdot}057 & 2{\cdot}211 & 3{\cdot}580 & 4{\cdot}856 & 5{\cdot}509 & 5{\cdot}703 & 5{\cdot}262 & 3{\cdot}649 \\ 0{\cdot}225 & 0{\cdot}919 & 2{\cdot}088 & 3{\cdot}679 & 5{\cdot}509 & 7{\cdot}165 & 7{\cdot}892 & 7{\cdot}483 & 5{\cdot}241 \\ 0{\cdot}182 & 0{\cdot}807 & 1{\cdot}945 & 3{\cdot}611 & 5{\cdot}703 & 7{\cdot}892 & 9{\cdot}490 & 9{\cdot}320 & 6{\cdot}608 \\ 0{\cdot}146 & 0{\cdot}683 & 1{\cdot}701 & 3{\cdot}247 & 5{\cdot}262 & 7{\cdot}483 & 9{\cdot}320 & 9{\cdot}714 & 7{\cdot}023 \\ 0{\cdot}096 & 0{\cdot}458 & 1{\cdot}156 & 2{\cdot}230 & 3{\cdot}649 & 5{\cdot}241 & 6{\cdot}608 & 7{\cdot}023 & 5{\cdot}347 \end{bmatrix}$$

34. In these examples the basic equation is such that the U elements settle down to constant values. For the matrix (67), as we have seen, they have periodic oscillation. In Tables 8 and 9 we show inverse matrices corresponding to (67), of orders 8 and 9 respectively, with boundary conditions $y = y' = 0$ at each end of the range. In the case of Table 8 the last U element is small, but there has been no change of sign: in the case of Table 9 there is one change of sign in the penultimate diagonal term. We note the rather large inverse elements in each case, and the fact that the nature of the matrix has changed completely, as in the corresponding second-order case. There is an eigensolution of the differential equation if the second boundary is at $x = 4{\cdot}7$ approximately, a position between those corresponding to Tables 8 and 9.

TABLE 8

A^{-1}

1·361	3·456	5·379	6·444	6·299	4·993	2·973	0·994
3·456	9·960	15·865	19·147	18·773	14·906	8·883	2·973
5·379	15·865	26·140	31·878	31·393	24·980	14·906	4·993
6·444	19·147	31·878	39·612	39·304	31·393	18·773	6·299
6·299	18·773	31·393	39·304	39·612	31·878	19·147	6·444
4·993	14·906	24·980	31·393	31·878	26·140	15·865	5·379
2·973	8·883	14·906	18·773	19·147	15·865	9·960	3·456
0·994	2·973	4·993	6·299	6·444	5·379	3·456	1·361

TABLE 9

A^{-1}

+0·072	−0·399	−1·099	−1·737	−2·092	−2·048	−1·626	−0·969	−0·324
−0·399	−1·571	−3·514	−5·326	−6·324	−6·158	−4·873	−2·898	−0·969
−1·099	−3·514	−6·430	−9·252	−10·787	−10·420	−8·212	−4·873	−1·626
−1·737	−5·326	−9·252	−12·332	−13·963	−13·313	−10·420	−6·158	−2·048
−2·092	−6·324	−10·787	−13·963	−15·012	−13·963	−10·787	−6·324	−2·092
−2·048	−6·158	−10·420	−13·313	−13·963	−12·332	−9·252	−5·326	−1·737
−1·626	−4·873	−8·212	−10·420	−10·787	−9·252	−6·430	−3·514	−1·099
−0·969	−2·898	−4·873	−6·158	−6·324	−5·326	−3·514	−1·571	−0·399
−0·324	−0·969	−1·626	−2·048	−2·092	−1·737	−1·099	−0·399	+0·072

General comments on precision. Nearness of eigensolutions

35. It is clear that the most important factor governing the precision is the possible existence of an eigensolution of the homogeneous problem. In this case the coefficients of the differential equation, those of the boundary conditions, and the positions of the boundaries, combine to give a finite-difference matrix which is singular, so that even if the non-homogeneous problem has a solution it is not unique; we can add a solution of the form $Af(x)$ where A is an indeterminate constant.

In this case the last diagonal element of the U matrix will be zero to our working number of figures. We shall get poor precision if this element is small, corresponding to the near existence of an eigensolution, that is, to a state of affairs in which very small changes in the conditions of the problem might provide an eigensolution.

A second factor governing precision is the order of the differential equation. As this increases, precision gets poorer.

A final factor is the form of the boundary conditions, even when there is no eigensolution. For example, for the equation $y^{\text{iv}} = 0$, Table 10 gives the inverse for a range of eleven points with y and y' specified at each end. Table 11 gives the corresponding results for y and y'' specified at each end (it is easy to see that this is the square of Table 1 for the second order case $y'' = 0$, y specified at each end). The inverse in the second case is much larger than that in the first. We know that the case y, y''' specified gives a singular matrix, with a rather trivial eigensolution, and we might say that y, y'' specified corresponds more nearly to this singular case than does y, y' specified, so that again we may associate this phenomenon, here rather loosely, with the near existence of an eigensolution.

TABLE 10

A^{-1}

0·401	0·635	0·731	0·718	0·625	0·482	0·319	0·165	0·049
0·635	1·647	2·141	2·224	2·000	1·576	1·059	0·553	0·165
0·731	2·141	3·449	3·871	3·625	2·929	2·001	1·059	0·319
0·718	2·224	3·871	5·012	5·000	4·188	2·929	1·576	0·482
0·625	2·000	3·625	5·000	5·625	5·000	3·625	2·000	0·625
0·482	1·576	2·929	4·188	5·000	5·012	3·871	2·224	0·718
0·319	1·059	2·001	2·929	3·625	3·871	3·449	2·141	0·731
0·165	0·553	1·059	1·576	2·000	2·224	2·141	1·647	0·635
0·049	0·165	0·319	0·482	0·625	0·718	0·731	0·635	0·401

TABLE 11

A^{-1}

2·85	4·80	5·95	6·40	6·25	5·60	4·55	3·20	1·65
4·80	8·80	11·20	12·20	12·00	10·80	8·80	6·20	3·20
5·95	11·20	15·05	16·80	16·75	15·20	12·45	8·80	4·55
6·40	12·20	16·80	19·60	20·00	18·40	15·20	10·80	5·60
6·25	12·00	16·75	20·00	21·25	20·00	16·75	12·00	6·25
5·60	10·80	15·20	18·40	20·00	19·60	16·80	12·20	6·40
4·55	8·80	12·45	15·20	16·75	16·80	15·05	11·20	5·95
3·20	6·20	8·80	10·80	12·00	12·20	11·20	8·80	4·80
1·65	3·20	4·55	5·60	6·25	6·40	5·95	4·80	2·85

Simultaneous differential equations

36. We remarked in Chapter 6 that it is often better to solve a fourth-order equation by transforming it into two simultaneous second-order equations. Consider, for example, the problem defined by the equations

$$\left.\begin{aligned} y^{\mathrm{iv}} &= f(x) \\ y = y'' &= 0 \text{ at each boundary} \end{aligned}\right\}. \tag{79}$$

A corresponding inverse matrix is that of Table 11. If we solve successively the equations

$$\left.\begin{aligned} \frac{d^2y''}{dx^2} &= f(x), \qquad y'' = 0 \text{ at each boundary} \\ \frac{d^2y}{dx^2} &= y'', \qquad y = 0 \text{ at each boundary} \end{aligned}\right\}, \tag{80}$$

we have to make a double application of the inverse of Table 1. The precision of y is effectively the same in the two methods, but that of y'', obtained from the first of (80), is clearly superior to that of y'' obtained by double differentiation of y calculated from (79); it is also superior to that of y obtained from (80).

37. When the fourth-order equation cannot be split in this convenient form but is treated as two simultaneous second-order equations rather than as two such successive equations, the precision of both y and y'' may still be superior to those obtained from the fourth-order equation. Consider, for example, the problem defined by

$$\left.\begin{array}{l} y^{\text{iv}}-y=f(x) \\ y=y'=1 \text{ at } x=0;\ y''=y,\ y'''=y' \text{ at } x=1 \end{array}\right\}, \tag{81}$$

solved in § 7 of Chapter 6. We can replace this by the problem defined by the equations

$$\left.\begin{array}{l} y''-z=0 \\ z''-y=f(x) \\ y=y'=1 \text{ at } x=0;\ y=z,\ y'=z' \text{ at } x=1 \end{array}\right\}, \tag{82}$$

solved in § 20 of Chapter 4.

If in the finite-difference replacements of (81) we include differences up to the fourth in the boundary conditions and differential equation, and eliminate pivotal values external to the range by our standard techniques, we have, at interval 0·2, the following matrix of coefficients for the homogeneous problem:

$$\begin{array}{ccccc} y(0\cdot2) & y(0\cdot4) & y(0\cdot6) & y(0\cdot8) & y(1\cdot0) \end{array}$$
$$\begin{bmatrix} 8\cdot998400 & -4\cdot500000 & 1 & & \\ -4 & 5\cdot998400 & -4 & 1 & \\ 1 & -4 & 5\cdot998400 & -4 & 1 \\ & 1 & -3\cdot856736 & 4\cdot706180 & -1\cdot816660 \\ & & 1\cdot719165 & -3\cdot506641 & 1\cdot718476 \end{bmatrix}. \tag{83}$$

Its inverse is shown in Table 12.

TABLE 12

A^{-1}

$$\begin{bmatrix} 0\cdot336 & 0\cdot827 & 1\cdot286 & 1\cdot694 & 1\cdot043 \\ 0\cdot686 & 2\cdot691 & 4\cdot591 & 6\cdot310 & 3\cdot999 \\ 1\cdot064 & 4\cdot665 & 9\cdot086 & 13\cdot149 & 8\cdot613 \\ 1\cdot485 & 6\cdot828 & 13\cdot950 & 21\cdot523 & 14\cdot635 \\ 1\cdot966 & 9\cdot267 & 19\cdot375 & 30\cdot764 & 21\cdot829 \end{bmatrix}.$$

For the simultaneous equations (82) we neglect differences higher than the second, and, using the same interval, can easily reduce the matrix to the following form:

$$\begin{matrix} y(0{\cdot}2) & y(0{\cdot}4) & y(0{\cdot}6) & y(0{\cdot}8) & y(1{\cdot}0) & z(0{\cdot}0) & z(0{\cdot}2) & z(0{\cdot}4) & z(0{\cdot}6) & z(0{\cdot}8) \end{matrix}$$
$$\begin{bmatrix} 2 & 0 & 0 & 0 & 0 & -0{\cdot}04 & 0 & 0 & 0 & 0 \\ -2 & 1 & 0 & 0 & 0 & 0 & -0{\cdot}04 & 0 & 0 & 0 \\ 1 & -2 & 1 & 0 & 0 & 0 & 0 & -0{\cdot}04 & 0 & 0 \\ 0 & 1 & -2 & 1 & 0 & 0 & 0 & 0 & -0{\cdot}04 & 0 \\ 0 & 0 & 1 & -2 & 1 & 0 & 0 & 0 & 0 & -0{\cdot}04 \\ -0{\cdot}04 & 0 & 0 & 0 & 0 & 1 & -2 & 1 & 0 & 0 \\ 0 & -0{\cdot}04 & 0 & 0 & 0 & 0 & 1 & -2 & 1 & 0 \\ 0 & 0 & -0{\cdot}04 & 0 & 0 & 0 & 0 & 1 & -2 & 1 \\ 0 & 0 & 0 & -0{\cdot}04 & 1 & 0 & 0 & 0 & 1 & -2 \\ 0 & 0 & 0 & -1 & 0 & 0 & 0 & 0 & 0 & 1 \end{bmatrix}. \tag{84}$$

Its inverse is shown in Table 13.

TABLE 13

A^{-1}

$$\begin{bmatrix} 0{\cdot}500 & -0{\cdot}020 & -0{\cdot}039 & -0{\cdot}057 & -0{\cdot}072 & 0{\cdot}020 & 0{\cdot}039 & 0{\cdot}057 & 0{\cdot}072 & 0{\cdot}085 \\ 1{\cdot}020 & 0{\cdot}959 & -0{\cdot}120 & -0{\cdot}194 & -0{\cdot}261 & 0{\cdot}041 & 0{\cdot}120 & 0{\cdot}194 & 0{\cdot}261 & 0{\cdot}317 \\ 1{\cdot}581 & 1{\cdot}977 & 0{\cdot}794 & -0{\cdot}380 & -0{\cdot}539 & 0{\cdot}063 & 0{\cdot}206 & 0{\cdot}380 & 0{\cdot}539 & 0{\cdot}676 \\ 2{\cdot}205 & 3{\cdot}073 & 1{\cdot}741 & 0{\cdot}420 & -0{\cdot}878 & 0{\cdot}088 & 0{\cdot}299 & 0{\cdot}580 & 0{\cdot}878 & 1{\cdot}140 \\ 2{\cdot}917 & 4{\cdot}293 & 2{\cdot}757 & 1{\cdot}236 & -0{\cdot}252 & 0{\cdot}117 & 0{\cdot}405 & 0{\cdot}804 & 1{\cdot}252 & 1{\cdot}690 \\ 0{\cdot}000 & -1{\cdot}000 & -1{\cdot}960 & -2{\cdot}842 & -3{\cdot}610 & 1{\cdot}000 & 1{\cdot}960 & 2{\cdot}842 & 3{\cdot}610 & 4{\cdot}233 \\ 0{\cdot}500 & -0{\cdot}020 & -1{\cdot}039 & -2{\cdot}017 & -2{\cdot}914 & 0{\cdot}020 & 1{\cdot}039 & 2{\cdot}017 & 2{\cdot}914 & 3{\cdot}694 \\ 1{\cdot}020 & 0{\cdot}959 & -0{\cdot}120 & -1{\cdot}194 & -2{\cdot}221 & 0{\cdot}041 & 0{\cdot}120 & 1{\cdot}194 & 2{\cdot}221 & 3{\cdot}159 \\ 1{\cdot}581 & 1{\cdot}977 & 0{\cdot}794 & -0{\cdot}380 & -1{\cdot}539 & 0{\cdot}063 & 0{\cdot}206 & 0{\cdot}380 & 1{\cdot}539 & 2{\cdot}636 \\ 2{\cdot}205 & 3{\cdot}073 & 1{\cdot}741 & 0{\cdot}420 & -0{\cdot}878 & 0{\cdot}088 & 0{\cdot}299 & 0{\cdot}580 & 0{\cdot}878 & 2{\cdot}140 \end{bmatrix}$$

It is clear that in the second method the precisions of y and z $(= y'')$ are much the same, and considerably better than that of y in the first method.

This, of course, is almost inevitable. In the fourth-order equation the non-homogeneous parts, forming in various combinations the right-hand sides of the original linear equations, have a factor h^4, while for the simultaneous second-order equations the factor is h^2. Since h is usually much less than unity the first factor is much smaller than the second, the inverse matrix must have correspondingly larger elements, and we must retain more decimals to achieve the same precision.

This ratio of h^2 in the two factors is, of course, implicit in the results of equations (19) and (24) for the effects of unit residuals in the relaxation process: in those equations n is inversely proportional to h.

Errors in the solution of linear equations

†38. So far we have been concerned with an assessment of the precision of our solutions of linear equations in relation to the precision of the given coefficients and right-hand sides. If m correct decimals are required in the answers, we must apparently use $m+n$ correct decimals in the coefficients of A and b, where n depends chiefly on the size of the coefficients of A^{-1}. In reaching this conclusion we assumed that any given set of equations was solved exactly so that, with coefficients of A and b assumed exact to $m+n$ decimals, and a perfect method of solution, the results rounded to m decimals give an accurate solution of our original unrounded equations.

In practice, however, we work in our solving process to a fixed number of decimals or, more rarely, significant figures, and commit rounding errors whenever any quantity is recorded. If we start with $(m+n)$ decimals, and the accumulation of rounding errors is so small that the first m decimals of our results are unaffected, no harm has been done, the rounding errors being less serious than the effects of imprecise coefficients. It is possible, however, for rounding errors to have a more serious effect and our solution, though theoretically of the required precision, may have errors in one or more of the first m decimals.

Much recent attention has been devoted to the analysis of this type of error accumulation, but the methods are lengthy and the results not easy to apply. It is found, however, that for direct methods of solution like 'elimination' or 'triangular resolution' the errors are greatest when the matrix is ill-conditioned, that is when the inverse has large elements.

39. Of more practical interest is the possibility of eliminating these rounding errors. One way is to keep guarding figures, adding one or two zeros after the $(m+n)$ decimals of the original coefficients, and working throughout with this extra number. This process, however, may be uneconomic and in any case we shall need to verify that the desired accuracy has been obtained.

There are two main methods for improving the accuracy of approximate solutions. Suppose that we are solving the linear equations by first inverting the matrix. The error of our solution is then caused by the errors in our alleged inverse. If the true inverse is A^{-1}, and we have calculated an approximation $B_0 = A^{-1}-C$, a better approximation is given by

$$B_1 = B_0(2I-AB_0), \tag{85}$$

and further corrections can be obtained by repetition of this process. We find

$$(I-AB_n) = (I-AB_{n-1})^2 = (I-AB_0)^{2^n} = (AC)^{2^n}, \tag{86}$$

so that convergence is assured if the latent roots of $AC = (I-AB_0)$ are all less than unity in modulus. The error $(I-AB_r)$ is squared at each stage, so that convergence of this iterative process is of second order. For computational purposes we can best write the iteration in the form

$$B_n = B_{n-1}+B_{n-1}(AC)^{2^{n-1}}, \tag{87}$$

successive powers of AC being calculable directly.

The considerable matrix multiplication involved may make this process prohibitively laborious. An alternative method is to use triangular resolution, obtaining approximations to L and U. By the usual back-substitutions we then calculate an approximate solution y_1 to the set of linear equations. Substitution in the original equations will give residuals

$$R_1 = Ay_1-b, \tag{88}$$

and, using the same L and U as before, we calculate a correction δy_1 given by

$$A\delta y_1 = -R_1, \tag{89}$$

repeating if necessary until there is no further change to the precision required. This iterative process has only first-order convergence, and might therefore be slow if L and U have large errors.

40. This method is closely connected with that indicated in § 5 for determining precision, and in practice the two processes will often be combined. The residual in (88) will then include terms originally neglected in truncating the elements of A and b, and the process can cease when further corrections are negligible to the required precision.

Though effectively equivalent to the retention of more and more accurate figures in A and b this process reduces to a minimum the most laborious part of the computation, the determination of L and U.

Effect of a small diagonal term of U

41. One topic that merits further discussion is the effect on accuracy of the incidence of a small diagonal element in the U matrix. This, as we have seen, has no catastrophic effect on precision unless the small element is the last or perhaps penultimate diagonal term of U. It might, however, have an effect on accuracy, since the latter depends on the chosen method of solution, unlike precision which is defined uniquely by the precision of the original coefficients.

The typical second-order difference equation has the form

$$a_r y_{r-1}+b_r y_r+c_r y_{r+1} = d_r, \tag{90}$$

with suitable modification at the ends of the range to satisfy the boundary conditions. With the notation of § 45 of Chapter 8, we determine the L and U matrices from equations typified by

$$\left.\begin{aligned} l_r u_{r-1} &= a_r \\ l_r v_{r-1}+u_r &= b_r \\ v_r &= c_r \end{aligned}\right\}, \tag{91}$$

the auxiliary variable z from the typical equation

$$l_r z_{r-1}+z_r = d_r, \tag{92}$$

and the required y from

$$u_r y_r+v_r y_{r+1} = z_r. \tag{93}$$

In (91) and (92) we work forwards with increasing r, in (93) we work backwards with decreasing r.

It is not necessary to record all these quantities, and a suitable method of calculation, with an easy form of presentation, is to eliminate l, calculating successive u_r^{-1} from the equation

$$u_r^{-1} = \frac{1}{b_r-c_{r-1}a_r u_{r-1}^{-1}}, \tag{94}$$

successive z_r from the equation

$$z_r = d_r-a_r u_{r-1}^{-1} z_{r-1}, \tag{95}$$

and successive y_r, in the reverse direction, from the equation

$$y_r = u_r^{-1}(z_r-c_r y_{r+1}). \tag{96}$$

If $u_r > 1$ everywhere, it is fairly clear from these equations that there is no substantial error accumulation, and that isolated errors will tend to die out. This is generally the case, as we have seen, with exponential-type solutions, and is connected with the fact that the solution of the homogeneous problem rarely passes through a zero, and at most once in the range.

Trouble may arise if u_r is anywhere very small, so that we are occasionally multiplying by quantities which are very large.

42. To illustrate this effect we consider the problem defined by the equation

$$y_{r-1}-1{\cdot}730y_r+y_{r+1} = 0{\cdot}100,$$

with y zero at the ends of a range of 11 points. We work throughout with three decimal places, recording all the significant figures which this may involve and stifling any doubts about final accuracy due to the one or two large numbers involved. The results of the calculations, using (94), (95), and (96) with $a = c = 1$, $b = -1{\cdot}730$, are shown in Table 14.

TABLE 14

r	1	2	3	4	5	6	7	8	9
u_r^{-1}	−0·578	−0·868	−1·160	−1·754	+41·667	− 0·023	−0·586	−0·874	−1·168
z_r	+0·100	+0·158	+0·237	+0·375	+ 0·758	−31·484	−0·624	−0·266	−0·132
y_r	0·235	0·507	0·742	0·877	0·875	0·737	0·581	0·367	0·154

A slightly different method of calculation, using equations (91), (92), and (93), gives a somewhat different final result, the detailed calculations being shown in Table 15.

TABLE 15

r	1	2	3	4	5	6	7	8	9
l_r		−0·578	−0·868	−1·160	−1·754	+41·667	−0·023	−0·586	−0·874
u_r	−1·730	−1·152	−0·862	−0·570	+0·024	−43·397	−1·707	−1·144	−0·856
z_r	+0·100	+0·158	+0·237	+0·375	+0·758	−31·484	−0·624	−0·266	−0·132
y_r	0·151	0·361	0·574	0·732	0·792	0·739	0·581	0·367	0·154

The results of Table 15 are more accurate than Table 14, but this is largely due to a fortuitous combination of rounding errors.

†43. Trouble starts with either method at stage 5, where u_5 is very small, and the following discussion is based on the first method. Clearly u_5^{-1} and z_6 may be considerably in error. Also, though u_6^{-1} will certainly be correct in all its decimals, it is multiplied by a large number to form z_7, which may also therefore be incorrect. Successive z_r and u_r will become more accurate. On the return journey, for the calculation of y_r, our first major source of error is in y_7, due to any error in z_7. For y_6, which again involves the large z_6, and y_5, involving the large u_5^{-1}, errors may be even greater, and do not decrease in successive y until u_r^{-1} becomes small again.

We could, of course, remove the difficulty by retaining more figures throughout, but this is a large price to pay. It is clear that there is no significant error in z_r as far as and including $r = 5$, and that if we could leave z_6 and calculate z_7 accurately by some process, we can proceed happily to the end. Similarly, on the return journey, y_r is accurate as far as y_7, if z_7 is correct, but we need alternative methods to calculate y_6 and y_5: if the latter are accurate our standard procedure gives no further major errors.

Special precautions are therefore needed in the calculation of y_r and y_{r+1}, when the corresponding u_r is small, and we proceed to develop formulae for this purpose.

44. If the small u_r is denoted by ϵ we have $u_r^{-1} = \epsilon^{-1}$, and

$$u_{r+1}^{-1} = -\frac{\epsilon}{c_r a_{r+1} - b_{r+1}\epsilon}, \tag{97}$$

from which we see that a small u_r is inevitably followed by a large u_{r+1}, their product being approximately $c_r a_{r+1}$, of the order of unity. Next we have

$$z_{r+1} = \epsilon^{-1}(d_{r+1}\epsilon - a_{r+1}z_r), \tag{98}$$

which is likely to be large, but we note the cancellation of ϵ in the product $z_{r+1}u_{r+1}^{-1}$, so that both

$$z_{r+2} = d_{r+2} - a_{r+2}u_{r+1}^{-1}z_{r+1} \quad \text{and} \quad y_{r+1} = u_{r+1}^{-1}(z_{r+1} - c_{r+1}y_{r+2})$$

are of reasonable size and accurately determinable. The formulae are given by

$$z_{r+2} = d_{r+2} + \frac{a_{r+2}(d_{r+1}\epsilon - a_{r+1}z_r)}{c_r a_{r+1} - b_{r+1}\epsilon}, \qquad y_{r+1} = \frac{a_{r+1}z_r + \epsilon(c_{r+1}y_{r+2} - d_{r+1})}{c_r a_{r+1} - b_{r+1}\epsilon}. \tag{99}$$

Finally, we have

$$y_r = \epsilon^{-1}(z_r - c_r y_{r+1}) = \frac{c_r(d_{r+1} - c_{r+1}y_{r+2}) - b_{r+1}z_r}{c_r a_{r+1} - b_{r+1}\epsilon}, \tag{100}$$

involving a further cancellation. Thus y_r can also be determined accurately, and we can then revert to the standard method.

The special formulae can be expressed more succinctly if we write

$$\left.\begin{aligned} c_r a_{r+1} - b_{r+1}\epsilon &= p \\ d_{r+1} - c_{r+1}y_{r+2} &= q \end{aligned}\right\}, \tag{101}$$

so that

$$\left.\begin{aligned} u_{r+1}^{-1} &= -\epsilon p^{-1} \\ z_{r+2} &= d_{r+2} + a_{r+2}(d_{r+1}\epsilon - a_{r+1}z_r)p^{-1} \\ y_{r+1} &= (a_{r+1}z_r - \epsilon q)p^{-1} \\ y_r &= (c_r q - b_{r+1}z_r)p^{-1} \end{aligned}\right\}. \tag{102}$$

Summarizing, we use the standard equations (94) and (95), calculating successive u^{-1} and z until we reach a small u_r. We calculate z_r by the standard method but ignore u_r^{-1} and z_{r+1}. At the next step u_{r+1}^{-1} is calculated from the first of (102) and z_{r+2} from the second of (102). We then revert to (94) and (95) to proceed to the end of the range or to the next small u. On the return journey we use the standard equation (96) as far as the calculation of y_{r+2}, then the special formulae, the third and fourth of (102) for y_{r+1} and y_r respectively, and revert finally to (96) to finish the calculation.

Repeating the calculations, with the use of the special formulae where relevant, we find the results shown in Table 16. Quantities obtained from the special formulae are underlined.

TABLE 16

r	1	2	3	4	5	6	7	8	9
u_r^{-1}	−0·578	−0·868	1·160	−1·754		−0·023	−0·586	−0·874	−1·168
z_r	+0·100	+0·158	+0·237	+0·375	+0·758		−0·625	−0·266	−0·132
y_r	0·155	0·369	0·583	0·740	0·797	0·739	0·581	0·367	0·154

The maximum error in y_r is less than 2 units in the third decimal and we have nowhere worked with more than three decimals or used many significant figures.

†**45.** It is clear that any error introduced by the small u_r is in a sense accidental, and caused by a method of computation in which recorded quantities are rounded off, introducing extra errors so that the cancellations in (99) and (100) are not complete. If we want, therefore, to use the standard method throughout, which is often desirable, at least some of the recorded quantities must have more correct figures. To ascertain this number we have to examine further the equations involved, and for simplicity we take the case $a_r = c_r = 1$, $b_r = -2$ for all r, which does not differ appreciably from any practical case.

Suppose then that we have just calculated the denominator of u_r^{-1}, and found it small. Any error in u_r, as we have seen, has negligible future effect, but in the recording and future use of u_r^{-1} we round off in some figure, thereby causing an error $\delta(u_r^{-1}) = \eta_1$ say. The quantity z_r has already been recorded with, in addition to any previous error which does not have major future effect, a rounding error of amount η_2, so that $\delta z_r = \eta_2$. As a result of these errors, and further rounding errors in the recordings, we find

$$\left.\begin{aligned} \delta z_{r+1} &= -z_r\,\eta_1 - u_r^{-1}\,\eta_2 + \eta_3 \\ \delta(u_{r+1}^{-1}) &= \eta_1\,u_{r+1}^{-2} + \eta_4 \\ \delta z_{r+2} &= -z_{r+1}(\eta_1\,u_{r+1}^{-2} + \eta_4) + u_{r+1}^{-1}(z_r\,\eta_1 + u_r^{-1}\eta_2 - \eta_3) + \eta_5 \end{aligned}\right\}. \tag{103}$$

If $u_r = \epsilon$, the magnitudes of these quantities, to the first order in ϵ, are given by

$$\left.\begin{aligned} \delta z_{r+1} &= -z_r\,\eta_1 - \eta_2\,\epsilon^{-1} + \eta_3 \\ \delta(u_{r+1}^{-1}) &= \eta_4 \\ \delta z_{r+2} &= \eta_4\,z_r\,\epsilon^{-1} - \eta_4\,d_{r+1} - \eta_2 + \eta_5 + \epsilon\eta_3 \end{aligned}\right\}. \tag{104}$$

At this stage we note that errors in u_{r+1}^{-1} and successive u do not accumulate or persist, but z_{r+1} and z_{r+2} have possible appreciable errors, of which the latter persists in all successive z. To prevent these errors we must take η_2 and η_4, that is round off z_r and u_{r+1}^{-1}, so that $\eta_2\,\epsilon^{-1}$ and $\eta_4\,z_r\,\epsilon^{-1}$ are at most 5 units in the first neglected place of the final result.

In the calculation of y_r we therefore assume no major error as far as

y_{r+2}, but make a rounding error in this quantity so that $\delta y_{r+2} = \eta_6$. Then

$$\left.\begin{aligned} \delta y_{r+1} &= (z_{r+1}-y_{r+2})\delta(u_{r+1}^{-1})+u_{r+1}^{-1}(\delta z_{r+1}-\eta_6)+\eta_7 \\ \delta y_r &= (z_r-y_{r+1})\delta(u_r^{-1})+u_r^{-1}(\delta z_r-\delta y_{r+1})+\eta_8 \end{aligned}\right\}, \tag{105}$$

which, on expansion and neglect of terms in ϵ and higher powers, become

$$\left.\begin{aligned} \delta y_{r+1} &= \eta_4(d_{r+1}-y_{r+2})-\eta_4 z_r \epsilon^{-1}+\eta_2+\eta_7 \\ \delta y_r &= \eta_4 z_r \epsilon^{-2}-\{\eta_7+\eta_4(d_{r+1}-y_{r+2})\}\epsilon^{-1}+\eta_8 \end{aligned}\right\}. \tag{106}$$

These equations show that η_4 must have a still smaller value than that indicated by (104). In fact we must have $\eta_4 z_r \epsilon^{-2}$ equal at most to 5 units in the first neglected place of the final result, and also $\eta_7 \epsilon^{-1}$ must have the same maximum value.

In our example $\epsilon = 0{\cdot}024$, and we attempt to get three accurate decimals in y and z. Thus we must record u_{r+1}^{-1} to at least six decimals and z_r and y_{r+1} to five, to be certain of preventing major error. Repeating the calculations under these conditions, we have the results of Table 17, finally identical with those of Table 16.

TABLE 17

r	1	2	3	4	5	6	7	8	9
u_r^{-1}	−0·578	−0·868	−1·160	−1·754	+41·667	− 0·023043	−0·586	−0·874	−1·168
z_r	+0·100	+0·158	+0·237	+0·375	+ 0·75775	−31·473	−0·625	−0·266	−0·132
y_r	0·155	0·369	0·583	0·740	0·797	0·73862	0·581	0·367	0·154

If y only is required, the absence of $\eta_2 \epsilon^{-1}$ in (106) and the third of (104) shows that z_r need not be recorded to extra decimals. Table 17 is then unchanged, except that $z_5 = 0{\cdot}758$, $z_6 = -31{\cdot}484$, and $y_6 = 0{\cdot}73887$.

46. The fourth-order problem is of course much more complicated, but it is still possible to work out formulae, similar to those of (102), for the safe negotiation of a small diagonal term of U. Again we wish to avoid the recording of large quantities and the multiplication of large and small quantities, and to take full advantage of any cancellations.

With the notation of (54), the general equations for the successive evaluation of m, l, u, v, and w are given by

$$\left.\begin{aligned} m_r u_{r-2} &= a_r \\ m_r v_{r-2}+l_r u_{r-1} &= b_r \\ m_r w_{r-2}+l_r v_{r-1}+u_r &= c_r \\ l_r w_{r-1}+v_r &= d_r \\ w_r &= e_r \end{aligned}\right\}, \tag{107}$$

and, with f_r on the right-hand side of (55), we determine the auxiliary z_r from

$$m_r z_{r-2}+l_r z_{r-1}+z_r = f_r, \tag{108}$$

and the required y_r from

$$u_r y_r + v_r y_{r+1} + w_r y_{r+2} = z_r, \tag{109}$$

working in order of increasing r in (107) and (108), decreasing r in (109).

Suppose now that $u_r = \epsilon$ is small: then, in the L and U matrices, m_r, l_r, v_r, and m_{r+1} are not affected, but

$$l_{r+1} = \epsilon^{-1}(b_{r+1} - m_{r+1} v_{r-1}) \tag{110}$$

and is large. Then

$$\left.\begin{aligned} u_{r+1} &= c_{r+1} - m_{r+1} e_{r-1} - \epsilon^{-1}(b_{r+1} - m_{r+1} v_{r-1}) v_r \\ \text{and} \quad v_{r+1} &= d_{r+1} - \epsilon^{-1}(b_{r+1} - m_{r+1} v_{r-1}) e_r \end{aligned}\right\} \tag{111}$$

are also large.

At the next step we have

$$m_{r+2} = \epsilon^{-1} a_{r+2}, \tag{112}$$

which is large, but

$$l_{r+2} = \frac{b_{r+2}\epsilon - a_{r+2} v_r}{(c_{r+1} - m_{r+1} e_{r-1})\epsilon - (b_{r+1} - m_{r+1} v_{r-1}) v_r} \tag{113}$$

is back to normal size, and can be calculated accurately from (113) and recorded. For u_{r+2} we have a combination of large terms in $m_{r+2} e_r$ and $l_{r+2} v_{r+1}$, but we can find a formula, from which the large terms cancel, given by

$$u_{r+2} = c_{r+2} + \\ + \frac{a_{r+2}\{v_r d_{r+1} - e_r(c_{r+1} - m_{r+1} e_{r-1})\} + b_{r+2}\{e_r(b_{r+1} - m_{r+1} v_{r-1}) - d_{r+1}\epsilon\}}{(c_{r+1} - m_{r+1} e_{r-1})\epsilon - (b_{r+1} - m_{r+1} v_{r-1}) v_r}. \tag{114}$$

The next quantity v_{r+2}, which is also of normal size, needs no special formula.

At the next stage we have

$$m_{r+3} = \frac{a_{r+3}\epsilon}{(c_{r+1} - m_{r+1} e_{r-1})\epsilon - (b_{r+1} - m_{r+1} v_{r-1}) v_r}, \tag{115}$$

which is small. Then l_{r+3} involves the product of the small m_{r+3} with the large v_{r+1}, so that we calculate it in the form

$$l_{r+3} = u_{r+2}^{-1}\left[b_{r+3} - \frac{a_{r+3}\{d_{r+1}\epsilon - (b_{r+1} - m_{r+1} v_{r-1}) e_r\}}{(c_{r+1} - m_{r+1} e_{r-1})\epsilon - (b_{r+1} - m_{r+1} v_{r-1}) v_r}\right]. \tag{116}$$

No further elements of L and U need special formulae.

Turning now to the auxiliary variable z, we note that the first quantity affected by the small u_r is z_{r+1}, given by

$$z_{r+1} = f_{r+1} - m_{r+1} z_{r-1} - \epsilon^{-1} z_r (b_{r+1} - m_{r+1} v_{r-1}), \tag{117}$$

which is large. For z_{r+2} we have another combination of large quantities which are found to cancel, and we obtain

$$z_{r+2} = f_{r+2} - l_{r+2}(f_{r+1} - m_{r+1} z_{r-1}) + \\ + \frac{\{a_{r+2}(m_{r+1} e_{r-1} - c_{r+1}) + b_{r+2}(b_{r+1} - m_{r+1} v_{r-1})\}z_r}{(c_{r+1} - m_{r+1} e_{r-1})\epsilon - (b_{r+1} - m_{r+1} v_{r-1})v_r}. \quad (118)$$

For z_{r+3} we can use the formula

$$z_{r+3} = f_{r+3} - l_{r+3} z_{r+2} - \\ - \frac{a_{r+3}\{(f_{r+1} - m_{r+1} z_{r-1})\epsilon - z_r(b_{r+1} - m_{r+1} v_{r-1})\}}{(c_{r+1} - m_{r+1} e_{r-1})\epsilon - (b_{r+1} - m_{r+1} v_{r-1})v_r}, \quad (119)$$

and we need no further special formulae for z.

Finally, for the required y, we need no special formulae as far as the calculation of y_{r+2}, but that of y_{r+1} involves large and small quantities, and we find the formula

$$y_{r+1} \\ = \frac{(e_r y_{r+2} - z_r)(b_{r+1} - m_{r+1} v_{r-1}) + \epsilon(f_{r+1} - m_{r+1} z_{r-1} - d_{r+1} y_{r+2} - e_{r+1} y_{r+3})}{(c_{r+1} - m_{r+1} e_{r-1})\epsilon - (b_{r+1} - m_{r+1} v_{r-1})v_r}. \quad (120)$$

The last quantity affected is y_r, and we find the expression

$$y_r = \frac{(z_r - e_r y_{r+2})(c_{r+1} - m_{r+1} e_{r-1}) - v_r(f_{r+1} - m_{r+1} z_{r-1} - d_{r+1} y_{r+2} - e_{r+1} y_{r+3})}{(c_{r+1} - m_{r+1} e_{r-1})\epsilon - (b_{r+1} - m_{r+1} v_{r-1})v_r}, \quad (121)$$

which involves several cancellations. Further values of y are unaffected by the disturbance, and can be calculated from the standard equation (109).

47. The special formulae can be expressed more simply if we write

$$\left.\begin{aligned} c_{r+1} - m_{r+1} e_{r-1} &= M \\ b_{r+1} - m_{r+1} v_{r-1} &= N \\ a_{r+2} v_r - b_{r+2} \epsilon &= P \\ f_{r+1} - m_{r+1} z_{r-1} &= Q \\ e_r y_{r+2} - z_r &= S \\ d_{r+1} y_{r+2} + e_{r+1} y_{r+3} &= T \\ \epsilon M - v_r N &= V \\ a_{r+2} M - b_{r+2} N &= W \end{aligned}\right\}. \quad (122)$$

For then

$$\left.\begin{aligned} l_{r+2} &= -V^{-1}P \\ u_{r+2} &= c_{r+2}+V^{-1}(d_{r+1}P-e_r W) \\ l_{r+3} &= u_{r+2}^{-1}\{b_{r+3}-a_{r+3}V^{-1}(d_{r+1}\epsilon-e_r N)\} \\ m_{r+3} &= a_{r+3}\epsilon V^{-1} \\ z_{r+2} &= f_{r+2}-l_{r+2}Q-V^{-1}Wz_r \\ z_{r+3} &= f_{r+3}-l_{r+3}z_{r+2}-a_{r+3}V^{-1}(Q\epsilon-z_r N) \\ y_{r+1} &= V^{-1}\{SN+\epsilon(Q-T)\} \\ y_r &= -V^{-1}\{SM+v_r(Q-T)\} \end{aligned}\right\}. \tag{123}$$

Summarizing the procedure, we use (107) and (108) to calculate l, m, u, v, w, and z until we reach a small u_r. This does not affect the use of these equations for l_r, v_r, w_r, z_r, and m_{r+1}. We do not record l_{r+1}, u_{r+1}, v_{r+1}, z_{r+1}, or m_{r+2}. For l_{r+2}, u_{r+2}, l_{r+3}, m_{r+3}, z_{r+2}, and z_{r+3} we use the special formulae of (123): v_{r+2} is calculated by the standard formula, as are all further elements of L, U, and z. On the return journey we use the standard (109) up to and including the calculation of y_{r+2}, but use the special expressions (123) for y_{r+1} and y_r, reverting finally to (109) to finish the determination of y.

Effects of the difference correction

48. In our examples we have generally used the simplest finite-difference equations. The difference correction involves differences of higher order than those retained in the original equations, and their inclusion is effected either by an iterative process, or by a direct process which often increases the order of the differential equation.

In one important case, however, involving a second-order equation without a first derivative, given by

$$y''+f(x)y = g(x), \tag{124}$$

for which we have in this chapter used the finite-difference equation

$$y_1+y_{-1}-(2-h^2f_0)y_0 = h^2g_0, \tag{125}$$

with a difference correction starting with $-\frac{1}{12}\delta^4y_0$, we know that we can derive the second-order difference equation

$$y_1(1+\tfrac{1}{12}h^2f_1)+y_{-1}(1+\tfrac{1}{12}h^2f_{-1})-y_0(2-\tfrac{5}{6}h^2f_0) = \tfrac{1}{12}h^2(g_1+g_{-1}+10g_0), \tag{126}$$

which has a much smaller difference correction starting with $\frac{1}{240}\delta^6y_0$.

The matrix corresponding to equation (126) is perhaps slightly less convenient for use with triangular resolution than that corresponding to (125), with its off-diagonal terms of unity. In many cases, however, the neglect of $\frac{1}{240}\delta^6 y$ is justified, and the first application of (126) will suffice when that of (125) needs the further use of a difference correction. The question arises whether the inverse matrix corresponding to (126) is substantially different from that corresponding to (125).

We must assume in each case that the interval is small enough to ensure that the accurate solution of the difference equations behaves in the same way as that of the differential equation. In this event the elements of the L and U matrices will have similar behaviour in the two cases, and only when the last diagonal term of U is small will there be substantial differences in the two inverses. In the worst case, it is possible for the inverse corresponding to (125) to be singular when that for (126) is not, and vice versa. The only dangerous event, therefore, is singularity or near singularity of an inverse, and in this case the results, as we have seen, must in any case be ill-defined.

49. Consider, for example, the matrices of Table 4, obtained from $y''+y = 0$ with fixed boundary values, and an interval $h = \frac{1}{2}$. The general finite-difference equation (125) reduces to $y_{-1}-1{\cdot}75y_0+y_1 = 0$, but that of (126) gives

$$\tfrac{49}{48}y_{-1}-\tfrac{43}{24}y_0+\tfrac{49}{48}y_1 = 0. \tag{127}$$

The elements of the inverse corresponding to A_4^{-1} of Table 4 are in this case smaller by about 5 per cent., those corresponding to A_5^{-1} are smaller by about 23 per cent., those corresponding to A_6^{-1} are greater by about 11 per cent., and we see that the differences are most marked in the neighbourhood of an eigensolution.

It would appear, however, that there is no appreciable difference in precision except near this exceptional case, for which it is then desirable to use from the outset the most accurate formulae, since otherwise we may not get a solution at all. This leads to the question of convergence of our standard iterative use of the difference correction.

Convergence of the iterative method

50. The original finite-difference equations can be represented by the matrix equation

$$Ay+Cy = b, \tag{128}$$

in which the matrix A is obtained from the truncated equations, b involves any non-homogeneous terms in differential equations and boundary conditions, and Cy is the vector of difference corrections. The matrix A is square, and we usually arrange that its order is the same as the number of unknown pivotal values in the range of integration. We can also regard C as a matrix but, if we use central differences, Cy will involve pivotal values outside the range, so that C is a rectangular matrix with more columns than rows, and the y on which it operates also includes some external values.

In the problem of § 48, however, if we take A corresponding to equation (125) and $(A+C)$ corresponding to (126), then both A and C are square and hence of easier form for analysis. We can then write (128) in the form

$$(A+C)y = b, \tag{129}$$

and the inclusion of C is equivalent to taking into account a very substantial part of the difference correction. Here, of course, we can solve (129) directly, but our general process is expressed by the iterative scheme

$$Ay_n = b - Cy_{n-1}. \tag{130}$$

This is equivalent to finding y from the infinite series

$$y = \{I - A^{-1}C + (A^{-1}C)^2 - (A^{-1}C)^3 + \ldots\}A^{-1}b, \tag{131}$$

which converges to $(A+C)^{-1}b$ if all the latent roots of $A^{-1}C$ are in modulus less than unity. This is the only criterion for the success of our iterative method of handling the difference correction.

51. If we take f constant and equal to unity, and $h = \frac{1}{2}$ in (125) and (126), we find the matrix C to have the form

$$\frac{1}{48}\begin{bmatrix} -2 & 1 & & & \\ 1 & -2 & 1 & & \\ & 1 & -2 & 1 & \\ & & \text{etc.} & & \end{bmatrix}, \tag{132}$$

and we can easily evaluate $A^{-1}C$ in typical cases. For example, from the results of Table 4 we find $A_4^{-1}C$ and $A_5^{-1}C$ as shown in Table 18. The latent roots are in each case much smaller than unity.

TABLE 18

$A_4^{-1}C$

$$\begin{bmatrix} 0{\cdot}029 & 0{\cdot}009 & 0{\cdot}008 & 0{\cdot}004 \\ 0{\cdot}009 & 0{\cdot}037 & 0{\cdot}013 & 0{\cdot}008 \\ 0{\cdot}008 & 0{\cdot}013 & 0{\cdot}037 & 0{\cdot}009 \\ 0{\cdot}004 & 0{\cdot}008 & 0{\cdot}009 & 0{\cdot}029 \end{bmatrix}$$

$A_5^{-1}C$

$$\begin{bmatrix} 0{\cdot}048 & 0{\cdot}043 & 0{\cdot}048 & 0{\cdot}040 & 0{\cdot}023 \\ 0{\cdot}043 & 0{\cdot}096 & 0{\cdot}083 & 0{\cdot}071 & 0{\cdot}040 \\ 0{\cdot}048 & 0{\cdot}083 & 0{\cdot}119 & 0{\cdot}083 & 0{\cdot}048 \\ 0{\cdot}040 & 0{\cdot}071 & 0{\cdot}083 & 0{\cdot}096 & 0{\cdot}043 \\ 0{\cdot}023 & 0{\cdot}040 & 0{\cdot}048 & 0{\cdot}043 & 0{\cdot}048 \end{bmatrix}$$

If, however, we take the equation

$$y'' + 1{\cdot}076y = g(x), \tag{133}$$

with y specified at $x = 0$ and $3{\cdot}0$, and $h = 0{\cdot}5$, we have, corresponding to the A matrix, the typical equation

$$y_{r-1} - 1{\cdot}731y_r + y_{r+1} = h^2 g(x), \tag{134}$$

the corresponding matrix C is that of (132) multiplied by the constant factor 1·076, and the matrices A^{-1} and $A^{-1}C$ are shown in Table 19. The largest latent root of $A^{-1}C$ is about 5·7, and we cannot use A to find a solution of (133).

TABLE 19

$$A^{-1} \qquad\qquad -A^{-1}C$$

$$\begin{bmatrix} 78{\cdot}647 & 137{\cdot}138 & 158{\cdot}739 & 137{\cdot}639 & 79{\cdot}514 \\ 137{\cdot}138 & 237{\cdot}386 & 274{\cdot}777 & 238{\cdot}253 & 137{\cdot}639 \\ 158{\cdot}739 & 274{\cdot}777 & 316{\cdot}900 & 274{\cdot}777 & 158{\cdot}739 \\ 137{\cdot}639 & 238{\cdot}253 & 274{\cdot}777 & 237{\cdot}386 & 137{\cdot}138 \\ 79{\cdot}514 & 137{\cdot}639 & 158{\cdot}739 & 137{\cdot}138 & 78{\cdot}647 \end{bmatrix} \qquad \begin{bmatrix} 0{\cdot}452 & 0{\cdot}827 & 0{\cdot}957 & 0{\cdot}830 & 0{\cdot}479 \\ 0{\cdot}827 & 1{\cdot}409 & 1{\cdot}657 & 1{\cdot}437 & 0{\cdot}830 \\ 0{\cdot}957 & 1{\cdot}657 & 1{\cdot}889 & 1{\cdot}657 & 0{\cdot}957 \\ 0{\cdot}830 & 1{\cdot}437 & 1{\cdot}657 & 1{\cdot}409 & 0{\cdot}827 \\ 0{\cdot}479 & 0{\cdot}830 & 0{\cdot}957 & 0{\cdot}827 & 0{\cdot}452 \end{bmatrix}$$

The inverse of the matrix corresponding to the more accurate equation

$$1{\cdot}02241\dot{6}y_{r-1} - 1{\cdot}77583\dot{3}y_r + 1{\cdot}02241\dot{6}y_{r+1} = h^2(1+\tfrac{1}{12}\delta^2)g(x) \qquad (135)$$

has much smaller elements, as shown in Table 20.

TABLE 20

$$-(A+C)^{-1}$$

$$\begin{bmatrix} 17{\cdot}448 & 29{\cdot}327 & 33{\cdot}490 & 28{\cdot}842 & 16{\cdot}605 \\ 29{\cdot}327 & 50{\cdot}938 & 58{\cdot}169 & 50{\cdot}095 & 28{\cdot}842 \\ 33{\cdot}490 & 58{\cdot}169 & 67{\cdot}543 & 58{\cdot}169 & 33{\cdot}490 \\ 28{\cdot}842 & 50{\cdot}095 & 58{\cdot}169 & 50{\cdot}938 & 29{\cdot}327 \\ 16{\cdot}605 & 28{\cdot}842 & 33{\cdot}490 & 29{\cdot}327 & 17{\cdot}448 \end{bmatrix}$$

This matrix can be used to obtain a solution if the remainder of the difference correction, C_1 say, is such that the latent roots of $(A+C)^{-1}C_1$ do not exceed unity in modulus.

Error introduced by the difference correction

52. We have now to consider the question of accuracy in the general case. Our first approximate solution of (128) is given by

$$y^{(1)} = A^{-1}b, \qquad (136)$$

and we have seen that, to get m correct decimals in $y^{(1)}$, we must have $(m+n)$ correct decimals in the elements of A and b, even if the solving process is exact, the number n depending chiefly on the size of the coefficients of A^{-1}. If $y^{(1)}$ is calculated to this precision, the last figure may have errors of as much as half a unit in that figure, and we must examine the effect of these errors on future calculations.

The next calculation is that of the difference corrections at the relevant pivotal points. Because the pivotal values have rounding errors, so also have the difference corrections: for example a correction like $-\frac{1}{12}\delta^4 y$ may have an error as large as $\frac{2}{3}$ of a unit in the last figure of y. This is greater than the error of the last figure of the original b, and suggests that there may be large errors in the correction $A^{-1}Cy^{(1)}$ to the first approximation $y^{(1)}$, large enough, indeed, to affect the last n places of $y^{(1)}$. Each successive application, moreover, might ruin the accuracy of a further n places.

We are saved from this unfortunate possibility by the fact that the errors in $Cy^{(1)}$ are not independent, even if they are in $y^{(1)}$: an error of $+\frac{2}{3}$ in one value of $Cy^{(1)}$ is inevitably followed by an almost equal negative value, so that when multiplied by A^{-1} the errors largely cancel, A^{-1} itself having rows which are 'smooth' as we go along the columns. The quantities multiplying the random errors in $y^{(1)}$ are indeed the elements of $A^{-1}C$, C now being regarded as a matrix, and the forming of $A^{-1}C$, as in § 51, involves some sort of differencing process on the rows of A^{-1}.

53. This is made clearer by examples. In the special case when C is square, we see from Tables 18 and 19 that $A^{-1}C$ is likely to have much smaller coefficients than those of A. The 95 per cent. error in the use of $A^{-1}C$ of Table 19 is at most 2 units in the last figure, if we can ignore the term $\sum y_s^2$ in (9), so that even in this extreme case the errors in $Cy^{(1)}$ cause almost negligible error in y.

For an example in which C is not square we take the problem leading to the rather large A_5^{-1} of Table 4, and we will examine the effect of the later inclusion of the major part $-\frac{1}{12}\delta^4 y$ of the difference correction. If we use central differences, this involves the use of the external values at points -1 and 7 of Fig. 34, in which 0 and 6 represent the boundary points and 1 to 5 the pivotal points at which y is calculated.

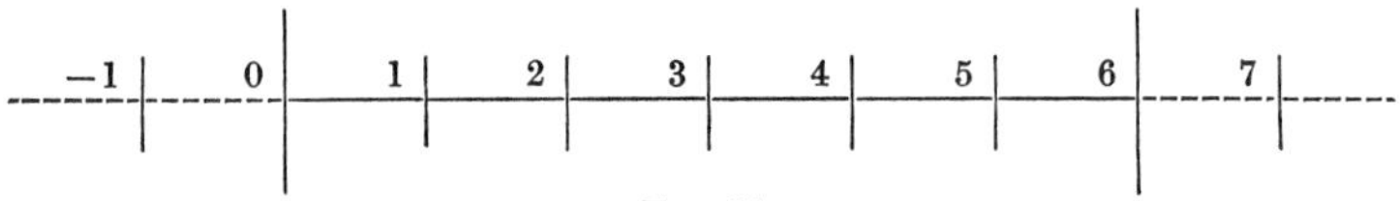

FIG. 34

The difference corrections are given approximately by

$$-12Cy_r = y_{r-2}-4y_{r-1}+6y_r-4y_{r+1}+y_{r+2}, \tag{137}$$

so that we can represent Cy by the product

$$-Cy = \frac{1}{12}\begin{bmatrix} 1 & -4 & 6 & -4 & 1 & 0 & 0 & 0 & 0 \\ 0 & 1 & -4 & 6 & -4 & 1 & 0 & 0 & 0 \\ 0 & 0 & 1 & -4 & 6 & -4 & 1 & 0 & 0 \\ 0 & 0 & 0 & 1 & -4 & 6 & -4 & 1 & 0 \\ 0 & 0 & 0 & 0 & 1 & -4 & 6 & -4 & 1 \end{bmatrix} \begin{bmatrix} y_{-1} \\ y_0 \\ y_1 \\ y_2 \\ y_3 \\ y_4 \\ y_5 \\ y_6 \\ y_7 \end{bmatrix}. \tag{138}$$

Then, using A_5^{-1} of Table 4, we can calculate $A^{-1}Cy$ in the form

$$-\frac{1}{12}\begin{bmatrix} -5\cdot281 & +12\cdot882 & -7\cdot861 & +0\cdot486 & -0\cdot572 & -0\cdot486 & -4\cdot709 & +9\cdot974 & -4\cdot433 \\ -8\cdot242 & +18\cdot544 & -7\cdot756 & -3\cdot152 & 0\cdot000 & -0\cdot848 & -8\cdot244 & +17\cdot456 & -7\cdot758 \\ -9\cdot143 & +20\cdot572 & -9\cdot715 & 0\cdot000 & -3\cdot428 & 0\cdot000 & -9\cdot715 & +20\cdot572 & -9\cdot143 \\ -7\cdot758 & +17\cdot456 & -8\cdot244 & -0\cdot848 & 0\cdot000 & -3\cdot152 & -7\cdot756 & +18\cdot544 & -8\cdot242 \\ -4\cdot433 & +9\cdot974 & -4\cdot709 & -0\cdot486 & -0\cdot572 & +0\cdot486 & -7\cdot861 & +12\cdot882 & -5\cdot281 \end{bmatrix}\begin{bmatrix} y_{-1} \\ y_0 \\ y_1 \\ y_2 \\ y_3 \\ y_4 \\ y_5 \\ y_6 \\ y_7 \end{bmatrix}. \tag{139}$$

We observe cancellations, particularly in the centre regions, but the process is not yet finished. The values y_{-1}, y_7 are not independent of the rest, but are calculated from internal pivotal values after rounding the latter, so that in this case, for example, we have

$$\left.\begin{aligned} y_{-1} &= 1\cdot75y_0 - y_1 \\ y_7 &= 1\cdot75y_6 - y_5 \end{aligned}\right\}. \tag{140}$$

Thus the errors in y_{-1} and y_7 are composed of final rounding errors η_{-1} and η_7, say, plus real errors dependent on those in y_0, y_{-1}, y_6, and y_5. If these are also denoted by η we have, for the final error expression in the correction to the first approximation $y^{(1)}$, the results

$$\begin{bmatrix} -0\cdot440 & +0\cdot303 & -0\cdot215 & +0\cdot040 & -0\cdot048 & -0\cdot040 & -0\cdot023 & +0\cdot185 & -0\cdot369 \\ -0\cdot687 & +0\cdot343 & +0\cdot040 & -0\cdot263 & 0\cdot000 & -0\cdot071 & -0\cdot040 & +0\cdot323 & -0\cdot646 \\ -0\cdot762 & +0\cdot381 & -0\cdot048 & 0\cdot000 & -0\cdot286 & 0\cdot000 & -0\cdot048 & +0\cdot381 & -0\cdot762 \\ -0\cdot646 & +0\cdot323 & -0\cdot040 & -0\cdot071 & 0\cdot000 & -0\cdot263 & +0\cdot040 & +0\cdot343 & -0\cdot687 \\ -0\cdot369 & +0\cdot185 & -0\cdot023 & -0\cdot040 & -0\cdot048 & +0\cdot040 & -0\cdot215 & +0\cdot303 & -0\cdot440 \end{bmatrix}\begin{bmatrix} \eta_{-1} \\ \eta_0 \\ \eta_1 \\ \eta_2 \\ \eta_3 \\ \eta_4 \\ \eta_5 \\ \eta_6 \\ \eta_7 \end{bmatrix}. \tag{141}$$

We must include η_0 and η_6, the errors in the boundary values, because even if these are known exactly we shall often record them to fewer figures, for the calculation of Cy, than were retained for the computation of $y^{(1)}$.

There is, as we see, substantial further cancellation, and we can now estimate the error from (9), where α_{rs} is now the (r, s) element of the rectangular matrix in (141). The maximum expected error is considerably less than 1 unit in the last place of $y^{(1)}$.

54. In practice, of course, we do not evaluate the matrix $A^{-1}C$, but multiply by A^{-1} the calculated vector Cy. To get the full effect of the cancellation we must record these quantities to more decimals than are apparently justified.

We have considered here only a fourth-difference correction. If sixth differences are significant they must of course be taken into account, and if A^{-1} is large a relatively small sixth difference can make an appreciable contribution to y.

If the difference correction of the first correction to $y^{(1)}$ is significant, we see that the further loss of accuracy is the same as that already lost, but if the finally alleged solution is inserted in the complete finite-difference equations, the residuals R calculated, and $A^{-1}R$ used to give a final correction, the final error is not greater than that caused by a single application of $A^{-1}C$.

55. We may also have to take account of difference corrections arising through conditions involving derivatives at the boundary, and our final example illustrates the same sort of cancellation as before in the error terms. In § 7 of Chapter 6 we discussed the problem defined by the equations

$$\left.\begin{array}{l} y^{\text{iv}}-y = 0, \\ y(0) = y'(0) = 1;\ y''(1) = y(1),\ y'''(1) = y'(1) \end{array}\right\}, \tag{142}$$

which produced the finite-difference equations given in that section and which, after eliminating external pivotal values, reduce to the following algebraic equations:

1	2	3	4	5	
$y(0{\cdot}2)$	$y(0{\cdot}4)$	$y(0{\cdot}6)$	$y(0{\cdot}8)$	$y(1{\cdot}0)$	
8·9984	−4·5	1			$= 6{\cdot}0996+\tfrac{1}{2}Cy(0)-Cy_1-3C_1\,y(0)$
−4	5·9984	−4	1		$= -1-Cy_2$
1	−4	5·9984	−4	1	$= -Cy_3$
	1	−3·856736...	4·706179...	−1·816660...	$= -Cy_4+0{\cdot}85958C_2\,y(1)+0{\cdot}14231C_3\,y(1)$
		1·719165...	−3·506641...	1·718475...	$= -Cy_5-1{\cdot}68501C_2\,y(1)+1{\cdot}70778C_3\,y(1)$

(143)

The inverse of the matrix on the left has already been given to a few figures in Table 12, and has fairly large coefficients. The difference-correction operators are given by

$$\left.\begin{array}{l} C = (-\tfrac{1}{6}\delta^6+\tfrac{7}{240}\delta^8-\ldots) \\ C_1 = (\tfrac{1}{30}\mu\delta^5-\tfrac{1}{140}\mu\delta^7+\ldots) \\ C_2 = (\tfrac{1}{90}\delta^6-\tfrac{1}{560}\delta^8+\ldots) \\ C_3 = \left\{-\left(\dfrac{1}{4}+\dfrac{0{\cdot}04}{30}\right)\mu\delta^5+\left(\dfrac{7}{120}+\dfrac{0{\cdot}04}{140}\right)\mu\delta^7+\ldots\right\} \end{array}\right\}. \tag{144}$$

The first approximation, obtained with neglect of difference corrections, should be exactly the same as that of Table 1 in Chapter 6, but we keep an extra figure in the calculations for the sake of accuracy and find the results of Table 21. We give also the differences and the calculated difference corrections, neglecting differences of order 9 and higher, and record the results to two extra decimals.

TABLE 21

x	$y^{(1)}$		δ^2		δ^4		δ^6		δ^8	10^6Cy	
−0·8	+0·455095										
		+ 96906									
·6	0·552001		+ 22801								
		119707		+ 4850							
·4	0·671708		27651		+1075						
		147358		5925		+ 236					
− ·2	0·819066		33576		1311		+ 53				
		180934		7236		289		+13			
+ ·0	1·000000		40812		1600		66		+0	−11·00	$10^6C_1\,y(0) = +\ 10{\cdot}64$
		221746		8836		355		13			
·2	1·221746		49648		1955		79		5	13·02	$10^6C_2\,y(1) = +\ 1{\cdot}93$
		271394		10791		434		18			
·4	1·493140		60439		2389		97		3	16·08	$10^6C_3\,y(1) = -218{\cdot}97$
		331833		13180		531		21			
·6	1·824973		73619		2920		118		4	19·55	
		405452		16100		649		25			
0·8	2·230425		89719		3569		143		7	23·63	
		495171		19669		792		32			
1·0	2·725596		109388		4361		175		7	28·96	
		604559		24030		967		39			
1·2	3·330155		133418		5328		214		+7	(−35·46)	
		737977		29358		1181		+46			
1·4	4·068132		162776		6509		+260				
		900753		35867		+1441					
1·6	4·968885		198643		+7950						
		1099396		+43817							
1·8	6·068281		+242460								
		+1341856									
+2·0	+7·410137										

We calculate $y(2{\cdot}0)$ and $Cy(1{\cdot}2)$ so that we can get a correction at $x = 1{\cdot}6$, as well as at previous pivotal points. This correction z, with its differences, is shown in Table 22, and differs substantially from that of Chapter 6, where the extra decimals were not retained in the difference corrections.

TABLE 22

x	z		δ^2		δ^4	
−0·4	−0·001393					
		+1056				
− ·2	·000337		−719			
		+ 337		+ 40		
+ ·0	·000000		679		+11	
		− 342		51		+1
·2	·000342		628		12	
		970		63		−1
·4	·001312		565		11	
		1535		74		+7
·6	·002847		491		18	
		2026		92		−2
0·8	·004873		399		16	
		2425		108		+1
1·0	·007298		291		17	
		2716		125		+2
1·2	·010014		166		+19	
		2882		+144		
1·4	·012896		− 22			
		−2904				
+1·6	−0·015800					

The major difference correction is clearly $C_3 y(1)$, and though the others are negligible for z, $C_3 z$ involves fifth differences at $x = 1{\cdot}0$, which are not quite oscillating about zero. Using the fifth differences, we find $10^6 C_3 z(1) = -0{\cdot}38$, and this causes further increments of -1, -3, -6, -11, and -16 at the respective pivotal points, the final solution differing from e^x by at most 1 unit in the sixth decimal.

56. This surprising result is of course better than reasonable expectation, but not to a remarkable extent. An investigation similar to that of § 53 shows that, in the calculation of the difference corrections, we use all the pivotal values from $y(1{\cdot}8)$ to $y(-0{\cdot}8)$, and, if the errors in these quantities were independent, our last row of $A^{-1}C$ would have the following appearance, with coefficients rounded to two decimals:

−0·8	−0·6	−0·4	−0·2	0·0	0·2	0·4
−0·01	−0·03	+0·72	−1·68	+0·79	+0·18	+5·13

0·6	0·8	1·0	1·2	1·4	1·6	1·8
−19·16	+29·46	−18·13	−4·21	+11·32	−4·99	+0·60.

These coefficients are smaller, but not substantially smaller, than those of the last row of A^{-1}. Again, however, the errors in values outside the range are not completely independent of those at internal points so that, for example, we have

$$\left.\begin{aligned}\epsilon(1\cdot2) &= 4\epsilon(1\cdot0)-5\cdot9984\epsilon(0\cdot8)+4\epsilon(0\cdot6)-\epsilon(0\cdot4)+\eta(1\cdot2)\\ \epsilon(1\cdot4) &= 4\epsilon(1\cdot2)-5\cdot9984\epsilon(1\cdot0)+4\epsilon(0\cdot8)-\epsilon(0\cdot6)+\eta(1\cdot4)\\ &\text{etc.}\end{aligned}\right\}, \tag{145}$$

where $\epsilon(x)$ is the error in $y(x)$, and η the extra rounding error. If we ignore the small coefficients of $y(-0\cdot8)$ to $y(0\cdot2)$, the remaining coefficients, multiplying *independent* errors, reduce to the following values:

0·4	0·6	0·8	1·0	1·2	1·4	1·6	1·8
+1·87	−3·54	+2·10	−0·41	+3·25	−2·60	−2·58	+0·60.

These are considerably smaller than those of A^{-1}, and clearly give rise to no great error.

57. As final experimental verification of the remark at the end of § 54, we might take values of e^x, rounded to five decimals, substitute them in the finite-difference equations (143), including difference corrections, and calculate residuals. We find the residual values −9·151, +7·149, −3·610, +1·288, and −0·885, in units of the fifth decimal, at the respective points 0·2, 0·4, 0·6, 0·8, and 1·0. Notwithstanding the large coefficients of A^{-1}, the values of $A^{-1}R$, giving the corrections to this alleged solution, are at most 1·4 units in the fifth decimal. With the 'first approximation' given to only five decimals, we would not expect to get the last figure completely accurate in the 'final result'. Similar calculations with six decimals produce a similar result.

58. It is not of course very satisfactory to have to rely on the small differences of Table 22 to produce an accurate solution. This is necessitated by the relatively large coefficients in the difference correction $C_3\,y(1)$, but could be avoided, the rather awkward right-hand sides of (143) simplified, and the first approximation made more accurate, by using the Taylor-series method to satisfy the boundary conditions. This is nearly always preferable whenever the labour of successively differentiating the differential equation is not prohibitive.

†59. All our methods of including the difference correction, except the special one of § 48 when $\delta^6 y$ is negligible, involve the use of central differences and the calculation of pivotal values external to the range. It is of course possible to use sloping-difference formulae near the boundaries to express the difference correction in terms of internal points only. This is not a very satisfactory process, for reasons of the relative inaccuracy of such formulae, but we might occasionally set up an original matrix, say, for a second-order equation, which already includes the fourth-difference correction, when sixth differences are expected to be negligible. One solution of the linear equations then gives the complete answer.

Suppose, for example, that for the simplest second-order equation (17) we use from the start the general formula

$$\delta^2 y_r - \tfrac{1}{12}\delta^4 y_r = h^2 f(x_r), \tag{146}$$

or

$$-\tfrac{1}{12}y_{r-2}+\tfrac{4}{3}y_{r-1}-\tfrac{5}{2}y_r+\tfrac{4}{3}y_{r+1}-\tfrac{1}{12}y_{r+2} = h^2 f(x_r). \tag{147}$$

If the boundary values are fixed we would need only one type of special formula, for the points adjacent to the boundaries, to replace (146), and we might take one of Table 7 in Chapter 2, to give

$$(\Delta^2 - \tfrac{1}{12}\Delta^4)y_0 = h^2 f(x_1), \tag{148}$$

or

$$-\tfrac{1}{12}y_4+\tfrac{1}{3}y_3+\tfrac{1}{2}y_2-\tfrac{5}{3}y_1+\tfrac{11}{12}y_0 = h^2 f(x_1), \tag{149}$$

where 0 is the boundary point and 1, 2,..., successive internal points.

The matrix would then have the following form:

$$\begin{array}{cccccccc} y_1 & y_2 & y_3 & y_4 & y_5 & y_6 & y_7 & \cdots \end{array}$$
$$\begin{bmatrix} -\frac{5}{3} & \frac{1}{2} & \frac{1}{3} & -\frac{1}{12} & 0 & 0 & 0 & \cdots \\ \frac{4}{3} & -\frac{5}{2} & \frac{4}{3} & -\frac{1}{12} & 0 & 0 & 0 & \cdots \\ -\frac{1}{12} & \frac{4}{3} & -\frac{5}{2} & \frac{4}{3} & -\frac{1}{12} & 0 & 0 & \cdots \\ 0 & -\frac{1}{12} & \frac{4}{3} & -\frac{5}{2} & \frac{4}{3} & -\frac{1}{12} & 0 & \cdots \end{bmatrix}. \tag{150}$$

We have here a five-term difference formula representing a second-order equation, and we have seen that the triangular decomposition process is equivalent to forward integration using the smaller factor of the operator and backward integration using the larger factor. The triangular matrices corresponding to (150) have the forms

$$\overset{L}{\begin{bmatrix} 1 & & & \\ \times & 1 & & \\ \times & \times & 1 & \\ & \times & \times & 1 \end{bmatrix}}, \quad \overset{U}{\begin{bmatrix} \times & \times & \times & \times & & \\ & \times & \times & \times & & \\ & & \times & \times & \times & \\ & & & \times & \times & \times \end{bmatrix}}, \tag{151}$$

corresponding to those of fourth-order equations. We have in fact introduced two unwanted solutions, and it is possible that these might dominate the true solutions so that incorrect results are obtained.

In the case of constant coefficients we know that the behaviour of the solution of the three-term difference equation depends on the roots $p_1 \geqslant p_2$ of a quadratic equation, while that of (147) has four roots $q_1 \geqslant q_2 \geqslant q_3 \geqslant q_4$, which determine its behaviour. In the three-term formula p_2 belongs to the L matrix, p_1 to the U matrix. For the five-term formula q_3 and q_4 belong to the L matrix, q_2 and q_1 to the U matrix. Now if our five-term formula is to give good results it is clear that q_3 must be approximately equal to p_2, and q_2 to p_1. If q_4 approximates to p_2, or q_1 to p_1, the 'wrong' roots will dominate the forward or backward integration, and we shall have instability.

For (146) without the δ^4 term, the roots belonging to the corresponding three-term formula are 1, 1 and for (146) with the δ^4 term, giving (147), the roots are those of the quartic equation

$$x^4-16x^3+30x^2-16x+1 = 0, \tag{152}$$

found to be $x = 7-\sqrt{48}$, 1, 1, $7+\sqrt{48}$. Here the roots corresponding to the quadratic separate the remaining roots of the quartic, and the situation is stable.

We infer that this process is generally stable. For desk-machine work, however, it is less satisfactory than our usual method, involving more work in the matrix processes, less precise formulae near the boundaries, and uncertainty, without further investigation, regarding the neglect of higher differences.

Eigenvalue problems

60. As before we would like to determine the precision of our results from that of the given coefficients. Suppose that y_r and λ_r are accurate solutions of the equation

$$(A-I\lambda_r)y_r = 0, \tag{153}$$

but that these are only approximations to the solution of the true equation

$$\{A+\delta A-I(\lambda_r+\delta\lambda_r)\}(y_r+\delta y_r) = 0. \tag{154}$$

We neglect small quantities of the second order and deduce the equation

$$(\delta A-I\delta\lambda_r)y_r+(A-I\lambda_r)\delta y_r = 0. \tag{155}$$

The transposed matrix A' has roots λ_r and vectors z_r, so that

$$A'z_r = \lambda_r z_r, \quad \text{or} \quad z'_r A = \lambda_r z'_r. \tag{156}$$

The known orthogonality relations, and our choice of normalization, are represented by the equations

$$z'_s y_r = y'_r z_s = 0, \qquad z'_r y_r = y'_r z_r = 1. \tag{157}$$

Premultiplying (155) by z'_r and using (156) and (157) we find, for the error $\delta\lambda_r$, the expression

$$\delta\lambda_r = z'_r\,\delta A y_r. \tag{158}$$

We can easily find the worst possible error, caused by maximum elements of δA equal to ϵ. If the components of the vectors y_r and z_r are denoted respectively by η_r and ζ_r we find, for the worst error, the expression

$$\delta\lambda_r = \epsilon(\sum|\zeta_r|)(\sum|\eta_r|). \tag{159}$$

In particular, if the matrix is symmetric so that $\eta_r = \zeta_r$, we find the simple result

$$\delta\lambda_r = \epsilon(\sum|\eta_r|)^2. \tag{160}$$

A more realistic statistical estimate can also be obtained. The standard deviation of $\delta\lambda_r$ is given by

$$\text{S.D. } (\delta\lambda_r) = \frac{\epsilon}{\sqrt{3}}\sqrt{(\sum\zeta_r^2\sum\eta_r^2)}, \tag{161}$$

and if the matrix is symmetric it reduces to

$$\text{S.D. } (\delta\lambda_r) = \frac{\epsilon}{\sqrt{3}} \tag{162}$$

for a unit vector. This shows that the 95 per cent. expectation of error in the case of a symmetric matrix is little more than ϵ, the maximum error in the coefficients, and is independent of the order of the matrix. For an unsymmetric matrix it may be considerably larger. We note also that the results hold for any one of the n latent roots λ_r.

61. From (155) we have

$$\delta y_r = -(A-I\lambda_r)^{-1}(\delta A-I\delta\lambda_r)y_r. \tag{163}$$

If λ is not an eigenvalue, we can expand $(A-I\lambda)^{-1}$ in the form

$$(A-I\lambda)^{-1} = -\sum_{s=1}^{n}\frac{y_s z_s'}{\lambda-\lambda_s} \tag{164}$$

so that, as $\lambda \to \lambda_r$, we can write

$$\delta y_r = \sum_{s\neq r}\frac{y_s z_s'}{\lambda_r-\lambda_s}(\delta A-I\delta\lambda_r)y_r + \lim_{p\to r}\frac{y_p z_p'(\delta A-I\delta\lambda_r)y_r}{\lambda_r-\lambda_p}. \tag{165}$$

The second term on the right is apparently indeterminate, but its limiting value is $y_r z_r' \delta y_r$. The first simplifies by the orthogonality rule so that we can write

$$(I-y_r z_r')\,\delta y_r = \left(\sum_{s\neq r}\frac{y_s z_s'}{\lambda_r-\lambda_s}\right)\delta A y_r. \tag{166}$$

There is little point in pursuing further this analysis, but we note that the main effect on precision of eigensolutions is the degree of separation of the eigenvalues.

62. We can see this phenomenon in an easier way by considering the residuals of an approximate solution. If λ_r and y_r are approximate solutions of (153) we can calculate the residuals, in general not zero, defined by the left-hand side of (153), and would like to make a pronouncement regarding their effect on λ_r and y_r. It is fairly safe to assume that λ_r is more correct than y_r, particularly if the matrix is symmetric and we have been able to use Rayleigh's Principle. We therefore consider the effect of small residuals on the eigenfunction only.

Suppose now that the y we use in (153) is given by

$$y = y_r + \sum_{r \neq s} \alpha_s y_s. \tag{167}$$

Then the residual vector is given by

$$R = \sum_{r \neq s} \alpha_s(\lambda_s - \lambda_r) y_s, \tag{168}$$

and, for λ_s near to λ_r, R is little changed by relatively large changes in α_s so that the precision of y_r, and hence also of y_s, is poor.

63. As in the solution of algebraic equations, so here errors introduced by inexact processes of solution are likely to be greatest when precision is poor, and it is then necessary to retain guarding figures in the calculations. The convergence of iterative processes, moreover, is likely to be slow in these circumstances, and this fact is the required warning that accuracy is poor and guarding figures necessary.

The retention of guarding figures causes no great inconvenience in the relaxation method or in the direct iterative processes, since this can be done at any stage in the proceedings, and with effect when the approximate answers are close to the required result. With the inverse iterative processes, however, that is iteration with A^{-1} or $(A-pI)^{-1}$ for calculating the smallest eigensolution or a particular one in the middle of the spectrum, we must use an accurate inverse at some stage, just as in the direct method we use the fully accurate A. The production of this accurate inverse involves the retention of guarding figures from the start.

64. In many eigenvalue problems involving discrete systems it may be required to calculate all or at least several of the eigensolutions. With methods of root removal, involving the formation of new matrices, it might be expected that there is a steady loss of accuracy as the work proceeds, though there is no satisfactory general theory on this matter. The iterative use of an accurate $(A-pI)^{-1}$, however, should guarantee the accuracy of each solution to the precision attainable, though it is not always possible in advance to specify suitable values of p.

For differential equations of eigenvalue type, however, it is uncommon to require more than one or two solutions, so that we are unaffected by possible steady deterioration in accuracy. In a large class of problems, moreover, the eigenvalues are reasonably well separated.

Non-linear equations

65. Questions of precision and accuracy in solving non-linear equations present no new problems. For linear equations, as we have seen, both accuracy and precision are determined in terms of the inverse matrix of the coefficients of the linear finite-difference equations. Non-linear equations are most often solved by iterative methods, in which corrections to a first approximation are obtained by linearizing the original equations. Precision and accuracy of the final solution depends on the matrix of these linear equations. Some of the coefficients of the linear equations will depend on the final solution, but this is a relatively minor complication whose main effect will be on the convergence of the iterative process of solution.

Additional notes and bibliography

§ 1. The term 'precision' we take to mean the number of figures or decimals to which a quantity is prescribed or might be expected to be correct. 'Accuracy' refers to the error in this quantity. For example, a table of $\sin x$, say, may be given to four decimals, which defines the precision of the table. The table may be accompanied by a remark that the end figure may be in error by as much as 1 unit, which defines its accuracy. Again, the precision of the solutions of linear equations, based on that of the coefficients in the equations, is a definite quantity, whereas the accuracy depends on the solving process used.

§ 12. Initial-value techniques applied to boundary-value problems are also capable of giving best precision when the solutions are of exponential type, but the methods are inconvenient, as illustrated in § 40 of Chapter 8.

§ 30. In examining the values of $|A|$ to determine the interval in n in which an eigensolution might occur, we must change the sign at each step, because the negative sign is involved in the ratio of $|A|$ and the last value y_n, which here vanishes in an eigensolution. See equations (95) and (114) of Chapter 8.

§ 32. The method of § 21, Chapter 5, for solving a first-order equation by deriving a second-order equation and using two-point boundary conditions, will sometimes suffer lack of precision through the introduction of a near eigensolution. For example, the system

$$y' - y + 1 = 0, \qquad y(0) = 1,$$

has the solution $y = 1$ everywhere. The derived system is given by

$$y'' - y + 1 = 0, \qquad y(0) = 1, \qquad y' - y + 1 = 0 \quad \text{at} \quad x = x_n,$$

where x_n is at the other end of the required range. The second-order equation has the general solution

$$y = Ae^x + Be^{-x} + 1,$$

and satisfies the first boundary condition if $B = -A$, so that

$$y = A(e^{x}-e^{-x})+1.$$

This satisfies the second boundary condition in theory only if $A = 0$, but in practice the term Ae^{x} satisfies $y'-y = 0$ for all x, and the term Ae^{-x} satisfies this condition approximately at $x = x_n$, provided x_n is large enough. There is 'almost' an eigensolution $A(e^{x}-e^{-x})$ if x_n is large, and the corresponding matrix of the finite-difference equations has large inverse elements. Our best hope of success is then bound up with a fairly small x_n.

In certain cases, however, x_n can be large provided that, while the particular integral is adequately represented by the finite-difference equations at the chosen interval, the interval is too large for adequate representation of the complementary functions. In this case the matrix method will give a good result, and there is no contaminating eigensolution. Such an example was given in the notes on Chapter 5.

§ 38. A discussion of errors in solving linear equations and allied problems has been given by

Turing, A. M., 1948, Rounding-off errors in matrix processes. *Quart. J. Mech. and Appl. Math.* **1**, 287–308.

§ 43, etc. The special formulae will be needed most often when the solutions are of oscillatory type. Boundary-value techniques, including relaxation because of the extra difficulties of liquidating residuals, may then have some inconvenience. Initial-value techniques are not adversely affected by internal zeros of complementary functions. Also the complementary functions in this case may not have rapid growth, so that initial-value techniques may be more convenient than those of boundary-value type. Compare, however, the additional comment on §§ 40 and 41 of Chapter 8.

§ 45. The method given in this section may be preferable to that of § 44 if ϵ is not very small, but for ϵ near zero the method of § 44 is clearly superior and, in particular, can be used when ϵ is exactly zero.

§ 59. In computation with high-speed machines it may be preferable to incorporate some of the difference correction into the original equations. It is only recommended, however, when this is known to be capable of providing the required precision. See the comment on § 2 in Chapter 4.

10

Miscellaneous Methods and Comments

1. In this chapter we discuss a variety of topics. Some of these have not had adequate treatment in previous chapters and others deal with methods which do not depend solely on the theory of finite differences.

Simultaneous differential equations

2. In previous chapters we have discussed the solution of simultaneous equations, mainly of the second order, but other forms of simultaneous equations arise which are best solved by boundary-value techniques. Perhaps the least obvious case is that of two first-order equations, each with one boundary condition but specified at different points. Consider, for example, the most general linear problem of this class, given by the equations

$$\left.\begin{aligned} y'+ay+bz &= f \\ z'+cy+dz &= g \end{aligned}\right\}, \tag{1}$$

with the conditions

$$\alpha_1 y+\beta_1 z = \gamma_1 \text{ at } x = x_0, \qquad \alpha_n y+\beta_n z = \gamma_n \text{ at } x = x_n. \tag{2}$$

The convenient use of three-term finite-difference formulae for the first-order equations would require extra boundary conditions, and the two-term formulae cannot be used in a step-by-step process, except indirectly after the manner of Chapter 8, since the required boundary conditions are at different points.

We might transform the problem by differentiation into the solution of two simultaneous second-order equations, both of which lack the first derivatives of y and z, with, as boundary conditions, equations (2) and (1) in which one of the latter is used at each boundary. As in §§ 19 and 20 of Chapter 5 we could then use powerful finite-difference equations, involving small difference corrections, both for differential equations and boundary conditions: by using the Taylor-series method we could even satisfy the latter exactly.

3. This is a useful method, and probably preferable when the differentiation involved is not difficult. We have noted, however, that precision generally deteriorates as the order of the equations increases, so

that the direct solution of the first-order equations should not be overlooked. Recalling the formula

$$\left.\begin{aligned} y_1 - y_0 &= \tfrac{1}{2}h(y_1' + y_0') + Cy_{\frac{1}{2}} \\ C &= -\tfrac{1}{12}\delta^3 + \tfrac{1}{120}\delta^5 - \tfrac{1}{840}\delta^7 + \ldots \end{aligned}\right\}, \tag{3}$$

we find that (1) can be replaced by the algebraic equations typified by

$$\left.\begin{aligned} -(1-\tfrac{1}{2}ha_r)y_r + \tfrac{1}{2}hb_r z_r + (1+\tfrac{1}{2}ha_{r+1})y_{r+1} + \tfrac{1}{2}hb_{r+1}z_{r+1} & \\ = \tfrac{1}{2}h(f_{r+1}+f_r) + Cy_{r+\frac{1}{2}} & \\ \tfrac{1}{2}hc_r y_r - (1-\tfrac{1}{2}hd_r)z_r + \tfrac{1}{2}hc_{r+1}y_{r+1} + (1+\tfrac{1}{2}hd_{r+1})z_{r+1} & \\ = \tfrac{1}{2}h(g_{r+1}+g_r) + Cz_{r+\frac{1}{2}} & \end{aligned}\right\}. \tag{4}$$

Equations (4) may be considered as the results of 'applying' the differential equations at a point half-way between the pivotal points r and $r+1$. If there are $(n-1)$ internal pivotal points there are altogether $(2n+2)$ unknowns. The use of (4) at every internal half-way point provides $2n$ equations, the remaining two equations being provided by the boundary conditions.

We note that each equation of type (4) involves two adjacent values of both y and z, while those for the boundary conditions involve one value of both y and z. The most general matrix of coefficients thus has the following appearance:

$$\begin{array}{cccccccccccccc} y_0 & z_0 & y_1 & z_1 & y_2 & z_2 & y_3 & z_3 & \cdot & \cdot & y_{n-1} & z_{n-1} & y_n & z_n \end{array}$$
$$\begin{bmatrix} \times & \times & & & & & & & & & & & & \\ \times & \times & \times & \times & & & & & & & & & & \\ \times & \times & \times & \times & & & & & & & & & & \\ & & \times & \times & \times & \times & & & & & & & & \\ & & \times & \times & \times & \times & & & & & & & & \\ & & & & \times & \times & \times & \times & & & & & & \\ & & & & \times & \times & \times & \times & & & & & & \\ & & & & & & \cdot & \cdot & \cdot & \cdot & \cdot & \cdot & & \\ & & & & & & & & & & \times & \times & \times & \times \\ & & & & & & & & & & \times & \times & \times & \times \\ & & & & & & & & & & & & \times & \times \end{bmatrix}. \tag{5}$$

This is probably the best arrangement of the matrix from the point of view of solution by triangular matrices, and the triangles L and U have the following simple and interesting forms:

$$\overset{L}{\begin{bmatrix} 1 & & & & & \\ \times & 1 & & & & \\ \times & \times & 1 & & & \\ & & \times & 1 & & \\ & & \times & \times & 1 & \\ & & & & \times & 1 \\ & & \cdot & \cdot & \cdot & \cdot \end{bmatrix}}, \quad \overset{U}{\begin{bmatrix} \times & \times & & & & & & \\ & \times & \times & \times & & & & \\ & & \times & \times & & & & \\ & & & \times & \times & \times & & \\ & & & & \times & \times & & \\ & & & & & \times & \times & \times \\ & & & & \cdot & \cdot & \cdot & \cdot \end{bmatrix}}. \tag{6}$$

In general there are alternately three and two non-zero terms in each row of L, and two and three alternately in corresponding rows of U. Formulae for the evaluation of these elements are easily obtained in the usual way from the rules for matrix multiplication.

The equations may also be solved by relaxation, though combined operations are usually necessary.

Other kinds of simultaneous differential equations, possibly with different orders, may occasionally present themselves for solution, but a combination of methods outlined here and in previous chapters should suffice for all cases.

Equations with discontinuous coefficients

†4. Another problem of not infrequent occurrence is presented by discontinuities in the coefficients of the differential equation or equations. For example, in the equation

$$y''+fy'+gy = k, \tag{7}$$

with specified two-point boundary conditions, the functions f, g, and k of x may change discontinuously at one or more values of x within the range of integration. Such a phenomenon, which may occur also in simultaneous equations, must be accompanied by extra conditions to provide a unique solution. The most common conditions in the case of (7) are that y and y' should be continuous at the critical points.

The only new point of interest is the method of establishing finite-difference equations for use near a critical point. We assume first that we can arrange this point to coincide with a pivotal point, which we label 0. To the left of this point the coefficients in (7) are denoted by f, g, and k, and to the right by F, G, and K respectively. The pivotal value to the left of 0 has the suffix -1, that to the right the suffix 1.

If we consider 0 to belong to the left-hand region, we have the finite-difference form of (7) given by

$$\bar{y}_1(1+\tfrac{1}{2}hf_0)+y_{-1}(1-\tfrac{1}{2}hf_0)-y_0(2-h^2g_0)+Cy_{0,L} = h^2k_0, \tag{8}$$

where the bar on y_1 denotes that this is not a pivotal value of the final result, but the value which that point would have in the absence of the discontinuity. The suffix L in the difference correction is explained later. Similarly, regarding 0 as belonging to the right-hand region, we have

$$y_1(1+\tfrac{1}{2}hF_0)+\bar{y}_{-1}(1-\tfrac{1}{2}hF_0)-y_0(2-h^2G_0)+Cy_{0,R} = h^2K_0, \tag{9}$$

with the bar having the same meaning as before, and R needing explanation. We have not given y_0 separate notations or values since the first continuity condition makes it the same whether belonging to the left-hand or right-hand region. The barred values must be eliminated, and for this purpose we use the second continuity condition involving the first derivative. This gives the finite-difference equation

$$\bar{y}_1-y_{-1}+C_1y_{0,L} = y_1-\bar{y}_{-1}+C_1y_{0,R}, \tag{10}$$

and, eliminating the barred quantities from (8), (9), and (10), we have a single equation connecting the three adjacent values, around the critical point, which actually represent the final solution. This equation is given by

$$\left.\begin{aligned} y_1(1+P_0/Q_0)+y_{-1}(1+q_0/p_0)-y_0(R_0/Q_0+r_0/p_0)& \\ = h^2(K_0/Q_0+k_0/p_0)+Cy_0& \\ Cy_0 = -p_0^{-1}Cy_{0,L}-Q_0^{-1}Cy_{0,R}+C_1y_{0,L}-C_1y_{0,R}& \end{aligned}\right\}, \tag{11}$$

where

$$\left.\begin{aligned} P_0 &= 1+\tfrac{1}{2}hF_0, & Q_0 &= 1-\tfrac{1}{2}hF_0, & R_0 &= 2-h^2G_0 \\ p_0 &= 1+\tfrac{1}{2}hf_0, & q_0 &= 1-\tfrac{1}{2}hf_0, & r_0 &= 2-h^2g_0 \end{aligned}\right\}. \tag{12}$$

The approximate equation, with Cy_0 neglected, is the required replacement for the usual formula at point 0 in the absence of discontinuities. The difference corrections can be included in a correcting process, but their calculation at this point involves a little more work. The correction $Cy_{0,L}$ is given by

$$Cy_{0,L} = \{hf_0(-\tfrac{1}{6}\mu\delta^3+\tfrac{1}{30}\mu\delta^5-\ldots)-\tfrac{1}{12}\delta^4+\tfrac{1}{90}\delta^6-\ldots\}y_{0,L}, \tag{13}$$

and in its calculation we must use values of y in the left-hand region and their continuation, including the term $\bar{y}_1$ in (8) and other values $\bar{y}_2, \bar{y}_3,\ldots,$ obtained in the usual step-by-step manner. The correction $Cy_{0,R}$ is defined in a similar way. The corrections of type $C_1 y$ in (10) also involve this extrapolation process, and are defined by

$$\left.\begin{aligned} C_1 y_{0,L} &= (-\tfrac{1}{3}\mu\delta^3+\tfrac{1}{15}\mu\delta^5-\ldots)y_{0,L} \\ C_1 y_{0,R} &= (-\tfrac{1}{3}\mu\delta^3+\tfrac{1}{15}\mu\delta^5-\ldots)y_{0,R} \end{aligned}\right\}. \tag{14}$$

In the case in which the first derivative is absent these equations are considerably simplified. For then $f = F = 0$, $P = Q = p = q = 1$, and (11), on division by two, becomes

$$y_1+y_{-1}-y_0\{2-\tfrac{1}{2}h^2(g_0+G_0)\}$$
$$= \tfrac{1}{2}h^2(K_0+k_0)+\tfrac{1}{2}\{(C_1 y_{0,L}-C_1 y_{0,R})-(Cy_{0,L}+Cy_{0,R})\}, \tag{15}$$

and the terms $Cy_{0,L}$ and $Cy_{0,R}$ involve only even differences.

5. Occasionally it will be convenient and desirable to use the Taylor-series method to satisfy exactly the conditions at the critical point, thus eliminating the corrections of type $C_1 y$. We can clearly express the $\bar{y}_1$, $\bar{y}_{-1}$ of (8) and (9) in terms of y_0 and y_0', since the second and higher derivatives in the Taylor series are thus expressible. We then have

$$\left.\begin{aligned} \bar{y}_1 &= \alpha_L+\beta_L y_0+\gamma_L y_0' \\ \bar{y}_{-1} &= \alpha_R+\beta_R y_0+\gamma_R y_0' \end{aligned}\right\}, \tag{16}$$

the y_0 and y_0' needing no extra suffix since the continuity conditions give them the same value in each region. From (8), (9), and (16) we now eliminate $\bar{y}_1$, $\bar{y}_{-1}$, and y_0' to produce a single equation of type (11), but in which difference corrections only of type $Cy_{0,L}$, $Cy_{0,R}$ appear.

6. If the critical point cannot be chosen to coincide with a pivotal point, the equations are necessarily more complicated. Consider the arrangement of Fig. 35, A denoting the critical point.

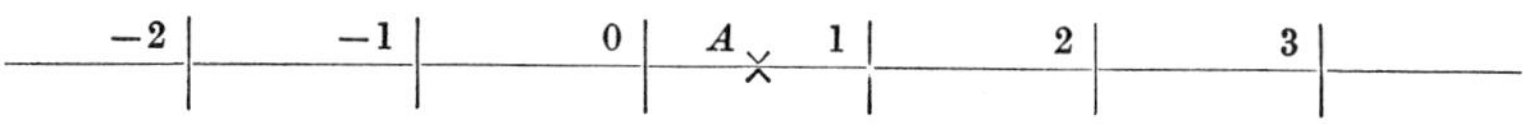

FIG. 35

With the same notation as before we can write

$$\left.\begin{aligned} \bar{y}_1(1+\tfrac{1}{2}hf_0)+y_{-1}(1-\tfrac{1}{2}hf_0)-y_0(2-h^2g_0)+Cy_0 &= h^2k_0 \\ y_2(1+\tfrac{1}{2}hF_1)+\bar{y}_0(1-\tfrac{1}{2}hF_1)-y_1(2-h^2G_1)+Cy_1 &= h^2K_1 \end{aligned}\right\}, \tag{17}$$

in which the difference corrections must use values belonging to their respective regions and their extrapolated barred values.

To make use by pure finite-difference processes of the continuity conditions at A we should have to use interpolation formulae to calculate

y_A and y'_A, regarding A as belonging separately to the two regions. For example, if $0A = h_1$, $A1 = h_2$, linear interpolation gives

$$\left.\begin{aligned} hy_{A,L} &= h_2 y_0 + h_1 \bar{y}_1 \\ hy_{A,R} &= h_2 \bar{y}_0 + h_1 y_1 \\ hy'_{A,L} &= h_2 y'_0 + h_1 \bar{y}'_1 \doteqdot \frac{h_2}{2h}(\bar{y}_1 - y_{-1}) + \frac{h_1}{2h}(\bar{y}_2 - y_0) \\ hy'_{A,R} &= h_2 \bar{y}'_0 + h_1 y'_1 \doteqdot \frac{h_2}{2h}(y_1 - \bar{y}_{-1}) + \frac{h_1}{2h}(y_2 - \bar{y}_0) \end{aligned}\right\}. \tag{18}$$

The new barred values $\bar{y}_2$ and $\bar{y}_{-1}$ can be calculated in terms of the old by further use of (17), so that, by equating the right-hand sides of the first two and last two equations of the set (18), we can find two equations containing only $\bar{y}_1$, $\bar{y}_0$, and unbarred quantities. From (17) we can then eliminate $\bar{y}_1$ and $\bar{y}_0$, producing two equations, involving the pivotal values actually required, which replace the standard equations at 0 and 1 in the absence of discontinuities.

Interpolation formulae of higher accuracy can of course be introduced, if (17) is used to extend the number of barred quantities, but the process becomes complicated.

7. The Taylor-series method is less cumbersome, unless the differentiation involved is prohibitive. We can write

$$\left.\begin{aligned} y_0 &= \left(y_A - h_1 y'_A + \frac{h_1^2}{2!} y''_A - \dots\right)_L \\ \bar{y}_1 &= \left(y_A + h_2 y'_A + \frac{h_2^2}{2!} y''_A + \dots\right)_L \end{aligned}\right\}, \tag{19}$$

and, using the differential equation and its derived forms in the left-hand region, these can be expressed in the form

$$\left.\begin{aligned} y_0 &= a_L + b_L y_{A,L} + c_L y'_{A,L} \\ \bar{y}_1 &= \alpha_L + \beta_L y_{A,L} + \gamma_L y'_{A,L} \end{aligned}\right\}. \tag{20}$$

We can similarly find the formulae

$$\left.\begin{aligned} y_1 &= a_R + b_R y_{A,R} + c_R y'_{A,R} \\ \bar{y}_0 &= \alpha_R + \beta_R y_{A,R} + \gamma_R y'_{A,R} \end{aligned}\right\}. \tag{21}$$

Equating $y_{A,L}$ with $y_{A,R}$ and $y'_{A,L}$ with $y'_{A,R}$ to satisfy the continuity, we deduce two equations connecting $\bar{y}_1$ and $\bar{y}_0$ with y_0 and y_1 and finally, using these equations in conjunction with (17), we find the two equations, involving only required quantities, which replace the standard forms at 0 and 1 in the absence of discontinuities. In these final equations only the corrections Cy_0 and Cy_1 appear, their effective definition being that of equation (13).

8. There is a slight variation of this problem which sometimes arises. Instead of being given the critical points x_A at which the differential equation changes, we are told that this change takes place in regions in which the solution y is smaller or greater than some critical value y_A. In these circumstances we must use an iterative method, calculating a first approximation assuming no discontinuities, finding by inverse interpolation the critical points x_A, and repeating the calculations using the techniques already described. For this new approximation the critical points will be in slightly different positions, and we must proceed iteratively, solving several problems of the type already described until there is no further change in the functional values and in the critical points.

On matters of detail we note that each approximation should give a good starting approximation to the next, so that the relaxation method is often the most convenient for solving the algebraic equations. Difference corrections need not be included until the last stages of the iteration.

The deferred approach to the limit

†9. This name was given by L. F. Richardson to a device for including the effects of the difference correction without its explicit evaluation. If a first approximation is calculated, as in all our methods, with the use of simple truncated finite-difference formulae, it is reasonable to suppose that the error in this approximation, at each pivotal point, is a function of the size of the interval h. If this function is expanded as a Taylor series we can write

$$y-y(h) = Bh+Ch^2+..., \tag{22}$$

there being no constant term if we assume that $y(h) \to y$ as $h \to 0$. This assumption is likely to be correct if the finite-difference equations themselves have meaning, that is in the absence of singularities and with sufficiently small values of h.

It is also reasonable to suppose that the terms on the right of (22) decrease in size, so that the term Bh is the major part of the error. If we calculate two approximate solutions, using the same simple equations but with different intervals h_1 and h_2, we can eliminate B and obtain

$$h_2\{y-y^{(1)}\}-h_1\{y-y^{(2)}\} = h_2(Ch_1^2+...)-h_1(Ch_2^2+...), \tag{23}$$

in which $y^{(r)}$ is written for $y(h_r)$. We therefore have an approximation, better than $y^{(1)}$ or $y^{(2)}$, given by

$$y = \frac{1}{h_2 - h_1}\{h_2 y^{(1)} - h_1 y^{(2)}\} - Ch_1 h_2 + \dots, \tag{24}$$

the next term being of order h^3.

We have thus reduced the error from $O(h)$ in (22) to $O(h^2)$ in (24). If we are prepared to compute a third approximation at interval h_3 we can eliminate both B and C and find a formula for y whose error is $O(h^3)$, and so on.

It is clear from (24) that for best precision the chosen h_2 should not be too close to h_1, and the usual method is to take $h_2 = \frac{1}{2}h_1$, a choice which has obvious advantages, for example, in the relaxation method. The better approximation is then given by

$$y = y^{(2)} + \{y^{(2)} - y^{(1)}\}. \tag{25}$$

10. In (22) we assumed that none of the coefficients $B, C, \dots,$ were zero, except possibly by accident. In practice it is very common for the coefficients of the odd powers to vanish, so that (22) becomes

$$y - y(h) = Ch^2 + Eh^4 + \dots . \tag{26}$$

If we now eliminate C from two approximate solutions, we find the better result

$$y = \frac{h_2^2 y^{(1)} - h_1^2 y^{(2)}}{h_2^2 - h_1^2}, \tag{27}$$

and if $h_2 = \frac{1}{2}h_1$ this reduces to

$$y = y^{(2)} + \tfrac{1}{3}\{y^{(2)} - y^{(1)}\}, \tag{28}$$

with error $O(h^4)$. The next approximation, obtained from three solutions with $h_2 = \frac{1}{2}h_1$, $h_3 = \frac{1}{2}h_2$, is given by

$$y = y^{(3)} + \tfrac{20}{45}\{y^{(3)} - y^{(2)}\} - \tfrac{1}{45}\{y^{(3)} - y^{(1)}\}, \tag{29}$$

the error in this case being $O(h^6)$.

We shall speak of (25) as an example of 'h-extrapolation', of (28) as an example of 'h^2-extrapolation', and so on, the power of the extrapolation depending on the first non-zero coefficient in (22). Mathematical investigation of the power of the extrapolation is complicated by the fact that the form of the differential equation and boundary conditions, and also the chosen method of solution, are all relevant factors. Examination of typical cases, however, will give some indication of the results to be expected in general.

11. Our object is to decide which, if any, of the coefficients in (22) are zero. If $y_r(h)$ is the approximate solution of the differential equation, that is the solution of the chosen difference equations, at pivotal point r and with interval h, then we can write (22) in the form

$$y_r(h) = y_r + h\frac{\partial y_r}{\partial h} + \frac{h^2}{2!}\frac{\partial^2 y_r}{\partial h^2} + \ldots, \tag{30}$$

y_r and its derivatives on the right being evaluated at $h = 0$, so that they refer to the exact solution of the differential equation and are independent of h. At the adjacent pivotal points $r+1$ and $r-1$ we have, by Taylor's series for a function of two variables, the equations

$$y_{r\pm1}(h) = y_r + h\left(\frac{\partial y_r}{\partial h} \pm \frac{\partial y_r}{\partial x}\right) + \frac{h^2}{2!}\left(\frac{\partial^2 y_r}{\partial x^2} \pm 2\frac{\partial^2 y_r}{\partial x \partial h} + \frac{\partial^2 y_r}{\partial h^2}\right) + \ldots. \tag{31}$$

Consider now the simple second-order equation

$$y'' + f(x)y' + g(x)y = k(x), \tag{32}$$

which we replace by approximate finite-difference equations typified by

$$(1+\tfrac{1}{2}hf_r)y_{r+1}(h) + (1-\tfrac{1}{2}hf_r)y_{r-1}(h) - (2-h^2 g_r)y_r(h) = h^2 k_r. \tag{33}$$

Substituting from (30) and (31) in (33) and equating to zero coefficients of successive powers of h, we find that the constant term and coefficient of h are identically zero, that of h^2 gives the relation

$$\frac{\partial^2 y_r}{\partial x^2} + f_r\frac{\partial y_r}{\partial x} + g_r y_r = k_r, \tag{34}$$

which is just the original differential equation, and successive later coefficients give the equations

$$\left.\begin{aligned}
\left(\frac{\partial^2}{\partial x^2} + f_r\frac{\partial}{\partial x} + g_r\right)\frac{\partial y_r}{\partial h} &= 0 \\
\left(\frac{\partial^2}{\partial x^2} + f_r\frac{\partial}{\partial x} + g_r\right)\frac{\partial^2 y_r}{\partial h^2} &= -\left(\frac{1}{6}\frac{\partial^4 y_r}{\partial x^4} + \frac{1}{3}f_r\frac{\partial^3 y_r}{\partial x^3}\right) \\
\left(\frac{\partial^2}{\partial x^2} + f_r\frac{\partial}{\partial x} + g_r\right)\frac{\partial^3 y_r}{\partial h^3} &= -\left(\frac{1}{2}\frac{\partial^4}{\partial x^4} + f_r\frac{\partial^3}{\partial x^3}\right)\frac{\partial y_r}{\partial h} \\
\ldots\ldots\ldots\ldots
\end{aligned}\right\}. \tag{35}$$

In these equations the differential operator on the left is always the same as that of the original differential equation, and on the right the terms are alternately functions of x only and homogeneous functions of some previous $\partial^s y_r/\partial h^s$. Their successive solution gives the required information about the coefficients $\partial y_r/\partial h$, $\partial^2 y_r/\partial h^2$, etc., of (30), the equivalents of B, C, etc., in (22).

The boundary conditions associated with (35) depend on those of the original problem. If in (32) y is specified at the two end points, then at these points it is independent of h, so that $\partial y/\partial h$, $\partial^2 y/\partial h^2$, etc., are all zero. We then see from (35) that in general $\partial y_r/\partial h$ and all odd derivatives $\partial^{2r+1}y/\partial h^{2r+1}$ are identically zero, $\partial^2 y_r/\partial h^2$ and all even derivatives $\partial^{2r}y/\partial h^{2r}$ are non-zero, and we have h^2-extrapolation.

The exceptional case arises when the homogeneous form of the original problem has an eigensolution, for then $\partial y_r/\partial h$ in (35) is not necessarily zero, and we might have only h-extrapolation. Since in these cases, however, any solution is not unique, the circumstance is hardly important.

12. As a simple example we take the problem defined by the equations

$$y''-y-x = 0, \qquad y(0) = 0, \qquad y(1) = 1, \tag{36}$$

solved by use of the difference correction in §§ 4–6 of Chapter 4. There we used the interval $h = \frac{1}{4}$ and produced the following first approximation:

$$y(\tfrac{1}{4}) = 0{\cdot}18023, \qquad y(\tfrac{1}{2}) = 0{\cdot}38735, \qquad y(\tfrac{3}{4}) = 0{\cdot}64993, \tag{37}$$

from difference equations typified by

$$y_1+y_{-1}-(2+h^2)y_0-h^2x_0 = 0. \tag{38}$$

If we had previously used the interval $h = \frac{1}{2}$, with only one internal point, we should have obtained the approximation $y(\frac{1}{2}) = 0{\cdot}38889$, and the assumption of h^2-extrapolation gives, from (28), a 'correction' of $-0{\cdot}00051$ to the value $0{\cdot}38735$, obtained at this point with the smaller interval, so that the extrapolated value is $0{\cdot}38684$.

If we use the smaller interval $h = \frac{1}{8}$ with equations (38), we find the following first approximation, from which the values at $x = \frac{1}{8}, \frac{3}{8}, \frac{5}{8}, \frac{7}{8}$ have been omitted:

$$y(\tfrac{1}{4}) = 0{\cdot}17999, \qquad y(\tfrac{1}{2}) = 0{\cdot}38695, \qquad y(\tfrac{3}{4}) = 0{\cdot}64957. \tag{39}$$

From (37) and (39), using h^2-extrapolation, we find the better results

$$y(\tfrac{1}{4}) = 0{\cdot}17991, \qquad y(\tfrac{1}{2}) = 0{\cdot}38682, \qquad y(\tfrac{3}{4}) = 0{\cdot}64945. \tag{40}$$

Finally, the use of (29), to produce an extrapolated value from the three approximations at $x = \frac{1}{2}$, gives the result $0{\cdot}38682$.

13. Returning to the general case, we must consider forms of boundary conditions other than 'y specified'. For example, with a general condition of the form

$$y'+ay = b, \tag{41}$$

we must examine the kinds of extrapolation resulting from our various ways of expressing this condition. A common method is to use the simplest central-difference expression for y', replacing (41) at the point 0 by

$$y_1(h)-y_{-1}(h)+2hay_0(h) = 2hb. \tag{42}$$

Substituting in (42) from (30) and (31) and equating to zero successive coefficients of powers of h, we find the equations

$$\left.\begin{aligned} \frac{\partial y_0}{\partial x}+ay_0 &= b \\ \left(\frac{\partial}{\partial x}+a\right)\frac{\partial y_0}{\partial h} &= 0 \\ \left(\frac{\partial}{\partial x}+a\right)\frac{\partial^2 y_0}{\partial h^2} &= -\frac{1}{3}\frac{\partial^3 y_0}{\partial x^3} \\ \cdots\cdots & \end{aligned}\right\}. \tag{43}$$

The first of these is the original boundary condition, and the remainder are now boundary conditions for use with (35). They show again that $\partial y_r/\partial h$ is identically zero unless the homogeneous form of the original problem has an eigensolution. The boundary conditions for all $\partial^{2r+1}y/\partial h^{2r+1}$ are also homogeneous, so that all odd derivatives vanish except in the critical case. Again, therefore, we have h^2-extrapolation.

14. A second favoured method of satisfying the boundary condition is to use Taylor's series, replacing (41) by

$$y_1(h) = A(h)y_0(h)+B(h), \tag{44}$$

where

$$\left.\begin{aligned} A &= 1-ah+\tfrac{1}{2}h^2(af_0-g_0)+\tfrac{1}{6}h^3(af'_0-af_0^2+f_0g_0+ag_0-g'_0)+\ldots \\ B &= bh+\tfrac{1}{2}h^2(k_0-bf_0)+\tfrac{1}{6}h^3(k'_0-bf'_0-f_0k_0+bf_0^2-bg_0)+\ldots \end{aligned}\right\}. \tag{45}$$

Substituting from (30), (31), and (45) in (44) and equating to zero coefficients of successive powers of h we find, after some reduction, the equations

$$\left.\begin{aligned} \left(\frac{\partial}{\partial x}+a\right)y_0 &= b \\ \left(\frac{\partial}{\partial x}+a\right)\frac{\partial y_0}{\partial h} &= 0 \\ \left(\frac{\partial}{\partial x}+a\right)\frac{\partial^2 y_0}{\partial h^2} &= -\left\{\frac{\partial^2}{\partial x^2}-(af_0-g_0)\right\}\frac{\partial y_0}{\partial h} \\ \cdots\cdots & \end{aligned}\right\}. \tag{46}$$

The first of (46) is just the original boundary condition, the remainder serving as one of the required boundary conditions for the successive equations (35). Again we find that $\partial y_r/\partial h$ is identically zero unless the homogeneous problem has an eigensolution, but $\partial^2 y_r/\partial h^2$ is not generally zero, and we have h^2-extrapolation.

15. We might also investigate the effects of other representations of the differential equation. Suppose, for example, that the differential equation lacks its first derivative in the form

$$y''+g(x)y = k(x), \tag{47}$$

and we use as approximations the finite-difference equations

$$y_{r+1}(1+\tfrac{1}{12}h^2g_{r+1})+y_{r-1}(1+\tfrac{1}{12}h^2g_{r-1})-(2-\tfrac{5}{6}h^2g_r)y_r = \tfrac{1}{12}h^2(k_{r+1}+k_{r-1}+10k_r). \tag{48}$$

Substituting for y_{r+1}, y_{r-1}, and y_r from (31) and (30), using Taylor's expansions for $g_{r\pm1}$ and $k_{r\pm1}$ and equating coefficients of powers of h, we find two identities, followed by

$$\frac{\partial^2 y_r}{\partial x^2}+g_r y_r = k_r, \tag{49}$$

which is just the original equation. Then comes

$$\left(\frac{\partial^2}{\partial x^2}+g_r\right)\frac{\partial y_r}{\partial h} = 0, \tag{50}$$

followed by
$$\left(\frac{\partial^2}{\partial x^2}+g_r\right)\frac{\partial^2 y_r}{\partial h^2} = \frac{1}{6}\frac{\partial^2}{\partial x^2}(k_r-y_r''-g_r y_r) = 0, \tag{51}$$

by virtue of the differential equation (47). Then comes

$$\left(\frac{\partial^2}{\partial x^2}+g_r\right)\frac{\partial^3 y_r}{\partial h^3} = -\frac{1}{2}\frac{\partial^2}{\partial x^2}\left(\frac{\partial^2}{\partial x^2}+g_r\right)\frac{\partial y_r}{\partial h}, \tag{52}$$

followed by

$$\left(\frac{\partial^2}{\partial x^2}+g_r\right)\frac{\partial^4 y_r}{\partial h^4} = \frac{1}{10}\frac{\partial^6 y_r}{\partial x^6}-\frac{\partial^2}{\partial x^2}\left(\frac{\partial^2}{\partial x^2}+g_r\right)\frac{\partial^2 y_r}{\partial h^2}. \tag{53}$$

Associated with the differential equations (50) to (53) for the quantities $\partial y_r/\partial h$, $\partial^2 y_r/\partial h^2$, $\partial^3 y_r/\partial h^3$, and $\partial^4 y_r/\partial h^4$, are boundary conditions depending on the original boundary conditions and on their representation. If the boundary conditions specify values of y, then all $\partial^r y/\partial h^r$ are zero at these points, and we see, from (50) to (53), that except in the presence of an eigensolution the first coefficient in (30) which does not vanish identically is that of h^4. In this case, therefore, we have h^4-extrapolation.

If a derivative is involved at the boundary, and we use the method of § 13, we find that $\partial^2 y_0/\partial h^2$ does not vanish at this boundary, so that we now have only h^2-extrapolation. If we use Taylor's series as in § 14, we find from (46) and its extensions that the boundary conditions are all homogeneous, and again we have h^4-extrapolation.

16. Before leaving second-order equations it is interesting to examine the sort of extrapolation resulting from the use of forward or backward differences for the satisfaction of derivative conditions at the boundary. Suppose, for example, that we use the equation

$$hy'_0 = (\Delta+\tfrac{1}{2}\Delta^2-\tfrac{1}{6}\Delta^3+\tfrac{1}{12}\Delta^4-\ldots)y_{-1} \qquad (54)$$

in its Lagrangian form

$$6hy'_0 = -2y_{-1}-3y_0+6y_1-y_2, \qquad (55)$$

in which the first neglected difference is $\Delta^4 y_{-1}$. If the boundary condition is

$$y'_0 = b, \qquad (56)$$

sufficiently general for our purpose, those for the functions $\partial y/\partial h$, $\partial^2 y/\partial h^2$, etc., are obtained by substitution of equations like (30) and (31) in (55) and (56), noting that, for y_2, equation (31) becomes

$$y_2(h) = y_0+h\left(\frac{\partial y_0}{\partial h}+2\frac{\partial y_0}{\partial x}\right)+\frac{h^2}{2!}\left(\frac{\partial^2 y_0}{\partial h^2}+2.2\frac{\partial^2 y_0}{\partial x\partial h}+2^2\frac{\partial^2 y_0}{\partial x^2}\right)+$$
$$+\frac{h^3}{3!}\left(\frac{\partial^3 y_0}{\partial h^3}+3.2\frac{\partial^3 y_0}{\partial x\partial h^2}+3.2^2\frac{\partial^3 y_0}{\partial x^2\partial h}+2^3\frac{\partial^3 y_0}{\partial x^3}\right)+\ldots. \qquad (57)$$

We then find, for the boundary conditions, the equations

$$\left.\begin{aligned} \frac{\partial y_0}{\partial x} &= b \\ \frac{\partial}{\partial x}\left(\frac{\partial y_0}{\partial h}\right) &= 0 \\ \frac{\partial}{\partial x}\left(\frac{\partial^2 y_0}{\partial h^2}\right) &= 0 \\ \frac{\partial}{\partial x}\left(\frac{\partial^3 y_0}{\partial h^3}\right) &= \frac{1}{2}\frac{\partial^4 y_0}{\partial x^4} \end{aligned}\right\}. \qquad (58)$$

These must be associated with the differential equations (35) or (50) to (53), according to the general finite-difference equations used. With equations (35), for example, we have h^2-extrapolation, but with (50) to (53) we have h^3-extrapolation, since, although the differential equation for $\partial^3 y/\partial h^3$ is homogeneous, its boundary condition is not.

17. Similar techniques can be used to establish the form of extrapolation for equations of higher order. In Chapter 6 we considered the linear equation

$$y^{\text{iv}}+f_1 y'''+f_2 y''+f_3 y'+f_4 y = f_5, \qquad (59)$$

associated with four boundary conditions, two at each end of the range, all of the form

$$g_0 y'''+g_1 y''+g_2 y'+g_3 y = g_4. \qquad (60)$$

For (59) we recorded the finite-difference equation

$$y_{r+2}(1+\tfrac{1}{2}hf_1-\tfrac{1}{12}h^2f_2-\tfrac{1}{12}h^3f_3)+y_{r+1}(-4-hf_1+\tfrac{4}{3}h^2f_2+\tfrac{2}{3}h^3f_3)+$$
$$+y_r(6-\tfrac{5}{2}h^2f_2+h^4f_4)+y_{r-1}(-4+hf_1+\tfrac{4}{3}h^2f_2-\tfrac{2}{3}h^3f_3)+$$
$$+y_{r-2}(1-\tfrac{1}{2}hf_1-\tfrac{1}{12}h^2f_2+\tfrac{1}{12}h^3f_3) = h^4f_5, \quad (61)$$

where y_s means $y_s(h)$, and all the functions f are evaluated at the pivotal point r. Substitution into (61) from equations like (30), (31), and (57) gives the results

$$\left.\begin{aligned}
&\left(\frac{\partial^4}{\partial x^4}+f_1\frac{\partial^3}{\partial x^3}+f_2\frac{\partial^2}{\partial x^2}+f_3\frac{\partial}{\partial x}+f_4\right)y_r = f_5\\
&\left(\frac{\partial^4}{\partial x^4}+f_1\frac{\partial^3}{\partial x^3}+f_2\frac{\partial^2}{\partial x^2}+f_3\frac{\partial}{\partial x}+f_4\right)\frac{\partial y_r}{\partial h} = 0\\
&\left(\frac{\partial^4}{\partial x^4}+f_1\frac{\partial^3}{\partial x^3}+f_2\frac{\partial^2}{\partial x^2}+f_3\frac{\partial}{\partial x}+f_4\right)\frac{\partial^2 y_r}{\partial h^2} = -\frac{1}{3}\frac{\partial^6 y_r}{\partial x^6}-\frac{f_1}{2}\frac{\partial^5 y_r}{\partial x^5}\\
&\cdots\cdots\cdots\cdots
\end{aligned}\right\}. \quad (62)$$

The first of (62) is just the original differential equation, and the rest give the differential equations for $\partial y_r/\partial h$, etc.

The boundary condition of type (60) is replaced in Chapter 6 by the finite-difference equation

$$y_2(h)(\tfrac{1}{2}g_0-\tfrac{1}{12}hg_1-\tfrac{1}{12}h^2g_2)+y_1(h)(-g_0+\tfrac{4}{3}hg_1+\tfrac{2}{3}h^2g_2)+$$
$$+y_0(h)(-\tfrac{5}{2}hg_1+h^3g_3)+y_{-1}(h)(g_0+\tfrac{4}{3}hg_1-\tfrac{2}{3}h^2g_2)+$$
$$+y_{-2}(h)(-\tfrac{1}{2}g_0-\tfrac{1}{12}hg_1+\tfrac{1}{12}h^2g_2) = h^3g_4. \quad (63)$$

Substitution for $y_2(h)$, etc., from (30), (31), and (57) provides boundary conditions for (62), given here by

$$\left.\begin{aligned}
&\left(g_0\frac{\partial^3}{\partial x^3}+g_1\frac{\partial^2}{\partial x^2}+g_2\frac{\partial}{\partial x}+g_3\right)y_0 = g_4\\
&\left(g_0\frac{\partial^3}{\partial x^3}+g_1\frac{\partial^2}{\partial x^2}+g_2\frac{\partial}{\partial x}+g_3\right)\frac{\partial y_0}{\partial h} = 0\\
&\cdots\cdots\cdots
\end{aligned}\right\}. \quad (64)$$

There are two conditions at each boundary point of type (64), and we see that they are homogeneous for $\partial y_0/\partial h$. The differential equation for $\partial y_r/\partial h$ is also homogeneous so that, with the usual exception involving an eigensolution, $\partial y_r/\partial h$ vanishes everywhere. On the other hand the differential equation for $\partial^2 y_r/\partial h^2$ is not homogeneous so that, whatever its boundary conditions may be, this function is not everywhere zero, and we have h^2-extrapolation.

18. This method is not confined to linear equations. As an illustrative example of its application to non-linear equations we consider the problem defined by the equations

$$\frac{d^2y}{dx^2}+\sin\left(\frac{dy}{dx}\right)+1 = 0, \qquad y(0) = 0, \qquad y(1) = 1, \tag{65}$$

solved in Chapter 4 by the effective use of a difference correction. The approximate finite-difference replacement of (65) is given by

$$y_{r+1}(h)+y_{r-1}(h)-2y_r(h)+h^2\sin\left\{\frac{y_{r+1}(h)-y_{r-1}(h)}{2h}\right\}+h^2 = 0, \tag{66}$$

and if we substitute in (66) from (30) and (31) we have the equation

$$\begin{aligned} h^2\frac{\partial^2 y_r}{\partial x^2}+h^3\frac{\partial^3 y_r}{\partial x^2\partial h}+\tfrac{1}{12}h^4\left(\frac{\partial^4 y_r}{\partial x^4}+\frac{6\partial^4 y_r}{\partial x^2\partial h^2}\right)+\dots+ \\ +h^2\sin\left\{\frac{\partial y_r}{\partial x}+h\frac{\partial^2 y_r}{\partial x\partial h}+\tfrac{1}{6}h^2\left(\frac{3\partial^3 y_r}{\partial x\partial h^2}+\frac{\partial^3 y_r}{\partial x^3}\right)+\right. \\ \left.+\tfrac{1}{24}h^3\left(\frac{4\partial^4 y_r}{\partial x\partial h^3}+\frac{4\partial^4 y_r}{\partial x^3\partial h}\right)+\dots\right\}+h^2 = 0. \end{aligned} \tag{67}$$

This equation can be expanded further to give the identity in h:

$$\begin{aligned} h^2\frac{\partial^2 y_r}{\partial x^2}+h^3\frac{\partial^3 y_r}{\partial x^2\partial h}+\tfrac{1}{12}h^4\left(\frac{\partial^4 y_r}{\partial x^4}+\frac{6\partial^4 y_r}{\partial x^2\partial h^2}\right)+\dots+ \\ +\left(\sin\frac{\partial y_r}{\partial x}\right)\left\{h^2-\tfrac{1}{2}h^4\left(\frac{\partial^2 y_r}{\partial x\partial h}\right)^2+\dots\right\}+ \\ +\left(\cos\frac{\partial y_r}{\partial x}\right)\left\{h^3\frac{\partial^2 y_r}{\partial x\partial h}+\tfrac{1}{6}h^4\left(\frac{3\partial^3 y_r}{\partial x\partial h^2}+\frac{\partial^3 y_r}{\partial x^3}\right)+\dots\right\}+h^2 = 0. \end{aligned} \tag{68}$$

Comparing coefficients of successive powers of h, we then find the equations

$$\left.\begin{aligned} &\frac{\partial^2 y_r}{\partial x^2}+\sin\left(\frac{\partial y_r}{\partial x}\right)+1 = 0 \\ &\left\{\frac{\partial^2}{\partial x^2}+\cos\left(\frac{\partial y_r}{\partial x}\right)\frac{\partial}{\partial x}\right\}\frac{\partial y_r}{\partial h} = 0 \\ &\left\{\frac{\partial^2}{\partial x^2}+\cos\left(\frac{\partial y_r}{\partial x}\right)\frac{\partial}{\partial x}\right\}\frac{\partial^2 y_r}{\partial h^2} = \sin\left(\frac{\partial y_r}{\partial x}\right)\left(\frac{\partial^2 y_r}{\partial x\partial h}\right)^2-\frac{1}{3}\cos\left(\frac{\partial y_r}{\partial x}\right)\frac{\partial^3 y_r}{\partial x^3}-\frac{1}{6}\frac{\partial^4 y_r}{\partial x^4} \\ &\cdots\cdots\cdots\cdots\cdots\cdots \end{aligned}\right\}. \tag{69}$$

The first of (69) is the original differential equation, the second is a homogeneous equation for $\partial y_r/\partial h$, the third a non-homogeneous equation for $\partial^2 y_r/\partial h^2$. The boundary conditions to be associated with (69) have already been discussed in some detail, and we conclude that the extrapolation is of order h^2. The exceptional event is the existence of an eigensolution of the equation

$$\frac{d^2z}{dx^2}+\cos\left(\frac{dy}{dx}\right)\frac{dz}{dx} = 0, \tag{70}$$

with the relevant boundary conditions. We see that, unlike the case of linear equations, this is not the homogeneous form of the original problem, but the homogeneous equation for small corrections to a first approximation (cf. Chapter 4, § 24).

19. We can also use the method for certain types of eigenvalue problems. Consider, for example, the equation

$$y''+f(x)y'+\{g(x)+\lambda\}y = 0, \tag{71}$$

with $y = 0$ at $x = x_0$ and x_n, the two ends of the range. It is required to find a value of λ for which y has non-trivial values. Here both y and λ are functions of h, though λ is independent of x, and we can write

$$\lambda(h) = \lambda+h\frac{\partial\lambda}{\partial h}+\frac{h^2}{2!}\frac{\partial^2\lambda}{\partial h^2}+\dots. \tag{72}$$

Using for (71) the finite-difference equation

$$(1+\tfrac{1}{2}hf_r)y_{r+1}+(1-\tfrac{1}{2}hf_r)y_{r-1}-\{2-h^2(g_r+\lambda)\}y_r = 0, \tag{73}$$

substituting from (30), (31), and (72), and equating to zero coefficients of successive powers of h, we find the equations

$$\left.\begin{aligned}
&\frac{\partial^2 y_r}{\partial x^2}+f_r\frac{\partial y_r}{\partial x}+(g_r+\lambda)y_r = 0\\
&\left(\frac{\partial^2}{\partial x^2}+f_r\frac{\partial}{\partial x}+g_r+\lambda\right)\frac{\partial y_r}{\partial h} = -y_r\frac{\partial\lambda}{\partial h}\\
&\left(\frac{\partial^2}{\partial x^2}+f_r\frac{\partial}{\partial x}+g_r+\lambda\right)\frac{\partial^2 y_r}{\partial h^2} = -y_r\frac{\partial^2\lambda}{\partial h^2}-2\frac{\partial\lambda}{\partial h}\frac{\partial y_r}{\partial h}-\frac{1}{3}f_r\frac{\partial^3 y_r}{\partial x^3}-\frac{1}{6}\frac{\partial^4 y_r}{\partial x^4}\\
&\dots\dots\dots\dots\dots\dots\dots\dots
\end{aligned}\right\}. \tag{74}$$

The boundary conditions on y may have various forms, but here we take the simplest case, $y = 0$ at each end of the range. The first of (74) is then the original problem, and the second gives a differential equation for $\partial y_r/\partial h$, with its boundary values zero.

Now a second-order equation of the form

$$f(D)y = k(x), \tag{75}$$

in which $f(D)z = 0$ has an eigenfunction z, and in which the boundary conditions for (75) are the same as those for the homogeneous equation, has a solution only if

$$\int_{x_0}^{x_n} \frac{1}{W} kz \, dx = 0, \tag{76}$$

where W is the Wronskian of two independent solutions of $f(D)z = 0$. In the case of (71) we have

$$W^{-1} = \exp \int f \, dx, \tag{77}$$

and, with $k = z\partial\lambda/\partial h$, equation (76) reduces to

$$\frac{\partial\lambda}{\partial h} \int_{x_0}^{x_n} \left(\exp \int f \, dx\right) z^2 \, dx = 0. \tag{78}$$

The integrand is always positive, so that (78) is satisfied only if $\partial\lambda/\partial h = 0$.

Returning to the second of (74), the solution for $\partial y_r/\partial h$ is then merely a multiple of the original y. Since, for any interval h, the solution $y(h)$ can be multiplied by any arbitrary constant, we cannot apply any extrapolation process unless the solutions are normalized in some way. Suppose, for example, that we give to y the value unity at some particular value of x, other than a node of the solution, and keep this value fixed for every interval h; then $\partial y/\partial h$ is zero at this point as well as at the two end points, and hence is zero everywhere.

We have now shown that both $\partial\lambda/\partial h$ and $\partial y/\partial h$ are zero. If we consider the third of (74) in relation to (76), we see no reason for the vanishing of $\partial^2\lambda/\partial h^2$ and therefore none for that of $\partial^2 y/\partial h^2$, so that the extrapolation is of order h^2 for both λ and y.

Equations with other types of boundary conditions and of higher orders can be treated similarly: h^2-extrapolation is the general rule.

20. The method of the deferred approach to the limit must be used with care, if full accuracy is desired in the solution. Even if the order of the extrapolation is known in theory, there is no information about the size of neglected coefficients in equations like (26), and no information about the chosen intervals as regards adequacy of finite-difference approximations to derivatives.

To be confident of the accuracy attained it is desirable to compute at least three solutions, at three different intervals, performing the extrapolation in two ways and comparing results. The method is then in general slower and less satisfactory than that using the difference correction, except possibly in the case of complicated non-linear equations in which the calculation of the difference corrections may be a heavy task.

Equations with eigensolutions

21. We have occasionally to solve equations like

$$y''+fy'+gy = k, \tag{79}$$

with boundary conditions of the form

$$\left.\begin{aligned} a_0 y(x_0)+b_0 y'(x_0) &= c_0 \\ a_n y(x_n)+b_n y'(x_n) &= c_n \end{aligned}\right\}, \tag{80}$$

when the homogeneous problem, defined by the equations

$$y''+fy'+gy = 0, \qquad a_0 y(x_0)+b_0 y'(x_0) = 0, \qquad a_n y(x_n)+b_n y'(x_n) = 0, \tag{81}$$

has a non-trivial eigensolution. If we use the matrix method we know that the structure of the matrix depends on the homogeneous problem, and in this case the matrix will be singular if the difference correction is zero, nearly singular in other cases. This singularity corresponds to the fact that if there is a solution to the problem it is not unique, for if $y^{(1)}$ is a particular integral of (79) which satisfies the boundary conditions (80), and $y^{(0)}$ is a solution of (81), then

$$y = Ay^{(0)}+y^{(1)} \tag{82}$$

is the general solution of (79) and (80), and the value of the constant A is indeterminate.

We assume, of course, that the problem has a solution, and in this connexion there must be suitable relations, of which (76) is a particular case, between the coefficients of the differential equation and the boundary conditions. For example the equation

$$y''+y = 1 \tag{83}$$

has the general solution

$$y = A\sin x+B\cos x+1, \tag{84}$$

and if the boundary conditions are $y(0) = 1, y(\pi) = 2$, they give the two incompatible equations

$$B+1 = 1, \qquad -B+1 = 2. \tag{85}$$

On the other hand, the conditions $y(0) = 2, y(\pi) = 0$ give the compatible equations

$$B+1 = 2, \qquad -B+1 = 0,$$

and the solution is

$$y = 1+\cos x+A\sin x. \tag{86}$$

The near-singularity of the finite-difference matrix would make it difficult to obtain a solution for any value of A. The indeterminacy in (82) and its particular case (86), however, allows us to find a particular solution by fixing the value of y at some point other than the boundaries: any point will do provided that $y^{(0)}$ in (82) or its particular form $\sin x$ in (86) does not vanish there, for otherwise A is still indeterminate.

Denoting this point by x_c, we have now two problems. The first is the solution of (79) with, as boundary conditions, the first of (80) at $x = x_0$ and a specified boundary value at $x = x_c$; the second is the solution of (79) with, as boundary conditions, the same specified value at $x = x_c$ and the second of (80) at $x = x_n$. Clearly none of the matrices involved are singular, and the solutions are easily obtainable and quite unique.

The remaining point of interest is whether the two solutions fit together, in the sense of being continuous, at the junction point x_c. The solutions of the differential equations certainly have this property, but the first approximations obtained with neglect of the difference corrections may not. We can, however, treat the two problems quite separately, obtaining the accurate solution for each, and their final combination will be continuous.

22. As an example we take the problem defined by (83) with $y(0) = 2$, $y(\pi) = 0$, and impose the extra condition that $y(\frac{1}{2}\pi) = 2$, giving $A = 1$ in (86) and the analytical solution

$$y = 1 + \cos x + \sin x. \tag{87}$$

In both regions the chosen finite-difference equations, with interval $h = \frac{1}{8}\pi$, are given by

$$\left.\begin{aligned} y_{r+1}+y_{r-1}-1{\cdot}8458y_r+Cy_r &= 0{\cdot}1542 \\ C &= -\tfrac{1}{12}\delta^4+\tfrac{1}{90}\delta^6-\dots \end{aligned}\right\}. \tag{88}$$

With neglect of Cy the first approximations in the two regions, and their extensions by step-by-step processes for the subsequent calculation of Cy, are given in Table 1. The values of Cy, using fourth and sixth differences, are also recorded.

Incorporation of Cy gives the corrections

$$y(\tfrac{1}{8}\pi) = -0{\cdot}0055, \quad y(\tfrac{1}{4}\pi) = -0{\cdot}0075, \quad y(\tfrac{3}{8}\pi) = -0{\cdot}0055,$$
$$y(\tfrac{5}{8}\pi) = -0{\cdot}0006, \quad y(\tfrac{3}{4}\pi) = +0{\cdot}0000, \quad y(\tfrac{7}{8}\pi) = +0{\cdot}0006,$$

and a second correction has a maximum of one unit. The full solution is assembled, together with its differences, on the left of Table 2. We see that the whole table differences perfectly smoothly. On the right of Table 2 are shown the differences of the assembled first approximations: the lack of continuity at the centre of the range is demonstrated.

TABLE 1

$8x/\pi$	y		δ^2		δ^4		δ^6	10^4Cy
−2	+0·9855							
		+5484						
−1	1·5339		− 823					
		4661		−719				
0	2·0000		1542		+238			
		3119		−481		+74		
+1	2·3119		2023		312		−48	−26·5
		+1096		−169		+26		
2	2·4215		2192		338		−52	−28·7
		−1096		+169		−26		
3	2·3119		2023		312		−48	−26·5
		3119		+481		−74		
4	2·0000		1542		+238			
		4661		+719				
5	1·5339		− 823					
		−5484						
+6	+0·9855							

$8x/\pi$	y		δ^2		δ^4		δ^6	10^4Cy
+2	+2·4069							
		−1029						
3	2·3040		−2011					
		3040		+469				
4	2·0000		−1542		+237			
		4582		706		−107		
5	1·5418		− 836		+130		−23	−11·1
		5418		836		130		
6	1·0000		0		0		0	0·0
		5418		836		130		
7	0·4582		+ 836		−130		+23	+11·1
		4582		706		−107		
8	+0·0000		+1542		−237			
		3040		+469				
9	−0·3040		+2011					
		−1029						
+10	−0·4069							

TABLE 2

$8x/\pi$	y		δ^2		δ^4		δ^6
+0	+2·0000						
		+3065					
1	2·3065		−1989				
		+1076		−163			
2	2·4141		2152		+326		
		−1076		+163		−23	
3	2·3065		1989		303		−47
		3065		466		70	
4	2·0000		1523		233		−38
		4588		699		108	
5	1·5412		−824		125		−17
		5412		824		−125	
6	1·0000		0		+0		
		5412		+824			
7	0·4588		+824				
		−4588					
+8	+0·0000						

$8x/\pi$	y		δ^2		δ^4		δ^6
+0	+2·0000						
		+3119					
1	2·3119		−2023				
		+1096		−169			
2	2·4215		2192		+338		
		−1096		+169		+53	
3	2·3119		2023		391		−377
		3119		560		−324	
4	2·0000		1463		67		+466
		4582		627		+142	
5	1·5418		−836		209		−351
		5418		836		−209	
6	1·0000		0		+0		
		5418		+836			
7	0·4582		+836				
		−4582					
+8	+0·0000						

The equation $y''+f(x)y = g(x)$, with one boundary at infinity

23. We have given several examples of the solution of this equation subject to given two-point boundary conditions, but most of them have involved functions $f(x)$ which were of one sign throughout the range of integration. In these circumstances the complementary functions had the same type of behaviour throughout the range, oscillatory for positive f, exponential for negative f. We have seen also, in Chapter 8, that step-by-step methods for boundary-value problems are more convenient in the former case than in the latter, while the converse is true of boundary-value techniques. We now consider some aspects of this equation, which has many physical contexts, when the sign of f varies in the range of integration and, in particular, when the range is of infinite extent.

Case $g(x) = 0, f(\infty) = k^2$

24. We consider first the case $g(x) = 0$, so that the equation becomes

$$y''+f(x)y = 0, \tag{89}$$

and we take $f(x)$ to have the form $k^2+O(x^{-2})$ as $x \to \infty$. Then there is a solution, everywhere well behaved, such that, for large values of x,

$$y = A\sin(kx+\epsilon), \tag{90}$$

and a common problem is the determination of the constant ϵ. We shall assume here that the well-behaved solution vanishes at the origin, and since the equation is homogeneous we can fix arbitrarily the value of y at any other non-nodal point.

There are at least three possible numerical methods of solution. First, we could use throughout a step-by-step method, with the initial conditions $y(0) = 0$, $y(h) = 1$, say. Second, we could use a boundary-value method throughout, taking $y(0) = 0$ and $y(X) = 1$, say, at a large value X of x. Third, we could use a combination of step-by-step and boundary-value techniques. The chosen method depends considerably on the behaviour of $f(x)$.

25. We take first the step-by-step method. If f is everywhere positive this is both convenient and accurate, and it is not difficult to decide how far to proceed so that y has settled down to the required form (90). One can, for example, assume that form at a certain stage, calculate ϵ as a function of x for several consecutive pivotal points, and stop when ϵ is constant to the required precision.

If f is initially negative, with one change of sign in the infinite range, the solution in the first region is of exponential type, and the step-by-step method is not quite so attractive, at least from the point of view of convenience. Full accuracy, however, can still be maintained. For if we start with $y(0) = 0$, $y(h) = 1$ and perform the step-by-step process, the errors introduced will not grow faster than the function itself, so that we can work throughout with a fixed number of significant figures, discarding end figures from time to time, and still have full accuracy in the significant figures.

Consider, in this respect, the solution (see § 40, Chapter 8) of the finite-difference equations

$$y_{r+1}+y_{r-1}-50{\cdot}0625y_r = 0, \tag{91}$$

with $y_0 = 0$, $y_5 = 1$. Table 29 of Chapter 8 gives, in the column headed $y^{(2)}$, a solution satisfying the first boundary condition, obtained by working with a constant number of decimals. If we repeat the calculation, working with a constant number of six significant figures, we find the values y_r shown in Table 3.

TABLE 3

x_r	y_r	y_r (*normalized*)
0	0·000000	0·000000
1	1·000000	·000000
2	50·0625	·000008
3	2505·25	·000399
4	125369	0·019983
5	6273780	1·000000

If we continue this process till we reach the end of the exponential region we can then normalize the results, dividing throughout by the largest pivotal value so that this becomes unity, and can then proceed, working with decimals rather than significant figures. All the values given in y_r (normalized) in Table 3 are certainly correct, and we could have recorded up to six significant figures without serious error.

This is not, however, a completely satisfactory method, at least for use with desk calculating machines, with which it is more convenient to work to a fixed number of decimals. Also it is not easy to decide when to discard a figure: for example in Table 3 we worked ruthlessly with six figures, but the value y_4 has only just this number, and y_5 has a consequent error in its sixth figure. This is rather an extreme example, however, and in many cases the step-by-step method is satisfactory.

26. We consider now the second possibility, the use of boundary-value methods throughout. If f is everywhere positive, giving oscillatory solutions, the relaxation method would hardly be satisfactory, and this applies also to any region of positive f. The same lack of attraction would seem to apply to the matrix method also, but further consideration yields compensating advantages.

In § 45 of Chapter 8 we found that, when the matrix corresponding to a second-order equation is decomposed into its triangles L and U, the ratio u_r/v_r of the elements of the U matrix is equal to the ratio $-y_{r+1}/y_r$ of successive values of the solution of the homogeneous equation whose initial value is zero. This is, of course, precisely our requirement, so that the triangular resolution alone effectively provides the required answer and there is no necessity, present in the use of the relaxation method, for fixing a second boundary value. In the triangles we need only the u_r, since v_r is equal to the corresponding element of the original matrix, and the u_r^{-1} is most easily found from the continued fraction (94) of Chapter 9.

The behaviour of the successive u_r is obvious. In exponential regions their values will have constant sign with magnitude probably decreasing monotonically. In oscillatory regions u_r will soon pass through a zero, and oscillate with varying period. In the case when f has one change of sign, from negative to positive, and is ultimately constant, the oscillation in u_r has ultimately a constant period. When this stage is reached, or soon after, we can easily calculate the ϵ of (90) from the equation

$$u_r^{-1}v_r = -\sin(kx_r+\epsilon)/\sin(kx_{r+1}+\epsilon), \tag{92}$$

and check that this stage has been reached by evaluating ϵ from (92) used at various values of r. For this purpose we can express (92) in the more convenient form

$$\tan\epsilon = -\frac{\sin kx_r + u_r^{-1}v_r \sin kx_{r+1}}{\cos kx_r + u_r^{-1}v_r \cos kx_{r+1}}. \tag{93}$$

Since u_r may pass close to zero in the oscillatory part, it would appear to be necessary to use the special methods and formulae of § 44 of Chapter 9, to avoid consequent loss of accuracy. We are not interested, however, in z_r or explicitly in y_r, and reference to that section shows that in fact we need no special formulae and need not record the small u_r.

This method is attractive provided there is no difference correction. The fact that we have nowhere recorded y_r or its differences, however, makes it difficult to estimate the adequacy of the chosen finite-difference formulae. We can, of course, use for (89) formulae involving a very small difference correction, and we could also use the method of the deferred approach to the limit for the final calculation of ϵ.

If a difference correction is applied in the matrix method much of its attraction disappears, for then we should have to record the first approximation y_r, and in the second approximation both z_r and y_r would have to be calculated, involving all the special formulae.

27. The third possibility uses both step-by-step methods and boundary-value techniques, and is most attractive when $f(x)$ has a single change of sign, from negative to positive. Any y_r can be given a specified value. Suppose this value is fixed at the pivotal point x_c nearest to the point at which $f(x)$ passes through its zero. In the first region we then have a pure boundary-value problem, with exponential-type solutions, and we can solve this by relaxation or matrix methods with full accuracy and convenience. From the solution so obtained we take in particular y_c and y_{c-h} and use them as initial values for a step-by-step method, extending the solution by this means as far as necessary in the oscillatory region.

In the exponential region we could calculate a first approximation, and correct it by inclusion of the difference correction, before any work was started in the oscillatory region. It is better, however, to calculate the first approximation in the second region before applying the difference correction in the first. Since the same finite-difference approximations are used throughout, in both regions, the complete solution will difference smoothly through the junction and the difference correction can be calculated everywhere. The latter can then be used in the exponential region, still keeping y_c fixed, to correct the first approximation and in particular to give a new starting value y_{c-h} for use with the step-by-step method, in which the difference correction in the second region is also included. If the complete solution were obtained in the first region before calculating the first approximation in the second, there would be a bump in the differences near the junction, and the calculation of the difference correction would be less satisfactory. This is the converse of the phenomenon of § 22.

28. As an illustration of the combined use of two techniques we take the equation

$$y''+\left(1-\frac{2}{x^2}\right)y = 0, \tag{94}$$

with $y = 0$ at $x = 0$. This is of the type already discussed, with one change of sign in f, giving a short exponential region followed by an oscillatory region. The change-over point is at $x = \sqrt{2}$, so that we fix $y = 1$ at $x = 1{\cdot}5$, use the interval $h = 0{\cdot}25$ and, to ensure for our illustrative purpose a non-negligible difference correction, we use throughout the finite-difference equation

$$\left.\begin{aligned} y_1+y_{-1}-\left\{2-h^2\left(1-\frac{2}{x_0^2}\right)\right\}y_0+Cy_0 = 0 \\ C = -\tfrac{1}{12}\delta^4+\tfrac{1}{90}\delta^6-\ldots \end{aligned}\right\}. \tag{95}$$

In the range $x = 0(0{\cdot}25)1{\cdot}5$ we use relaxation methods, ignore the difference correction for the moment, and obtain the values shown on the left of Table 4. These are extended by step-by-step methods into the oscillatory region, as far as the convenient stopping point $x = 3{\cdot}5$, and the values are also given on the left of Table 4, together with all the significant differences and the calculated values of the difference corrections Cy.

TABLE 4

x	y		δ^2		δ^4	$10^4 Cy$	η
+0·00	+0·0000		+692		−44		+0·0000
		+ 346		− 22			
0·25	·0346		670		52	+4·3	·0002
		1016		74			
0·50	·1362		596		47	3·9	·0005
		1612		121			
0·75	·2974		475		37	3·1	·0007
		2087		158			
1·00	·5061		317		28	2·3	·0007
		2404		186			
1·25	0·7465		+131		14	1·2	·0004
		2535		200			
1·50	1·0000		− 69		− 2	+0·2	+ ·0000
		2466		202			
1·75	1·2466		271		+15	−1·3	− ·0004
		2195		187			
2·00	1·4661		458		25	2·1	·0006
		1737		162			
2·25	1·6398		620		38	3·2	·0006
		1117		124			
2·50	1·7515		744		46	3·8	− ·0002
		+ 373		78			
2·75	1·7888		822		52	4·3	+ ·0005
		− 449		− 26			
3·00	1·7439		848		+56	−4·7	·0015
		−1297		+ 30			
3·25	1·6142		−818				+0·0029
		−2115					
+3·50	+1·4027						

In the exponential region the difference corrections become new residuals, whose liquidation gives the increments η at $x = 0{\cdot}25$, $0{\cdot}50$, $0{\cdot}75$, $1{\cdot}00$, and $1{\cdot}25$, and with the last value and the condition $\eta = 0$ at $x = 1{\cdot}5$ we can restart the step-by-step method, now incorporating the rest of the difference correction. The corrections η are shown on the right of Table 4.

We can clearly extend the solution indefinitely by this process. In this example, of course, the exponential region is quite short, and the step-by-step method could be used throughout without great inconvenience.

If $f(x)$ has more than one change of sign it may be preferable to use only one technique throughout the range. The chosen one will depend on the relative lengths of the exponential and oscillatory regions.

Case $g(x) = 0, f(\infty) = -k^2$

29. In this case the solution for large values of x is of exponential form and it is usually required to find a solution which is zero at infinity. This means that the equation and initial condition are such that the increasing exponential solution is suppressed, and we either have to find an eigenvalue and eigenfunction, or just the latter when the eigenvalue is already known.

In Chapter 7 we used relaxation methods to find an eigenvalue λ of the problem defined by the equations

$$y''+\left(\frac{2}{x}-\lambda\right)y = 0, \qquad y(0) = y(\infty) = 0. \tag{96}$$

Instead of the second boundary condition we suggested two more convenient possibilities. In the first we fixed $y(X) = 0$ at some large value X of x, and in the second we used a less severe condition at X based on an estimate of the behaviour of the solution for large x.

It is also of interest to examine the use of step-by-step and matrix methods for the determination of the eigenvalue when the range is of infinite extent.

30. The step-by-step method depends on finding a value of λ such that successive values of y ultimately decrease numerically and never again increase. The introduction of rounding errors at various stages of the process ensures that, working to a finite number of figures, solutions for all λ will ultimately diverge to $\pm\infty$, and a refined technique, not properly falling within the scope of this volume, is necessary to find an accurate solution.

31. The matrix method has similar difficulties. If the increasing exponential solution is to be excluded, the diagonal terms u_r of the U matrix, being the ratios $-y_{r+1}/y_r$ of the step-by-step solution, must never exceed unity, and should ultimately tend to a constant value if a fixed number of decimals is used in the calculation. This value is the smaller of the two roots of an equation corresponding to (40) of Chapter 9. For all values of λ other than the eigenvalues of the finite-difference equations, some multiple of the increasing exponential solution is present, and the u_r will soon begin to approach the larger root.

In previous discussions of eigenvalue problems we decided that the criterion for an eigensolution was the singularity of the corresponding finite-difference matrix. If the range and therefore also the matrix is finite, there is singularity only if the last diagonal term u_n of U is zero. Here, however, the matrix is infinite, and singular only if there is an infinity of u_r all less than unity. This corresponds to the suppression of the increasing exponential solution.

With the use of the eigenvalue in the finite-difference equations we should in theory reach the desirable state of an infinity of u_r all less than unity, but in practice, through the necessity of working with a finite number of figures, we shall introduce rounding errors which grow in exponential regions, and the u_r will ultimately and inevitably change from the smaller to the larger root of the relevant quadratic equation. In the circumstances we have to deduce the eigenvalue from a consideration of the relative behaviour of the u_r for various values of λ.

As an example we take the problem of equation (96), use the interval $h = \frac{1}{2}$ and the finite-difference equation

$$y_{r+1}+y_{r-1}-\left\{2-\frac{1}{2x}+\tfrac{1}{4}\lambda\right\}y_r = 0, \tag{97}$$

and record successive u_r for various values of λ. For $\lambda = 0{\cdot}9$ we find the following values of $-u_r$:

1·2250, 0·9087, 0·7912, 0·7111, 0·6187, 0·4420, −0·1803, 7·6463, 1·9831, 1·6207, 1·5171,... .

If the corresponding solution y is initially positive, it clearly passes through a zero and then diverges to $-\infty$, so that an increasing exponential solution with negative coefficient is undoubtedly present. With $\lambda = 1{\cdot}0$ the values of $-u_r$ are as follows:

1·2500, 0·9500, 0·8641, 0·8427, 0·8633, 0·9250, 1·0260, 1·1503, 1·2696, 1·3624, 1·4251,... .

Here the solution y is everywhere positive, and diverges to $+\infty$: the increasing solution has a positive coefficient, and we infer that there is an eigenvalue of the finite-difference equations between $\lambda = 0{\cdot}9$ and $\lambda = 1{\cdot}0$.

We can bracket the eigenvalue still more closely and obtain the following results:

($\lambda = 0{\cdot}9443$)

$-u$: 1·2361, 0·9271, 0·8241, 0·7727, 0·7419, 0·7215, 0·7072, 0·6971, 0·6905, 0·6879, 0·6915,...,

($\lambda = 0{\cdot}9441$)

$-u$: 1·2360, 0·9269, 0·8238, 0·7721, 0·7408, 0·7195, 0·7033, 0·6891, 0·6737, 0·6517, 0·6107, 0·5152.

For both these λ's we have a fairly long string of small u_r: in the first case the solution is clearly diverging to $+\infty$, and in the second will soon pass through a zero and then diverge to $-\infty$. We infer that the solution lies between the two, and with $\lambda = 0{\cdot}9442$, and working with an extra figure, we find the following values of $-u$:

1·23605, 0·92702, 0·82399, 0·77244, 0·74145, 0·72067, 0·70559, 0·69380, 0·68360, 0·67321, 0·65972, 0·63693, 0·58910, 0·46712, 0·02860....

This solution will also diverge to $-\infty$, but has the required form for a larger range, and we can assert with some confidence that the eigenvalue is 0·9442 to four decimals.

We cannot, however, be very sure of the accuracy of our eigenfunction over the latter part of the range, and this uncertainty is inevitable in exponential regions with techniques which are essentially of a step-by-step variety. The matrix method, of course, is in this category when it is not supplied with a specific (numerical) boundary condition at both ends of a finite range.

32. The difficulty disappears if we can apply this second boundary condition. In the use of the relaxation method we suggested two alternatives, of which the second, involving an estimate of the behaviour of the solution for large x, is more representative of the true state of affairs than the ruthless specification of a zero at large x. We cannot use with this the boundary condition at $x = 0$, for the matrix is still effectively singular. We can, however, use the method of § 27, fixing y at some point, preferably at the junction of the oscillatory and exponential regions, solve separate problems in the two regions, and examine the continuity at the junction to determine the accuracy of the alleged eigenvalue.

In the exponential region we can use matrix methods with no worry regarding accuracy. For example, if for equations (96) and (97) we fix $y = 1$ at the junction point $x = 2$, corresponding roughly to a suspected eigenvalue of about 0·9442, and use at $x = 10$ the condition

$$y(10) = 0{\cdot}61517y(9{\cdot}5), \tag{98}$$

corresponding to equation (137) of Chapter 7 for this value of λ, we then have to solve the equations

$$\left.\begin{aligned} -2{\cdot}03605y(2{\cdot}5)+y(3{\cdot}0) \qquad\qquad\qquad\qquad &= -1 \\ y(2{\cdot}5)-2{\cdot}06938y(3{\cdot}0)+y(3{\cdot}5) \qquad\qquad &= 0 \\ \cdot\quad\cdot\quad\cdot\quad\cdot\quad\cdot\quad\cdot\quad\cdot\quad\cdot\quad\cdot\quad\cdot\quad\cdot\quad\cdot \qquad & \\ y(8{\cdot}5)-2{\cdot}18049y(9{\cdot}0)+y(9{\cdot}5) &= 0 \\ y(9{\cdot}0)-1{\cdot}56825y(9{\cdot}5) &= 0 \end{aligned}\right\} \tag{99}$$

in the last of which the condition (98) is incorporated.

The behaviour of the diagonal terms of the U matrix is now all that could be desired. The matrix is not singular, and since the u_r are now ratios of successive terms of the solution which starts at zero at $x = 2$, and in which there is no reason to expect the vanishing of the increasing solution, their values are all greater than unity except the very last term, which is as large as 0·9. The increasing solution is suppressed, of course, by the last equation of (99) and not, as previously when starting from $x = 0$, by the structure of the differential equation. By the usual matrix method we calculate the solution shown in Table 5.

TABLE 5

x	2·0	2·5	3·0	3·5	4·0	4·5	5·0	5·5	6·0
y	1·00000	0·77256	0·57298	0·41315	0·29182	0·20290	0·13934	0·09473	0·06387

x	6·5	7·0	7·5	8·0	8·5	9·0	9·5	(10·0)
y	0·04276	0·02845	0·01883	0·01239	0·00810	0·00525	0·00335	(0·00206)

In the range $x = 0{\cdot}0(0{\cdot}5)2{\cdot}0$ we have two possibilities. The first is to carry out a step-by-step process starting with $y(0) = 0$, $y(\frac{1}{2}) = 1$, normalizing the solution so obtained so that $y(2) = 1$. The results of this procedure are shown in Table 6.

TABLE 6

x	y	y (*normalized*)
+0·0	+0·00000	+0·00000
0·5	1·00000	1·05914
1·0	1·23605	1·30915
1·5	1·14584	1·21360
+2·0	+0·94416	+1·00000

If $\lambda = 0{\cdot}9442$ were a correct eigenvalue of the difference equations, the solution of Table 6 would fit smoothly with that of Table 5. The easiest way to test this is to find the residual of the finite-difference equation at $x = 2{\cdot}0$, given by

$$R(2{\cdot}0) = y(1{\cdot}5)+y(2{\cdot}5)-1{\cdot}98605y(2{\cdot}0) = 0{\cdot}00011. \qquad (100)$$

We decide that both eigenvalue and eigenfunction are accurate to four decimals.

The second possibility is to carry the solution back from $x = 2{\cdot}0$ to the origin by a step-by-step process, starting with the values $y(2{\cdot}5)$ and $y(2{\cdot}0)$ of the matrix solution. The results of this procedure are shown in Table 7.

TABLE 7

x	2·5	2·0	1·5	1·0	0·5	0·0
y	0·77256	1·00000	1·21349	1·30893	1·05888	−0·00010

With the true eigenvalue we should have $y(0{\cdot}0) = 0$, but again we can assert that our approximate λ is fairly close to the truth.

Repetitions of these techniques for various values of λ will enable the eigenvalue to be obtained as accurately as desired. We can also speed the convergence of this method by solving a variational equation, as in Chapter 8, but using matrix methods in appropriate regions rather than step-by-step methods throughout. The process of matching at $x = 2{\cdot}0$ is an example of 'matching in the middle' mentioned in Chapter 8.

33. Several other points deserve comment. First, the difference correction can be included at any convenient stage, though in this example the interval is too large to permit its accurate computation near the origin. If the variational method is used in the calculation, the inclusion of the difference correction is effected by suitable modifications of the method given in §§ 25 and 26 of Chapter 8. Second, in this example the first approximation in the exponential region could be calculated from (99) by a backward step-by-step method, starting with $y(10) = 1$, say. The next value $y(9{\cdot}5)$ comes from (98), and successive values by successive use of (99), starting from the bottom. As in the example of § 25, end figures could be discarded from time to time, and five-figure accuracy achieved everywhere. When the difference correction is incorporated, however, this method is unsatisfactory since it would be necessary, as in § 40 of Chapter 8, to calculate two trial solutions and use throughout five decimals, involving perhaps many significant figures. Third, we may note that we did not use the correct boundary condition at our representation of infinity: we assumed that the solution behaved there like $e^{-x\sqrt{\lambda}}$, whereas the true behaviour is like $f(x)e^{-x\sqrt{\lambda}}$, where $f(x)$ is a slowly varying function (the actual eigenfunction is xe^{-x}). This would have a small effect on the accuracy of our eigenfunction for large x, but the eigenvalue is hardly affected, as can be seen by comparing our solution with that of Chapter 7, in which the second boundary condition was quite different.

Finally, we mention an extension of these techniques to another problem of practical significance, in which $f(x)$ changes sign twice in the range 0 to ∞. Here $f(x)$ is initially negative, goe through zero at $x = x_1$, stays positive as far as $x = x_2$, then goes negative and remains negative up to $x = \infty$. We thus have an oscillatory region sandwiched between two exponential regions. In this case the technique needs only slight amendments. We fix y at $x = x_2$, solving the equation between x_2 and X, representing infinity, by matrix methods. This solution is then continued backwards to $x = x_1$ by a step-by-step method, and with the value so obtained at $x = x_1$, and the given value zero at the origin, we calculate the solution in $(0, x_1)$ by matrix methods. The matching point is at $x = x_1$. The points x_1 and x_2 may of course be functions of λ, but their precise position is not of great importance.

34. If the eigenvalue of the differential equation is already known, and we merely want the eigenfunction, the problem is simplified. We must remember, however, that the eigenvalue of the difference equations is not identical with that of the differential equation, so that with our first approximation, neglecting the difference correction, our continuity or other matching conditions are violated. The simplification lies in the knowledge that the lack of continuity has nothing to do with λ, and will disappear when the difference corrections are included.

Case $g(x) \neq 0$

35. Little comment is needed in the case $g(x) \neq 0$. If the complementary function is of oscillatory type at infinity we are not likely to be embarrassed by eigensolutions, and $g(x)$ might have any form. Where possible, we would use matrix methods in exponential regions, step-by-step methods in oscillatory regions.

If the complementary function is of decreasing exponential type at infinity, there is likely to be an eigensolution so that $g(x)$ will have special form to permit solution of the complete equation. The technique of §§ 21 and 22, suitably modified by the remarks of this section, will give a solution without difficulty.

Tabulation of functions for large arguments

†36. Many mathematical functions behave at infinity in a way which makes their direct tabulation difficult or even impossible. For example the exponential integrals, defined by the definite integrals

$$\mathrm{Ei}(x) = \int_{-\infty}^{x} t^{-1}e^{t}\,dt, \qquad -\mathrm{Ei}(-x) = \int_{x}^{\infty} t^{-1}e^{-t}\,dt, \tag{101}$$

have for large argument the forms

$$\left.\begin{aligned} \mathrm{Ei}(x) &= x^{-1}e^{x}S(x) \\ -\mathrm{Ei}(-x) &= x^{-1}e^{-x}S(-x) \end{aligned}\right\}, \tag{102}$$

where

$$S(x) \sim 1 + \frac{1!}{x} + \frac{2!}{x^2} + \cdots. \tag{103}$$

This series is divergent, but for large x its terms first decrease to a minimum before beginning to increase, and it is asymptotic in the sense that if we stop at the smallest term the value so obtained differs from $S(x)$

by a quantity of the order of magnitude of that term: in this case it can be shown, moreover, that the error is certainly less than that term.

The tabulation of $S(x)$ is clearly the main necessity before that of $\mathrm{Ei}(x)$ and $\mathrm{Ei}(-x)$ can be regarded as complete. If we are to take the tabulation 'all the way to infinity', however, we must use an auxiliary argument. The most obvious choice is $z = x^{-1}$, so that the range $x = 10$ to ∞ is covered by $z = 0{\cdot}1$ to 0. We note, moreover, that with this argument $S(z^{-1})$ and $S(-z^{-1})$ are continuous through $z = 0$, the series

$$S(z^{-1}) = 1+1!\,z+2!\,z^2+... \tag{104}$$

being applicable for both positive and negative z. This is effectively the Taylor's series, but it is divergent for all $z \neq 0$; its radius of convergence is zero.

If (104) is used for the calculation of S, the precision obtainable is limited by the size of the smallest term. At $z = 0{\cdot}1$, for example, the smallest term in (104) is $10!(0{\cdot}1)^{10} = 0{\cdot}00036...$ so that, with the use of (104) only, four decimals in S could hardly be guaranteed at this point.

37. An alternative and attractive method is to solve a differential equation for S. With $z = x^{-1}$ as independent variable, the equation is given by

$$S''+\frac{2}{z}S'-\left(\frac{1}{z^4}-\frac{2}{z^3}\right)S+\left(\frac{1}{z^4}-\frac{1}{z^3}\right) = 0. \tag{105}$$

If we change the dependent variable to $T = z(S-1)$ we remove the first-derivative term and have for T the differential equation

$$T''-\left(\frac{1}{z^4}-\frac{2}{z^3}\right)T+\frac{1}{z^2} = 0, \tag{106}$$

and the 'Taylor's series' for T is given by

$$T = z^2+2!\,z^3+3!\,z^4+... . \tag{107}$$

If we can calculate $T(z)$ by other means, say by ascending series or quadrature, at $z = -0{\cdot}1$ and $z = 0{\cdot}1$, corresponding to $x = -10$ and 10, we have a boundary-value problem and can calculate T at intermediate pivotal points. The simplest finite-difference equation corresponding to (106) is given by

$$\left.\begin{aligned} T_{r+1}+T_{r-1}-\left\{2+h^2\left(\frac{1}{z_r^4}-\frac{2}{z_r^3}\right)\right\}T_r+\frac{h^2}{z_r^2}+CT_r &= 0 \\ C = -\tfrac{1}{12}\delta^4+\tfrac{1}{90}\delta^6-... & \end{aligned}\right\}. \tag{108}$$

The coefficient of T_r is always greater than 2 in the range $z = -0{\cdot}1$ to $+0{\cdot}1$, so that the set of linear equations is ideal for relaxation or matrix methods. At the origin of z we can overcome apparent difficulties by using (107) in the equation (108) for $r = 0$, obtaining the simple equation

$$T_1+T_{-1}-2h^2+CT_0 = 0. \tag{109}$$

This method has been used to calculate T for $z = -0{\cdot}10(0{\cdot}01)0{\cdot}10$. The first approximation differences perfectly, the difference corrections are easily calculable and two applications of the process give results correct to nine decimal places.

We could use instead of (108) the more powerful formula

$$\left.\begin{aligned}(1-\tfrac{1}{12}h^2f_{r+1})T_{r+1}+(1-\tfrac{1}{12}h^2f_{r-1})T_{r-1}-\\ -(2+\tfrac{5}{6}h^2f_r)T_r+\tfrac{1}{12}h^2(g_{r+1}+g_{r-1}+10g_r)+CT_r = 0\\ C = \tfrac{1}{240}\delta^6-\tfrac{13}{15120}\delta^8+\ldots\end{aligned}\right\}, \tag{110}$$

where

$$f = \frac{1}{z^4}-\frac{2}{z^3}, \qquad g = \frac{1}{z^2}. \tag{111}$$

In this case the difference correction has to be applied only once, but the arithmetic is heavier both by relaxation and matrix methods.

The asymptotic behaviour of many other functions can be evaluated in a similar manner. One example has already been given at the end of Chapter 5.

Non-linear eigenvalue problems

38. By a non-linear eigenvalue problem we mean here that the parameter whose critical value or values is sought occurs non-linearly in the differential equation which might, for example, have the form

$$y''+(a+b\lambda+c\lambda^2)y = 0, \tag{112}$$

where a, b, and c are functions of x. If homogeneous boundary conditions are given at each end of the range, we can still set up finite-difference equations, themselves homogeneous, but the matrix equivalent of these equations no longer has one of the simple forms

$$(A-I\lambda)y = 0, \qquad (A-B\lambda)y = 0, \tag{113}$$

but is represented by

$$(A+\lambda B+\lambda^2C)y = 0, \tag{114}$$

where B and C are diagonal matrices in (114).

The condition for a non-trivial solution y is again that the determinant $|A+\lambda B+\lambda^2C|$ should vanish, and the corresponding values of λ are the eigenvalues. We could calculate these values by expansion of the determinant as a polynomial in λ, then evaluating its roots, or preferably by evaluation of the determinant for various values of λ with inverse interpolation for the zeros. Methods for the calculation of eigensolutions based on the iterative processes of Chapter 7, however, are no longer applicable, and though relaxation can still be used for this purpose the technique is much more difficult and has no help from Rayleigh's Principle or other similar devices.

Methods based on matching techniques, either at one end of the range or somewhere in the middle, preferably with the simultaneous solution of the variational equation used in Chapter 8 can, however, still be used, and the only remaining interest is to examine whether convergence to a solution can be accelerated by initial investigations to determine an approximate or plausible value of λ.

39. In certain cases simple considerations can be used to provide the required information. Consider, for example, the equation

$$y''-f(x,\lambda)y = 0, \tag{115}$$

for which $y = 0$ at $x = 0$ and 1. We can postulate that y' is positive at the origin without losing generality. Now if $f(x,\lambda)$ is positive at all points in $0 \leqslant x \leqslant 1$, for all values of λ, there is clearly no real solution. This we can see both geometrically and arithmetically. From the former point of view, it is clear that the curvature of y is everywhere positive so that y increases at an increasing rate and cannot pass through a zero within the range. Arithmetically, it is clear that, using the finite-difference equations

$$y_{r+1}+y_{r-1}-(2+h^2f_r)y_r = 0, \tag{116}$$

the coefficient of y_r is everywhere greater than 2, and the solution is therefore ever increasing and cannot return to zero.

For a real solution, then, it is essential that an originally positive $f(x,\lambda)$ should become zero, and in fact change sign and become negative, at some point in the range. In a practical problem $f(x,\lambda)$ had the form

$$f(x,\lambda) = \frac{6}{x^2}-\lambda(7-13x)(1-x)+\lambda^2x^2(1-x)^4, \tag{117}$$

which is initially positive for all finite λ, and is positive also in the neighbourhood of $x = 1$. We therefore look for possible changes of sign in f. Regarded as a quadratic in λ, f vanishes when

$$\lambda = \frac{(7-13x)\pm\sqrt{\{145(x-0{\cdot}66)(x-0{\cdot}26)\}}}{2x^2(1-x)^3}. \tag{118}$$

We notice first that f does not vanish in $0{\cdot}26 < x < 0{\cdot}66$ for any real value of λ. Since f is positive at $x = 0$, and certainly positive at $x = \frac{7}{13}$ for all values of λ, we reach the following conclusions.

(i) If f vanishes in $0 < x < 0{\cdot}26$ it must vanish twice in this region. From (117) the corresponding values of λ must be positive. We see that any positive eigenvalues are associated with eigenfunctions which have a short exponential region, followed by a short oscillatory region, followed by a long exponential region.

(ii) If f vanishes in $0{\cdot}66 < x < 1{\cdot}00$ it must vanish twice in this region. From (117) the corresponding values of λ must be negative. Any negative eigenvalues are then associated with eigenfunctions which have a long exponential region, followed by a short oscillatory region, followed by a short exponential region. Case (ii) clearly gives a better chance of obtaining an eigensolution.

It is not easy to decide from these simple considerations how large the numerical value of λ should be, and it is probably quickest at this stage to try tentative negative values of λ in (116), perhaps obtaining rough solutions to a few figures and at a reasonably large interval. If we take $h = 0{\cdot}2$, the function $h^2f(x)$ has the following values at the pivotal points:

$$x = 0{\cdot}2, \qquad h^2f = 6{\cdot}000-0{\cdot}1408\lambda+0{\cdot}00065536\lambda^2,$$
$$x = 0{\cdot}4, \qquad h^2f = 1{\cdot}500-0{\cdot}0432\lambda+0{\cdot}00082944\lambda^2,$$
$$x = 0{\cdot}6, \qquad h^2f = 0{\cdot}66\dot{6}+0{\cdot}0128\lambda+0{\cdot}00036864\lambda^2,$$
$$x = 0{\cdot}8, \qquad h^2f = 0{\cdot}375+0{\cdot}0272\lambda+0{\cdot}00004096\lambda^2.$$

For all λ the function which starts with $y(0) = 0$, $y(0{\cdot}2) = 1{\cdot}0$ is still increasing rapidly at $x = 0{\cdot}8$, so that for an eigensolution there must be a catastrophic decrease in the last interval. Thus $2+h^2f(0{\cdot}8)$ must be considerably less than unity. This suggests that $-\lambda$ is more of the order of 10^2 rather than of unity, and with this information we could use standard techniques, at a smaller interval, to produce an accurate solution. There is an eigenvalue of about -186.

Semi-analytical methods

†40. We end this chapter with a short description of three related methods for solving linear differential equations which make no use of finite differences. All these methods are based on the representation of the solution in a series of the form

$$y = f_0(x)+\sum_{r=1}^{s} c_r f_r(x), \tag{119}$$

in which the $f_r(x)$ are prescribed functions of x and the c_r are constants to be determined. The choice of the f_r is to some extent arbitrary, except that $f_0(x)$ must satisfy all the boundary conditions and $f_r(x)$ $(r \geqslant 1)$ the homogeneous form of those conditions. Thus the expression (119) satisfies the complete boundary conditions of the problem for all values of the c_r, but it does not necessarily satisfy the differential equation, and the three methods are based on different approximate techniques for meeting this requirement.

Collocation

41. In the method of collocation we satisfy the differential equation at s different points, not necessarily equidistant, in the range of integration. As a result we have s linear algebraic equations for the s constants $c_1, c_2,..., c_s$.

If the differential equation is given by

$$\phi(D)y - g(x) = 0, \tag{120}$$

and if $\phi(D)f_r(x_i)$ is denoted by ϵ_{ri}, the linear equations are given by

$$\sum_{r=1}^{s} c_r \epsilon_{ri} + \epsilon_{0i} = g(x_i), \qquad i = 1, 2,..., s. \tag{121}$$

Their solution, substituted in (119), gives an approximate answer to the required solution of the differential equation.

Least squares

42. In the method of least squares we no longer select s discrete points, but consider all points of the range x_0 to x_n. If $\phi(D)f_r(x)$, at a general point x, is denoted by $\epsilon_r(x)$, then the substitution of (119) in (120) gives an error

$$\sum_{r=1}^{s} c_r \epsilon_r(x) + \epsilon_0(x) - g(x). \tag{122}$$

The constants c_r are here chosen so that

$$\int_{x_0}^{x_n} \left\{\sum_{r=1}^{s} c_r \epsilon_r(x) + \epsilon_0(x) - g(x)\right\}^2 dx \tag{123}$$

is a minimum, the conditions for which are given by

$$\frac{\partial}{\partial c_i} \int_{x_0}^{x_n} \left\{ \sum_{r=1}^{s} c_r \epsilon_r(x) + \epsilon_0(x) - g(x)\right\}^2 dx = 0, \quad i = 1, 2,..., s. \tag{124}$$

Again this gives s linear algebraic equations for the s constants $c_1, c_2,..., c_s$.

The method of Galerkin

43. In Galerkin's method, also, we consider an integral over the whole range, and the equations for the c_r are given by

$$\int_{x_0}^{x_n} f_i(x)\left\{ \sum_{r=1}^{s} c_r \epsilon_r(x) + \epsilon_0(x) - g(x)\right\} dx = 0, \quad i = 1, 2,..., s. \tag{125}$$

Eigenvalue problems

44. If the differential equation and boundary conditions are homogeneous we omit the term $f_0(x)$ in (119), so that ϵ_{0i} in (121) and ϵ_0 in (124) and (125) are all zero. The linear equations for the constants c_r are then homogeneous, and have non-trivial solutions only for the discrete values of a parameter, giving the eigenvalues, for which the determinant of the matrix vanishes.

General comments

45. We shall not give detailed discussion or examples of these methods, which have been described fully in the literature. We remark, however, that the accuracy of the results can be guaranteed only by repeating the techniques for different values of s, and stopping when successive numerical values agree to the required precision. In the papers quoted in the bibliography attention is also given to considerations governing the choice of the functions $f_r(x)$, the connexion between the methods for large values of s, the relation between Galerkin's method and Rayleigh's Principle, and the application of the methods to non-linear equations.

Additional notes and bibliography

§ 4. If there is only one critical point it is possible to choose the argument interval so that the critical point is a pivotal point, though for a convenient h we may have unequal intervals at both boundaries. With two critical points we can also arrange to have the unequal intervals at the boundaries. For more than two such points the situation of § 6 is generally unavoidable, as also in the case of § 8 when the position of the critical point or points depends on the solution.

§ 9. This device, which has been used in many contexts, was first suggested by

RICHARDSON, L. F., and GAUNT, J. A., 1926, The deferred approach to the limit. *Phil. Trans.* A, **226,** 299–361.

§ 36. Tables of the function $T(z)$, calculated by solving its differential equation, have been given by

FOX, L., and MILLER, J. C. P., 1951, Notes on Numerical Analysis—5. Table-making for large arguments. The exponential integral. *Mathematical Tables and Other Aids to Computation,* **5,** 163–6.

§§ 40–44. The methods of collocation and least squares and Galerkin's method have been described in detail in the following papers:

FRAZER, R. A., JONES, W. P., and SKAN, Sylvia W., 1937, Approximations to functions and to the solutions of differential equations. *Aeronautical Research Committee: Reports and Memoranda No.* 1799. London, H.M.S.O.

DUNCAN, W. J., 1937, Galerkin's method in mechanics and differential equations. *Aeronautical Research Committee: Reports and Memoranda No.* 1798. London, H.M.S.O.

DUNCAN, W. J., 1938, The principles of the Galerkin method. *Aeronautical Research Committee: Reports and Memoranda No.* 1848. London, H.M.S.O.

Somewhat similar methods have been developed by

DENNIS, S. C. R., and POOTS, G., 1955, The solution of linear differential equations. *Proc. Camb. Phil. Soc.* **51**, 422–32,

and

CLENSHAW, C. W., 1957, The numerical solution of linear differential equations in Chebyshev series. *Proc. Camb. Phil. Soc.* **53**, 134–49.

Here solutions are sought in terms of convergent expansions of the form

$$y = \sum_{0}^{\infty} a_r f_r(x),$$

where the $f_r(x)$ are orthogonal functions, not satisfying the boundary conditions as in the previous methods. Dennis and Poots choose $f_r(x)$ to be either $\sin r\pi x/l$ or $\cos r\pi x/l$, where l is the total range of x under consideration, so that the series is the Fourier series for y. The derivatives of y and any functions of x occurring in the differential equation are expressible in the same form, and the equating to zero of the coefficients of each $f_r(x)$ gives an infinite set of linear algebraic equations in the coefficients a_r. Extra equations are provided by the boundary conditions, and the complete set is usually easily solved by either iterative or direct methods.

Clenshaw chooses $f_r(x)$ to be the Chebyshev polynomial, normalized to the range $-1 \leqslant x \leqslant 1$, or $0 \leqslant x \leqslant 1$, according to the range of integration involved. Again there are simultaneous equations for the coefficients a_r. Both methods can be used to determine eigenvalues by a trial and error process, but they are convenient only for linear differential equations.

Index

A CATALOG OF SELECTED

DOVER BOOKS

IN SCIENCE AND MATHEMATICS

A CATALOG OF SELECTED
DOVER BOOKS
IN SCIENCE AND MATHEMATICS

QUALITATIVE THEORY OF DIFFERENTIAL EQUATIONS, V.V. Nemytskii and V.V. Stepanov. Classic graduate-level text by two prominent Soviet mathematicians covers classical differential equations as well as topological dynamics and erqodic theory. Bibliographies. 523pp. 5⅜ × 8½. 65954-2 Pa. $10.95

MATRICES AND LINEAR ALGEBRA, Hans Schneider and George Phillip Barker. Basic textbook covers theory of matrices and its applications to systems of linear equations and related topics such as determinants, eigenvalues and differential equations. Numerous exercises. 432pp. 5⅜ × 8½. 66014-1 Pa. $8.95

QUANTUM THEORY, David Bohm. This advanced undergraduate-level text presents the quantum theory in terms of qualitative and imaginative concepts, followed by specific applications worked out in mathematical detail. Preface. Index. 655pp. 5⅜ × 8½. 65969-0 Pa. $12.95

ATOMIC PHYSICS (8th edition), Max Born. Nobel laureate's lucid treatment of kinetic theory of gases, elementary particles, nuclear atom, wave-corpuscles, atomic structure and spectral lines, much more. Over 40 appendices, bibliography. 495pp. 5⅜ × 8½. 65984-4 Pa. $11.95

ELECTRONIC STRUCTURE AND THE PROPERTIES OF SOLIDS: The Physics of the Chemical Bond, Walter A. Harrison. Innovative text offers basic understanding of the electronic structure of covalent and ionic solids, simple metals, transition metals and their compounds. Problems. 1980 edition. 582pp. 6⅛ × 9¼. 66021-4 Pa. $14.95

BOUNDARY VALUE PROBLEMS OF HEAT CONDUCTION, M. Necati Özisik. Systematic, comprehensive treatment of modern mathematical methods of solving problems in heat conduction and diffusion. Numerous examples and problems. Selected references. Appendices. 505pp. 5⅜ × 8½. 65990-9 Pa. $11.95

A SHORT HISTORY OF CHEMISTRY (3rd edition), J.R. Partington. Classic exposition explores origins of chemistry, alchemy, early medical chemistry, nature of atmosphere, theory of valency, laws and structure of atomic theory, much more. 428pp. 5⅜ × 8½. (Available in U.S. only) 65977-1 Pa. $10.95

A HISTORY OF ASTRONOMY, A. Pannekoek. Well-balanced, carefully reasoned study covers such topics as Ptolemaic theory, work of Copernicus, Kepler, Newton, Eddington's work on stars, much more. Illustrated. References. 521pp. 5⅜ × 8½. 65994-1 Pa. $11.95

PRINCIPLES OF METEOROLOGICAL ANALYSIS, Walter J. Saucier. Highly respected, abundantly illustrated classic reviews atmospheric variables, hydrostatics, static stability, various analyses (scalar, cross-section, isobaric, isentropic, more). For intermediate meteorology students. 454pp. 6⅛ × 9¼. 65979-8 Pa. $12.95

THE ELECTROMAGNETIC FIELD, Albert Shadowitz. Comprehensive undergraduate text covers basics of electric and magnetic fields, builds up to electromagnetic theory. Also related topics, including relativity. Over 900 problems. 768pp. 5⅜ × 8¼. 65660-8 Pa. $15.95

FOURIER SERIES, Georgi P. Tolstov. Translated by Richard A. Silverman. A valuable addition to the literature on the subject, moving clearly from subject to subject and theorem to theorem. 107 problems, answers. 336pp. 5⅜ × 8½. 63317-9 Pa. $7.95

THEORY OF ELECTROMAGNETIC WAVE PROPAGATION, Charles Herach Papas. Graduate-level study discusses the Maxwell field equations, radiation from wire antennas, the Doppler effect and more. xiii + 244pp. 5⅜ × 8½. 65678-0 Pa. $6.95

DISTRIBUTION THEORY AND TRANSFORM ANALYSIS: An Introduction to Generalized Functions, with Applications, A.H. Zemanian. Provides basics of distribution theory, describes generalized Fourier and Laplace transformations. Numerous problems. 384pp. 5⅜ × 8½. 65479-6 Pa. $9.95

THE PHYSICS OF WAVES, William C. Elmore and Mark A. Heald. Unique overview of classical wave theory. Acoustics, optics, electromagnetic radiation, more. Ideal as classroom text or for self-study. Problems. 477pp. 5⅜ × 8½. 64926-1 Pa. $10.95

CALCULUS OF VARIATIONS WITH APPLICATIONS, George M. Ewing. Applications-oriented introduction to variational theory develops insight and promotes understanding of specialized books, research papers. Suitable for advanced undergraduate/graduate students as primary, supplementary text. 352pp. 5⅜ × 8½. 64856-7 Pa. $8.50

A TREATISE ON ELECTRICITY AND MAGNETISM, James Clerk Maxwell. Important foundation work of modern physics. Brings to final form Maxwell's theory of electromagnetism and rigorously derives his general equations of field theory. 1,084pp. 5⅜ × 8½. 60636-8, 60637-6 Pa., Two-vol. set $19.90

AN INTRODUCTION TO THE CALCULUS OF VARIATIONS, Charles Fox. Graduate-level text covers variations of an integral, isoperimetrical problems, least action, special relativity, approximations, more. References. 279pp. 5⅜ × 8½. 65499-0 Pa. $7.95

HYDRODYNAMIC AND HYDROMAGNETIC STABILITY, S. Chandrasekhar. Lucid examination of the Rayleigh-Benard problem; clear coverage of the theory of instabilities causing convection. 704pp. 5⅜ × 8¼. 64071-X Pa. $12.95

CALCULUS OF VARIATIONS, Robert Weinstock. Basic introduction covering isoperimetric problems, theory of elasticity, quantum mechanics, electrostatics, etc. Exercises throughout. 326pp. 5⅜ × 8½. 63069-2 Pa. $7.95

DYNAMICS OF FLUIDS IN POROUS MEDIA, Jacob Bear. For advanced students of ground water hydrology, soil mechanics and physics, drainage and irrigation engineering and more. 335 illustrations. Exercises, with answers. 784pp. 6⅛ × 9¼. 65675-6 Pa. $19.95

NUMERICAL METHODS FOR SCIENTISTS AND ENGINEERS, Richard Hamming. Classic text stresses frequency approach in coverage of algorithms, polynomial approximation, Fourier approximation, exponential approximation, other topics. Revised and enlarged 2nd edition. 721pp. 5⅜ × 8½.
65241-6 Pa. $14.95

THEORETICAL SOLID STATE PHYSICS, Vol. I: Perfect Lattices in Equilibrium; Vol. II: Non-Equilibrium and Disorder, William Jones and Norman H. March. Monumental reference work covers fundamental theory of equilibrium properties of perfect crystalline solids, non-equilibrium properties, defects and disordered systems. Appendices. Problems. Preface. Diagrams. Index. Bibliography. Total of 1,301pp. 5⅜ × 8½. Two volumes. Vol. I 65015-4 Pa. $12.95
Vol. II 65016-2 Pa. $12.95

OPTIMIZATION THEORY WITH APPLICATIONS, Donald A. Pierre. Broad-spectrum approach to important topic. Classical theory of minima and maxima, calculus of variations, simplex technique and linear programming, more. Many problems, examples. 640pp. 5⅜ × 8½. 65205-X Pa. $13.95

THE MODERN THEORY OF SOLIDS, Frederick Seitz. First inexpensive edition of classic work on theory of ionic crystals, free-electron theory of metals and semiconductors, molecular binding, much more. 736pp. 5⅜ × 8½.
65482-6 Pa. $15.95

ESSAYS ON THE THEORY OF NUMBERS, Richard Dedekind. Two classic essays by great German mathematician: on the theory of irrational numbers; and on transfinite numbers and properties of natural numbers. 115pp. 5⅜ × 8½.
21010-3 Pa. $4.95

THE FUNCTIONS OF MATHEMATICAL PHYSICS, Harry Hochstadt. Comprehensive treatment of orthogonal polynomials, hypergeometric functions, Hill's equation, much more. Bibliography. Index. 322pp. 5⅜ × 8½. 65214-9 Pa. $9.95

NUMBER THEORY AND ITS HISTORY, Oystein Ore. Unusually clear, accessible introduction covers counting, properties of numbers, prime numbers, much more. Bibliography. 380pp. 5⅜ × 8½. 65620-9 Pa. $8.95

THE VARIATIONAL PRINCIPLES OF MECHANICS, Cornelius Lanzcos. Graduate level coverage of calculus of variations, equations of motion, relativistic mechanics, more. First inexpensive paperbound edition of classic treatise. Index. Bibliography. 418pp. 5⅜ × 8½. 65067-7 Pa. $10.95

MATHEMATICAL TABLES AND FORMULAS, Robert D. Carmichael and Edwin R. Smith. Logarithms, sines, tangents, trig functions, powers, roots, reciprocals, exponential and hyperbolic functions, formulas and theorems. 269pp. 5⅜ × 8½. 60111-0 Pa. $5.95

THEORETICAL PHYSICS, Georg Joos, with Ira M. Freeman. Classic overview covers essential math, mechanics, electromagnetic theory, thermodynamics, quantum mechanics, nuclear physics, other topics. First paperback edition. xxiii + 885pp. 5⅜ × 8½. 65227-0 Pa. $18.95

ORDINARY DIFFERENTIAL EQUATIONS, Morris Tenenbaum and Harry Pollard. Exhaustive survey of ordinary differential equations for undergraduates in mathematics, engineering, science. Thorough analysis of theorems. Diagrams. Bibliography. Index. 818pp. 5⅜ × 8½. 64940-7 Pa. $15.95

STATISTICAL MECHANICS: Principles and Applications, Terrell L. Hill. Standard text covers fundamentals of statistical mechanics, applications to fluctuation theory, imperfect gases, distribution functions, more. 448pp. 5⅜ × 8½. 65390-0 Pa. $9.95

ORDINARY DIFFERENTIAL EQUATIONS AND STABILITY THEORY: An Introduction, David A. Sánchez. Brief, modern treatment. Linear equation, stability theory for autonomous and nonautonomous systems, etc. 164pp. 5⅜ × 8¼. 63828-6 Pa. $4.95

THIRTY YEARS THAT SHOOK PHYSICS: The Story of Quantum Theory, George Gamow. Lucid, accessible introduction to influential theory of energy and matter. Careful explanations of Dirac's anti-particles, Bohr's model of the atom, much more. 12 plates. Numerous drawings. 240pp. 5⅜ × 8½. 24895-X Pa. $5.95

ORDINARY DIFFERENTIAL EQUATIONS, I.G. Petrovski. Covers basic concepts, some differential equations and such aspects of the general theory as Euler lines, Arzel's theorem, Peano's existence theorem, Osgood's uniqueness theorem, more. 45 figures. Problems. Bibliography. Index. xi + 232pp. 5⅜ × 8½. 64683-1 Pa. $6.95

GREAT EXPERIMENTS IN PHYSICS: Firsthand Accounts from Galileo to Einstein, edited by Morris H. Shamos. 25 crucial discoveries: Newton's laws of motion, Chadwick's study of the neutron, Hertz on electromagnetic waves, more. Original accounts clearly annotated. 370pp. 5⅜ × 8½. 25346-5 Pa. $8.95

INTRODUCTION TO PARTIAL DIFFERENTIAL EQUATIONS WITH APPLICATIONS, E.C. Zachmanoglou and Dale W. Thoe. Essentials of partial differential equations applied to common problems in engineering and the physical sciences. Problems and answers. 416pp. 5⅜ × 8½. 65251-3 Pa. $9.95

BURNHAM'S CELESTIAL HANDBOOK, Robert Burnham, Jr. Thorough guide to the stars beyond our solar system. Exhaustive treatment. Alphabetical by constellation: Andromeda to Cetus in Vol. 1; Chamaeleon to Orion in Vol. 2; and Pavo to Vulpecula in Vol. 3. Hundreds of illustrations. Index in Vol. 3. 2,000pp. 6⅛ × 9¼. 23567-X, 23568-8, 23673-0 Pa., Three-vol. set $38.85

ASYMPTOTIC EXPANSIONS FOR ORDINARY DIFFERENTIAL EQUATIONS, Wolfgang Wasow. Outstanding text covers asymptotic power series, Jordan's canonical form, turning point problems, singular perturbations, much more. Problems. 384pp. 5⅜ × 8½. 65456-7 Pa. $9.95

AMATEUR ASTRONOMER'S HANDBOOK, J.B. Sidgwick. Timeless, comprehensive coverage of telescopes, mirrors, lenses, mountings, telescope drives, micrometers, spectroscopes, more. 189 illustrations. 576pp. 5⅜ × 8¼. 24034-7 Pa. $8.95

GEOMETRY OF COMPLEX NUMBERS, Hans Schwerdtfeger. Illuminating, widely praised book on analytic geometry of circles, the Moebius transformation, and two-dimensional non-Euclidean geometries. 200pp. 5⅜ × 8¼. 63830-8 Pa. $6.95

MECHANICS, J.P. Den Hartog. A classic introductory text or refresher. Hundreds of applications and design problems illuminate fundamentals of trusses, loaded beams and cables, etc. 334 answered problems. 462pp. 5⅜ × 8½. 60754-2 Pa. $8.95

TOPOLOGY, John G. Hocking and Gail S. Young. Superb one-year course in classical topology. Topological spaces and functions, point-set topology, much more. Examples and problems. Bibliography. Index. 384pp. 5⅜ × 8¼. 65676-4 Pa. $7.95

STRENGTH OF MATERIALS, J.P. Den Hartog. Full, clear treatment of basic material (tension, torsion, bending, etc.) plus advanced material on engineering methods, applications. 350 answered problems. 323pp. 5⅜ × 8½. 60755-0 Pa. $7.50

ELEMENTARY CONCEPTS OF TOPOLOGY, Paul Alexandroff. Elegant, intuitive approach to topology from set-theoretic topology to Betti groups; how concepts of topology are useful in math and physics. 25 figures. 57pp. 5⅜ × 8½. 60747-X Pa. $2.95

ADVANCED STRENGTH OF MATERIALS, J.P. Den Hartog. Superbly written advanced text covers torsion, rotating disks, membrane stresses in shells, much more. Many problems and answers. 388pp. 5⅜ × 8½. 65407-9 Pa. $9.95

COMPUTABILITY AND UNSOLVABILITY, Martin Davis. Classic graduate-level introduction to theory of computability, usually referred to as theory of recurrent functions. New preface and appendix. 288pp. 5⅜ × 8½. 61471-9 Pa. $6.95

GENERAL CHEMISTRY, Linus Pauling. Revised 3rd edition of classic first-year text by Nobel laureate. Atomic and molecular structure, quantum mechanics, statistical mechanics, thermodynamics correlated with descriptive chemistry. Problems. 992pp. 5⅜ × 8½. 65622-5 Pa. $18.95

AN INTRODUCTION TO MATRICES, SETS AND GROUPS FOR SCIENCE STUDENTS, G. Stephenson. Concise, readable text introduces sets, groups, and most importantly, matrices to undergraduate students of physics, chemistry, and engineering. Problems. 164pp. 5⅜ × 8½. 65077-4 Pa. $5.95

THE HISTORICAL BACKGROUND OF CHEMISTRY, Henry M. Leicester. Evolution of ideas, not individual biography. Concentrates on formulation of a coherent set of chemical laws. 260pp. 5⅜ × 8½. 61053-5 Pa. $6.00

THE PHILOSOPHY OF MATHEMATICS: An Introductory Essay, Stephan Körner. Surveys the views of Plato, Aristotle, Leibniz & Kant concerning propositions and theories of applied and pure mathematics. Introduction. Two appendices. Index. 198pp. 5⅜ × 8½. 25048-2 Pa. $5.95

THE DEVELOPMENT OF MODERN CHEMISTRY, Aaron J. Ihde. Authoritative history of chemistry from ancient Greek theory to 20th-century innovation. Covers major chemists and their discoveries. 209 illustrations. 14 tables. Bibliographies. Indices. Appendices. 851pp. 5⅜ × 8½. 64235-6 Pa. $17.95

CHALLENGING MATHEMATICAL PROBLEMS WITH ELEMENTARY SOLUTIONS, A.M. Yaglom and I.M. Yaglom. Over 170 challenging problems on probability theory, combinatorial analysis, points and lines, topology, convex polygons, many other topics. Solutions. Total of 445pp. 5⅜ × 8½. Two-vol. set.
Vol. I 65536-9 Pa. $6.95
Vol. II 65537-7 Pa. $6.95

FIFTY CHALLENGING PROBLEMS IN PROBABILITY WITH SOLUTIONS, Frederick Mosteller. Remarkable puzzlers, graded in difficulty, illustrate elementary and advanced aspects of probability. Detailed solutions. 88pp. 5⅜ × 8½.
65355-2 Pa. $3.95

EXPERIMENTS IN TOPOLOGY, Stephen Barr. Classic, lively explanation of one of the byways of mathematics. Klein bottles, Moebius strips, projective planes, map coloring, problem of the Koenigsberg bridges, much more, described with clarity and wit. 43 figures. 210pp. 5⅜ × 8½. 25933-1 Pa. $4.95

RELATIVITY IN ILLUSTRATIONS, Jacob T. Schwartz. Clear non-technical treatment makes relativity more accessible than ever before. Over 60 drawings illustrate concepts more clearly than text alone. Only high school geometry needed. Bibliography. 128pp. 6⅛ × 9¼. 25965-X Pa. $5.95

AN INTRODUCTION TO ORDINARY DIFFERENTIAL EQUATIONS, Earl A. Coddington. A thorough and systematic first course in elementary differential equations for undergraduates in mathematics and science, with many exercises and problems (with answers). Index. 304pp. 5⅜ × 8¼. 65942-9 Pa. $7.95

FOURIER SERIES AND ORTHOGONAL FUNCTIONS, Harry F. Davis. An incisive text combining theory and practical example to introduce Fourier series, orthogonal functions and applications of the Fourier method to boundary-value problems. 570 exercises. Answers and notes. 416pp. 5⅜ × 8½. 65973-9 Pa. $8.95

THE THOERY OF BRANCHING PROCESSES, Theodore E. Harris. First systematic, comprehensive treatment of branching (i.e. multiplicative) processes and their applications. Galton-Watson model, Markov branching processes, electron-photon cascade, many other topics. Rigorous proofs. Bibliography. 240pp. 5⅜ × 8½. 65952-6 Pa. $6.95

AN INTRODUCTION TO ALGEBRAIC STRUCTURES, Joseph Landin. Superb self-contained text covers "abstract algebra": sets and numbers, theory of groups, theory of rings, much more. Numerous well-chosen examples, exercises. 247pp. 5⅜ × 8½. 65940-2 Pa. $6.95
